# Port and Harbour Structures

Sundaravadivelu Ranganathan
and Sakthivel Sundaravadivelu

CRC Press
Taylor & Francis Group
Boca Raton London New York

CRC Press is an imprint of the
Taylor & Francis Group, an **informa** business

Designed cover image: © Sundaravadivelu Ranganathan and Sakthivel Sundaravadivelu

First edition published 2026
by CRC Press
2385 NW Executive Center Drive, Suite 320, Boca Raton FL 33431

and by CRC Press
4 Park Square, Milton Park, Abingdon, Oxon, OX14 4RN

*CRC Press is an imprint of Taylor & Francis Group, LLC*



ISBN: 978-1-032-42118-6 (hbk)
ISBN: 978-1-032-42119-3 (pbk)
ISBN: 978-1-003-36127-5 (ebk)

DOI: 10.1201/9781003361275

Typeset in Times
by Apex CoVantage, LLC

# Contents

# Authors' Biographies

**Sundaravadivelu** Ranganathan was Emeritus Professor (Retd.) from the Department of Ocean Engineering at IIT Madras and currently serves as Director of Prof. Sundaravadivelu Ranganathan Advisory Services Pvt. Ltd. He holds a B.E. in Civil Engineering from Annamalai University, an M. Tech in Structural Engineering, and a Ph.D. in Ocean Engineering from IIT Madras.

With over four decades of experience, he has contributed extensively to the development of greenfield and brownfield ports in India and abroad, consulting on nearly 1,000 projects. His research spans marine structures, soil-structure-wave interaction, geotextile applications, sea water intake and outfalls, sea walls, groynes, deep-water ports, offshore platforms, shipyards Dredging and physical and numerical modelling of port and harbour structures.

Prof. Sundaravadivelu has published around 200 papers in international journals and conferences and guided 23 Ph.D. scholars. He has held key positions at IIT Madras, including Professor & Head of the Department, Chair Professor, Member Board of IC and SR, Chairman of the Estate & Works Committee, and Member of the Board of Governors.

His accolades include being a Fellow of the Indian National Academy of Engineering and serving on numerous national committees such as the Indian Maritime University, Central Road Research Institute, Chilika Development Authority, and Chairperson CED47, Ports, Harbours and Offshore Installations, Bureau of Indian Standards, Government of India.

**Sakthivel** Sundaravadivelu is the Managing Director and Principal Marine Consultant at Prof. Sundaravadivelu Ranganathan Advisory Services Pvt. Ltd. With over two decades of experience in waterborne transport infrastructure, he has successfully led more than 500 projects involving Ports, Shipyards, Dredging, offshore, sea water intake and Coastal protection structures.

He holds a B.E. in Civil Engineering from Madras University, an M. Tech in Ocean Engineering from IIT Madras, and a Postgraduate Certification in Strategic Port Management from IIM Ahmedabad. His expertise spans FEED engineering, contract documentation, analysis, design and detailed engineering, Techno-economic feasibility report (TEFR), Master planning, Detailed Project Reports (DPR) and Project Management Consultancy.

Sakthivel is proficient in structural analysis software such as SACS and STAAD. Pro, wave flume and basin model studies and is well-versed in applying Indian and international codes for design of marine structures. He is an active member of professional societies including ASME, ASCE, and The Indian Society of Hydraulics, and is an alumnus of IIT Madras.

# Preface

The harbours of the modern civilization in the 20th century are semi-natural or artificial, and the port and harbour structures are developed in deep waters based on research and development, modern software, new technology, construction techniques and materials. This gave birth to a specialized form of civil engineering which is known today as marine civil infrastructure. It is the aim of the authors to present to students and practicing engineers the development in ports and harbours, shipyards and environmental impacts of these developments. Dr. Sundaravadivelu Ranganathan, FNAE and Mr. Sakthivel Sundaravadivelu present this book based on their experience in design of ports in India and abroad. This book details various Indian codes in Indian context and comparison to highlight the difference between other international codes. The Indian codes were recently modified by Bureau of Indian Standards under active participation by Prof. Sundaravadivelu Ranganathan (Chairperson BID Committee CED 47 on Ports, Harbours and Offshore Installations) and his team, and the details are incorporated in this book. An entire chapter is dedicated to estimate load acting on the system. The design of slipway, marine travel lift, and ship lift is also presented. The environmental issues, physical and numerical model studies, non-linear soil structure interaction, etc. are illustrated with typical examples for green port development. The brownfield development of existing berthing structures for deeper draft is presented as case studies based on published literature by the authors The type of development with capital dredging, maintenance dredging and disposal is highlighted. The various procedures and documentation required for environmental clearance along with EIA and EMP are also highlighted. The case studies on rehabilitation and retrofitting to make the infrastructure operation with less down time are given in detail.

# Acknowledgement

We would like to express our sincere gratitude to all those who have supported the completion of this book, *Port and Harbour Structures*.

We are deeply indebted to Mrs. Vanaja for her unwavering encouragement and moral support during this work.

We also acknowledge with appreciation the contributions of leading industry partners whose expertise and collaboration added immense value to this publication:

1. Liebherr Mobile Harbour Crane
2. HiTech Fenders
3. Hitech Civil Engineers Pvt Ltd.
4. Fosroc Chemicals
5. Vizhinjam International Seaport
6. Adani Ports and SEZ Ltd
7. ITD CEM
8. TNEB and TANGEDCO

The contribution of Dr K. Balasubramanian Managing Director, Hitech Concrete Solutions Chennai Pvt. Ltd. (NABL Accredited Laboratory as per ISO/IEC 17025: 2005) has contributed immensely for the retrofitting and rehabilitation techniques.

Special thanks to Shri G. V. Satyanarayana, Deputy Chief Engineer (Retd.), Visakhapatnam Port Trust and Dr. M. V. Ramanamurthy Director (Retd) National Centre for Coastal Research, Ministry of Earth Sciences, Govt of India for their valuable technical input drawn from their distinguished career in the Indian port sector.

We also like to express our thanks to Dr. P. Kavitha, Mrs. S. Poornima Devi and the project staff and students of Dept. of Ocean Engineering, IIT Madras and Prof. R. Sundaravadivelu Advisory Services Pvt Ltd.

**Dr. Sundaravadivelu Ranganathan FNAE**
**Mr. Sakthivel Sundaravadivelu**

# 1 Introduction

## 1.1 SHIP DIMENSION

### 1.1.1 General

Ships are large floating vessels used for transporting cargoes and passengers across open seas. The key ways to describe a ship are by its length (LoA: length overall), width (beam), draft (the depth which is below the water level), dead weight tonnage (DWT) and kind of cargo the vessel carries. Ships are generally classified during the design stage based on route of operation and the purpose of the ship. A variety of parameters such as draft, beam, length overall, gross tonnage, deadweight tonnage etc. are taken into consideration while designing and constructing a seagoing ship. Based on these details the ships are registered. Ship registration is the process by which a ship is documented under a nationality, which allows the ship to travel internationally. Figure 1.1 shows the size of the ship in a side view.

### 1.1.2 Classification of Vessels

Vessels are classified as follows

- Container, liquid and dry bulk
- Handy, Panamax, Capsize, VLCC
- Barges
- Offshore supply vessel
- Floating drydock
- Tugs
- Fishing vessels

### 1.1.3 Vessel Motion

The motion of the vessel includes:

- Heave, yaw, surge, sway, pitch and roll (Figure 1.2)

One must also check the motions during the following conditions, for example,

- Tranquility for loading/unloading
- Swell and sea for berthing
- Orientation in the direction of current
- Quick release hook
- Tidal effects to fix the top level

DOI: 10.1201/9781003361275-1

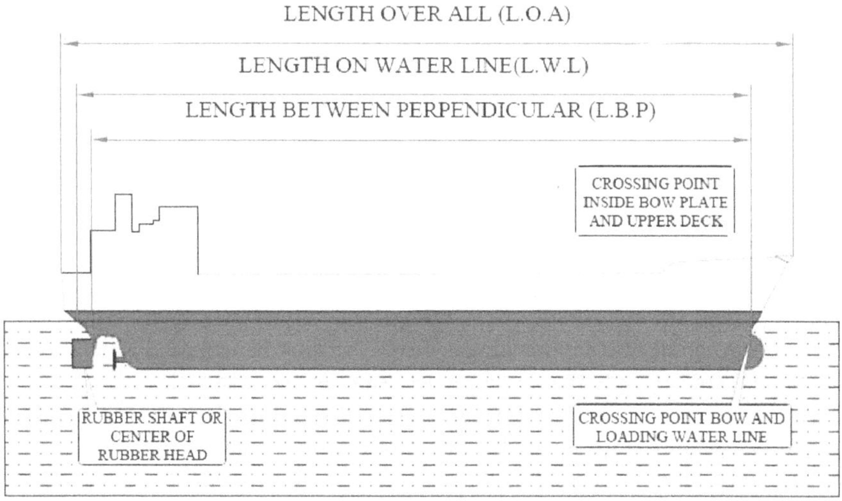


**FIGURE 1.1** Ship size (side view).

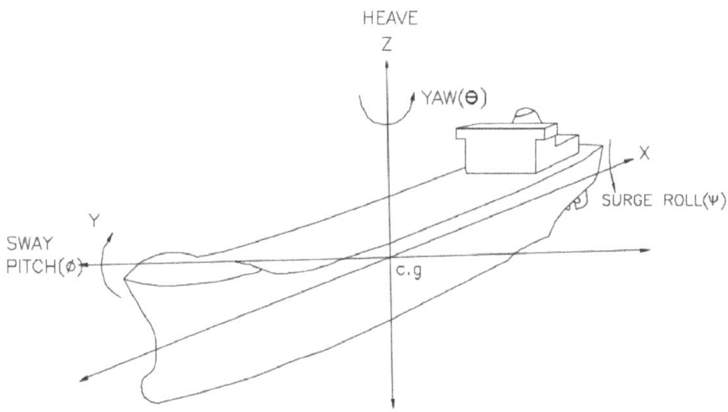


**FIGURE 1.2** Vessel motion.

Bow, stern, breast and spring lines are provided to secure the vessels as shown in Figure 1.3.

Permissible ship motions as per PIANC WG 24 1995 ‘Criteria for movements of moored ships in harbours’ for general cargo vessels and bulk carriers are as follows:

- Surge: 1.00 m
- Sway: 0.75 m
- Yawing: 1.5°
- Rolling: 2.5°
- Pitching: 1°
- Heave: 0.50 m

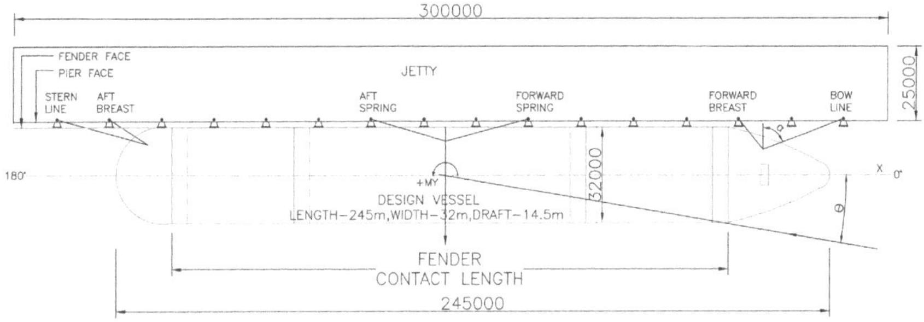


**FIGURE 1.3** Vessel mooring lines and berth layout.

The permissible current velocity in the head and beam sea are 1.50 m/s and 0.40 m/s, respectively. The acceptable significant wave height in head sea is 1.50 m for loading and 1.00 m for unloading. The acceptable significant wave height in head sea for is 2.50 m and 1.20 m in quarter and beam sea.

### 1.1.4 Vessel Dimensions

| | | |
|---|---|---|
| Length overall (LOA) | - | Length from the tip of the bow to stern |
| Beam (b) | - | Vessel width at its widest point |
| Draft (d) | - | The vertical distance between the bottom of the hull and the waterline |
| Freeboard (f) | - | Distance from the waterline to the first deck |

#### Length overall (LOA):

It is the distance measured parallel to the summer load waterline between the extreme points at the forward and the aftermost point of the stern. The extreme point at the forward can be taken on the bulbous bow.

#### Length between perpendiculars (LBP):

The distance measured along the summer load water plane from fore to aft perpendicular is the length between perpendiculars.

#### Length on the waterline (LWL):

It is the distance measured between the intersection points of the bow and the after end with the summer load waterline. In case it is not mentioned, the length of the summer load waterline is considered as the length of the waterline.

#### Width or beam or extreme breadth:

The width of the ship is also known as the *extreme breadth*. It is generally measured in the middle of the vessel. It is the distance between the two sides of the ship at the greatest width.

**Freeboard:**

Freeboard of the ship is essential in determining the stability of the ship. It is the difference between the depth at the side and the draught, i.e., the height of the deck above the waterline. Freeboard is generally less at amidships than at bow and stern.

**Depth:**

The depth of the ship varies along the length. But the depth of the ship is taken as the distance between the undersides of the deck amid the ship to the bottom of the keel. The exact depth varies throughout the ship. For example, the depth is greater at the stern than at mid-ship.

**Draught:**

Draught or draft is the distance between the keel and the waterline at any point along the length of the ship. Moulded draught is measured from the inside of the keel plating.

### 1.1.5 Ship Tonnages

Gross registered tonnage (GRT) – It is broadly the capacity in cubic feet of the spaces within the hull and of the enclosed spaces above the deck provided for cargo, stores, passengers and crew, with certain exceptions, divided by 100. Thus 100 cubic feet of capacity is equivalent to 1 gross registered tonnage. It is the ship's internal volume given in the GRT measure, each unit of which is equal to 100 cubic feet.

Net registered tonnage (NRT) – It is derived from the gross tonnage by deducting spaces used for the accommodation of the master, officers, crew, navigation, propelling machinery and fuel.

Dead weight tonnage (DWT) – It is the weight in tons (of 2240 lb) of 1016 kg cargo, stores, fuel, passengers and crew carried by the ship when loaded to the maximum summer load line.

Displacement tonnage (DT) – The actual weight of the vessel, or the weight of water, the ship displaces when afloat and may be either 'loaded' or 'light'.

Displacement 'loaded' – The weight, in long tons, of the ship and its contents when fully loaded with cargo to the plimsoll mark or load line.

Displacement 'light' – The weight, in long tons, of the ship without cargo, fuel and stores.

### 1.1.6 Ship Characteristics

Ship characteristics are as per IS 4651 (Part 3) (2023) (Planning and Design of Ports and Harbours – Loading). The relationship between the type of ships and their corresponding gross registered weight, net weight, deadweight and displacement weight in tonnages is given in Chapter 2.

Ship dimensions: For preliminary design purpose, corresponding loading may be taken for the ship dimensions given in Annexure A of IS 4651 (Part 3) (2023), Planning and Design of Ports and Harbours – Loading). For detailed design, ship dimensions appropriate to the type of service required may be obtained from the Register of Shipping. Lloyds Register of Shipping has divided the design vessels into eight types. PIANC WG 49 – 'Horizontal and Vertical Dimensions of Harbour Approach Channels for Container and Bulk Ships' (2014) has prepared tables for the eight types of vessels with ship dimensions corresponding to 50%, 75% and 90% confidence limits.

## 1.2 HARBOUR

Port and harbour structures are constructed to provide berthing facilities to the ships for loading and unloading cargo or for embarking and disembarking passengers. The harbour is a protected water area that facilitates safe and suitable accommodation for ships. The port is a place where vessels may discharge or receive cargo. It may be the entire harbour, including its approaches and anchorage or only the commercial part of a harbour where the quays, wharves, facilities for transfer of cargo, docks and repair shops are situated.

### 1.2.1 Types of Harbour

The types of harbour are classified based on safety, location and usage and are detailed as follows:

Based on protection – natural, semi-natural and artificial
Based on locations – sea, river, creek, canal and lagoon
Based on usage – commercial, fishery, military, marine and harbour of refuge

A fishing vessel could obtain shelter during bad weather within a track or area of water close to the shore, providing a good hold for anchoring, protected by natural or artificial harbour walls against the fury of storms. However, commercial vessels need to go out of the harbour to open sea, as such vessels may hit the sea bed due to harbour oscillations inside the harbour during cyclones.

**Natural harbour:**

Natural formations affording safe discharge facilities for ships on sea-coasts, in the form of creeks and basins, are called natural harbours. With the rapid development of naval operations engaged either in commerce or war, improved accommodation and facilities for repairs, storage of cargo and connected amenities are to be provided in natural harbours. The size and draft of present-day vessels have necessitated the improvement measures for natural harbours. The factors such as local geographical features, growth of population and development of the area have made natural harbours attractive. Mumbai and Kandla are examples of natural harbours.

**Semi-natural/semi-artificial harbour:**

The semi-natural type of harbour is protected on sides by headlands protection, and it requires man-made protection only at the entrance. Vishakhapatnam is a semi-natural harbour.

**Artificial harbour:**

Where such natural facilities as sheltered basin are not available, shelters making use of engineering skills and methods are constructed, and such harbours are called artificial or man-made harbours. Chennai is an artificial harbour.

**Sea harbour:**

A sea harbour is a harbour along the seashore. These harbours are of large area, and loading and unloading of goods and passengers are done in larger quantities. These ports connect different countries.

**River harbour:**

A river harbour is also known as an estuary harbour (Haldia Dock Complex) and is constructed along the banks of a river which finally connects to the sea. Possibilities of navigation inside a country have non-coastal areas.

**Canal harbour:**

A canal harbour is similar to a river harbour, but a canal is connected to the sea through a river.

**Lagoon harbour:**

A lagoon, with good natural shelter, has outlets which are maintained through the coast, even when the tidal rise is relatively small, aided by the discharge of rivers falling into it. These outlets afford navigable access to the lagoon in which, moreover, natural channels are formed by the action of the currents. Lagoons are also created adjacent to the sea-coast, where large land area is available and breakwater are provided for safe navigation (Paradip and New Mangalore Port).

**Commercial harbour:**

The commercial harbour is where loading and unloading of cargoes are done. It requires areas with facilities to carry out repair work. Some commercial harbours are limited to loading and unloading of single commodities like coal, crude petroleum etc.

**Fishery harbour:**

A fishery harbour carries seafood or marine animals. They are always open to fishing ships for loading and unloading. Sufficient space is provided near the harbour for preserving seafood. The tranquility for fishery harbour is recommended with wave height less than 0.3 m.

**Military harbour:**

These harbours are used as naval bases, where war ships are rested while there is no work in the sea or during repair works. Private ships are restricted in military harbours.

**Marine harbour:**

Marine harbours are small harbours in which moorings are provided for yachts and small ships and are not suitable for large cargoes. The boats, ship repair works, and fuelling are done in this type of harbour. Small marinas have fewer than 100 berths, and large marinas have more than 200 berths. Resort facilities, yacht clubs, sports, and fishing facilities are provided in this type of harbour.

**Harbours of refuge:**

Harbours are used to shelter ships during emergency situations like storms or any other natural calamities and provide good anchorage. A harbour of refuge is constructed with a large entrance, which is easily accessible.

### 1.2.2 Harbour Components

The following are the main components of a harbour:

- Breakwaters
- Entrance channel
- Turning circle
- Dock basin
- Berth and berthing structures
- Sand traps

**Breakwaters:**

A breakwater is a structure protecting the shore area, harbour, anchorage or basin from wave disturbance. A breakwater protects the harbour against wave action, which reduces the intensity of wave action. A breakwater may be constructed parallel or perpendicular to the coast to maintain tranquility condition. It can be either a rubble mound or a concrete caisson. It is connected to the shoreline and extends out into the water. Breakwater reduces the intensity of wave action in inshore waters, providing safe harbourage. The breakwater and its types are explained in Chapter 4.

**Entrance channel/navigational channel:**

Waterways should be laid out in proper configuration and designed for good control and safe manoeuvrability of ships under all conditions, including wind, currents, waves, visibility and adverse weather conditions. The entrance channel should be aligned in such a way that the expected design ships can navigate with reasonable safety under adverse conditions of tide, current, wave and wind action. Channels can be classified as unrestricted, semi-restricted and fully restricted.

The width of a restricted channel measured at the bottom of the dredged bed may be taken as the sum of the following three zones:

- The manoeuvring lane (single lane) in a straight channel should be 180% to 200% of the vessel beam and in curved channels widened suitably.

- Bank clearance on each side should be 75% to 150% of the beam of the largest vessel.
- Passing clearance, the distance between adjacent manoeuvring lanes in a two-lane channel, should not be less than the beam of the largest vessel.

The depth of the channel depends on vessel size and draft, number of lanes, type of channel, wind, wave and current action and dredging. The current in the channel section at the harbour entrance should not be too strong or sluggish to maintain the required sediment movement. When the current velocity is more than 2 m/s, it may be necessary to adjust the channel cross-section to maintain an optimum flow.

### Turning circle:

The size or diameter of the turning circle depends on the geometry of the water area available and berth arrangement. The diameter should be a minimum of 1.70 to 2.00 times the length of the largest vessel to be turned with the assistance of tugs as per IS 4651 part 5 (1980) Planning and design of ports and harbours – Layout and functional requirement. The diameter may be four times the length of the design ship without tug assistance. Generally two or more tugs (40T–60T capacity) are used for bigger vessels (greater than 60,000 DWT).

### Dock basin:

The depth of the harbour basin below the chart datum is determined based on design vessel size, tranquility and salinity conditions. The minimum depth should not be less than the loaded draft of the largest vessel and an allowance of 0.60 m to 0.75 m for under keel clearance. Water area requirements are:

- Berthing area
- Passage and manoeuvring area
- Turning basin
- Anchorage and offshore moorings

In the berth pockets adjacent to the berth, the depth should be sufficient with adequate under-keel clearance under all tidal conditions.

### Berth and berthing structures:

The berth is a ship station at an anchor or at a wharf, or alongside a structure. Berth occupancy is the duration over which a ship lies anchored at the berth for loading and unloading operations in a specified period of time. It is usually expressed as a ratio of the actual time the berth is occupied by the ship to the total time available in the specified period.

### Types of Berth:

- Passenger berth
- General cargo berth

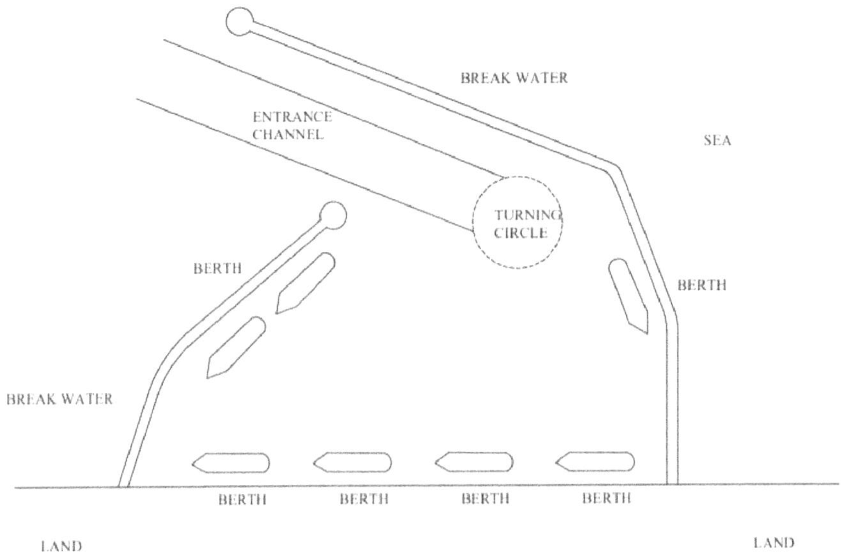


**FIGURE 1.4** Layout of harbour.

- Bulk cargo or ore berth
- Container berth
- Liquid cargo berth

Berthing structures are constructed for the berthing and mooring of vessels to enable the loading and unloading of cargo and for embarking and disembarking of passengers and vehicles. A berthing structure is a facility where the vessel may be safely moored. The different types of berthing structures are explained in Chapter 5. The planning and design of berthing structures depend upon various factors. A berthing structure is an intensive project; hence optimum use of both space and capital by proper planning of the different units of the structure for the present and future demands is to be considered.

The typical layout of a harbour is shown in Figure 1.4, and Figure 1.5 shows the section of berthing structure with ship.

### 1.2.3 Accessibility and Size of Harbour

Accessibility depends on the location of the harbour. The harbour entrance should be designed and located for quick, easy navigation by ships. At the same time, it should be narrow enough not to expose the harbour to the effects of the stormy sea. The entrance is generally placed at the sea, with a passage to the interior of the harbour so arranged as to minimize the effect of rough sea. Thus, the size is determined by accommodation required for convenience manoeuvring, navigation and adaptability to the natural features.

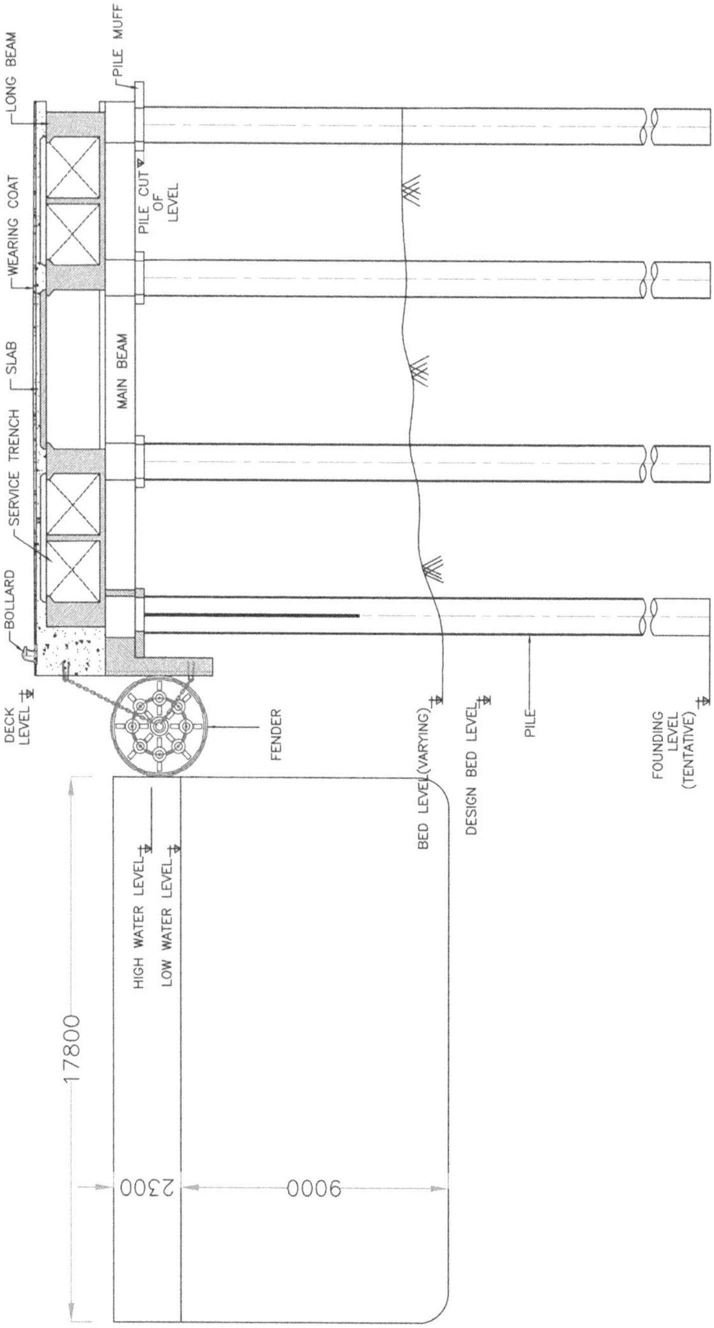


**FIGURE 1.5** Berthing structure.

### 1.2.4 Site Selection

The guiding factors which play a significant role in choosing the site for a harbour are as follows:

- Availability of economic land and construction materials
- Transport and communication facilities
- Natural protection from winds and waves
- Industrial development of the locality
- Sea-bed subsoil and foundation conditions
- Traffic potential
- Availability of electrical energy and freshwater
- Favorable marine conditions
- Defence and strategic aspects

### 1.2.5 Harbour Planning

The following points should be enumerated for planning the harbour:

- It is necessary to carry out a thorough survey of the neighbourhood, including the foreshore and the depths of water in the vicinity.
- The borings on land should also be made to know the probable subsurface conditions on land. It will be helpful in locating the harbour works appropriately.
- The nature of the harbour, whether sheltered or not, should be studied.
- The sea bed profile determines the foundation.
- The problem of silting or erosion of coastline should be carefully studied.
- The natural meteorological phenomena should be studied at the site, especially with respect to intensity/frequency of storms, rainfall, range of tides, maximum and minimum temperature and winds, humidity, direction and velocity of currents.

## 1.3 OCEAN ENVIRONMENT

Some of the natural and meteorological phenomena which primarily affect the location and design of the harbour are given as follows:

- Wind, waves and their combined effect on harbour structures
- Tide and tidal range
- Coastal currents and evidence of silting, including littoral drift, coastal erosion etc.

### 1.3.1 Waves and Wind

The *sea wave* is the most powerful force acting on harbour barriers and against which the engineer must contend. Waves are caused by the wind blowing on the surface of the sea. Operational winds must be specified beyond which the vessel has to be demoored. The extreme wave and wind shall be obtained for a 100-year return period.

### 1.3.2 Tides

Tides on the coastline are caused by the gravitational attraction of the sun and moon and the rotation of the earth. The effect of tides is to raise and lower the water level during stated periods. This apparent variation of sea level is known as the tidal range. In India, most ports have diurnal variation.

Spring tides and neap tides:

- Waves at new and full moon or rather a day or two after (or twice in each lunar month). The tides rise higher and fall lower than at other times, and these are called spring tides.
- One or two days after the moon is in the quarter, i.e., about seven days from new and full moons (twice in a lunar month), the tides rise and fall less than at other times and are then called neap tides.

### 1.3.3 Currents

Currents are caused by winds, temperature differences and the topography of the bottom of the ocean. The current in the channel section at the harbour entrance should not be too strong or too sluggish to maintain the required sediment movement. Strong currents and winds have a significant effect on the alignment, width, depth, turning circles, etc., of the proposed channel and can alter the requirements appreciably.

## 1.4 MARINE SURVEY

Bathymetric and topographic surveys of the sea-floor require the utilization of sonic depth recorders. The echo sounder works by transmitting brief pulsed acoustic signals at a high frequency from a source mounted beneath the ship's hull. Interpretation of echogram characteristics obtained from the simple sonic depth recorders helps to identify and trace depth.

Undersea mapping by side-scan sonar has now developed to the point where techniques can obtain pseudo-photographs of the sea-floor covering more than half a kilometre. Towed side scan sonars suitable for deep-sea conditions are relatively new devices that are more widely used for the more detailed features, obstacles and to evaluate the nodule also.

Pingers and transponders are two essential sonar tools that emit periodical impulses of a fixed frequency and are standard aids to exploration. They are used in connection with towed sampling equipment.

Remote sensing aircraft and satellites, using several fractions of the electromagnetic spectrum, are now capable of measuring sea surface temperature, current boundaries, sea state, sea ice, evidence of sea life, pollution and volcanoes under shallow water at a rate as much as 1000 times faster than a ship.

## REVIEW QUESTIONS

1. Write the classification of vessels and their respective DWT.
2. Define LOA and free board.
3. Describe the permissible motion of vessels.
4. Define DWT and DT. Write the relationship between DWT and GRT for bulk carriers.
5. What is the difference between a port and a harbour?
6. How are harbours classified? Describe with an example.
7. Name the components of a harbour.
8. Write the guiding factors for selection of site for a harbour.
9. Describe in brief about harbour planning.
10. Define spring tides and neap tides.

## REFERENCES

**IS 4651 (Part 1) (2020)** *Code of Practice for Planning and Design of Ports and Harbours, PART I: Site Investigation* (CED 47: Ports and Harbours)

**IS 4651 (Part 3) (2020)** *Code of Practice for Planning and Design of Ports and Harbours, PART III: Loading* (CED 47: Ports and Harbours)

**IS 4651 (Part 5) (1980)** *Code of Practice for Planning and Design of Ports and Harbours, PART V: Layout and Functional Requirements* (CED 47: Ports and Harbours)

**PIANC WG 24 (1995)** *Criteria for Movements of Moored Ships in Harbours*

**PIANC WG 33 (2002)** *Guidelines for the Design of Fender Systems*

**PIANC WG 49 (2014)** *Horizontal and Vertical Dimensions of Harbour Approach Channels for Container and Bulk Ships*

**PIANC WG 211 (2020)** *Mooring of Large Ships at Quay Walls*

**Carl A. Thoresen (2014)** *Port Designer Handbook* (3rd Edition), ISBN 978-0-7277-6004-3, http://dx.doi.org/10.1680/pdh.60043.001

# 2 Indian Codes

This chapter mainly deals with the Indian Standard codes related to port and harbour structures. The information and detailed explanation of the codes for designing port and harbour structures are given in this chapter. The recommended codes and standards are given below.

| | |
|---|---|
| IS 1893 (Part 1) (2016) | Criteria for Earthquake Resistant Design of Structures |
| IS 4651 (Part 1) (2020) | Code of Practice for Planning and Design of Ports and Harbours – Site Investigation |
| IS 4651 (Part 2) (2020) | Code of Practice for Planning and Design of Ports and Harbours – Geotechnical Engineering |
| IS 4651 (Part 3) (2023) | Code of Practice for Planning and Design of Ports and Harbours – Loadings |
| IS 4651 (Part 4) (2020) | Code of Practice for Planning and Design of Ports and Harbours – General Design Considerations |
| IS 4651 (Part 5) 1980 | Code of Practice for Planning and Design of Ports and Harbours – Layout and Functional Requirements |
| IS 2911 (Part 1/Sec 2) (2010) | Code of Practice for Design and Construction of Pile Foundations – Bored Cast In-Situ Piles |

Other IS Codes used for design are IS 456 (2000) Code of Practice for Plain and Reinforced Concrete, SP16 – Design Aids to IS 456, SP34 – Handbook on Concrete Reinforcement and Detailing, and IS 800 (2007) Code of Practice for General Construction in Steel.

## 2.1 IS 4651 – CODE OF PRACTICE FOR PORT AND HARBOUR STRUCTURES

This code of practice for port and harbour structures consists of five parts. Part 1 is for site investigation, Part 2 is for geotechnical engineering, Part 3 is for loadings, Part 4 is for general design considerations, and Part 5 is for layout and functional requirements. All these codes are given in detail in this section.

### 2.1.1 IS 4651 (Part 1) (2020) – Code of Practice for Planning and Design of Ports and Harbours (Site Investigation)

This Part I deals with site investigation and collection of data necessary for the planning, design and construction of marine structures of ports and harbours. The information required is grouped under the following headings as per IS 4651(Part 1) (2020).


DOI: 10.1201/9781003361275-2

#### 2.1.1.1 General

A brief description of the site, including its latitude, longitude, geographical location, accessibility, etc., along with a map should be given. The historical background of the area, purpose of the project, type of hinterland and communications facilities available may also be briefly described.

Wherever possible, advantage should be taken of existing local data on tides, storms, waves, littoral drifts, mud banks, etc. and records of previous investigations in the vicinity and information compiled. The behaviour of existing structures which may be of similar nature to the ones proposed and the influence of the soil and water on the materials of construction should be studied and recorded.

The primary data collection is grouped into the following categories:

- Survey and investigations
- Met-ocean observations

#### 2.1.1.2 Survey

Topography survey maps of scale 1:50,000 and contour intervals of 20 m for general planning and for detailed planning, 1:5000, 1:2500 and 1:1250 with a contour interval of 1 m are required. Survey maps for detailed planning shall be of scale less than 1:5000 with a contour interval of 1 m, depending on the topography of the area.

Hydrographic survey charts required for detailed planning shall not be less than 1:5000. The recommended scales are 1:2500 and 1:1250.

The choice between the selection of multi-beam echo sounding (MBES) or single-beam echo sounding (SBES) is determined by economic considerations. Where the highest possible coverage is needed, such as in dredging, rocky bottom, pipe routes, etc., the multi-beam survey is preferable.

#### 2.1.1.3 Meteorological Data

*Winds*: For preliminary studies, information may be obtained from the available meteorological records of the area. Data on the velocity and direction of the wind at the proposed site shall be obtained by installing continuous and self-recording anemometers. The data shall be collected for at least a period of 1 year and shall also be correlated with the data available at places nearest to the site.

*Cyclones*: Details of the velocity of maximum winds, radius of maximum wind velocity, duration, pressure drop at the cyclone centre and speed of movement of cyclone centre are required. From this, a design cyclone is adopted and waves that could be incident at a place are computed.

*Rainfall*: The annual average rainfall, months at which the maximum rainfall occurs, maximum intensity of rainfall and its duration, and average number of wet days in a year should be collected from the India Meteorological Department for a minimum period of 3 years.

*Relative humidity*: Data on the maximum, mean and minimum relative humidity for every month shall be obtained for a minimum period of 5 years.

*Temperature*: The normal ambient air temperatures with emphasis on daily and seasonal variation shall be collected from the India Meteorological Department.

*Barometric pressures*: Data on monthly average barometric pressures should be collected for the nearest site from the India Meteorological Department.

#### 2.1.1.4 Oceanographic Data

Considering the time requirements and logistical constraints for oceanographic data for large projects, the preliminary assessment and feasibility reports may be prepared with secondary data. Reports for statutory clearance purposes (for example, detailed project report (DPR), environmental impact assessment (EIA), etc.) shall make use of validated numerical model results, while a comprehensive observation program shall be set in motion at the start of these activities.

The numerical model after due calibration with field observation shall be used to derive various design parameters and evaluate different scenarios of outcome caused by the project elements on the environment. The components of the model and its details are presented in Table 2.1.

**TABLE 2.1**
**Numerical Simulation Model Components**

| Sl. No. | Components | Input | Output | Purpose |
|---|---|---|---|---|
| 1 | Hydrodynamics model | Bathymetry, tide, currents at boundaries | Elevation and current field for an entire domain | a) Basic component to enable other components<br>b) Define elevation and current variation at any location within domain<br>c) Inundation on coast |
| 2 | Wave model | Wave (random/regular) at boundaries | Wave parameters like Hs, Tp, mean direction | a) To derive design wave height, direction and period at any location<br>b) To derive a sediment transport model<br>c) To derive design water levels<br>d) Wave setup, run up<br>e) Hind casting wave parameters to derive return period statistics |
| 3 | Sediment transport model | Sediment characteristics grain size, critical shear stress, suspended and bed load | Sediment transport rate in the domain | Morphological evaluation of sea bed and coast with and without the project elements |
| 4 | Water quality model | Water quality parameters and pollutants | Dispersion characteristics | Evaluate the environmental changes due to project elements |

NOTE: Information as given in 'Handbook of Cyclonic Storms', 'Tracks of Storms and Depressions in the Bay of Bengal and Arabian Sea – 1964' and 'Synoptic Charts', issued by the India Meteorological Department, may also be referred.

Oceanographic data to be collected should cover the following:

1. Tides
2. Waves (wind waves and swells)
3. Storm surges
4. Currents
5. Salinity
6. Sea water temperature
7. Suspended load

*Tides*: Record of the tidal information over as long a period as possible from port authorities or the Geodetic and Research Branch, Survey of India or the hydrographic department of the Indian Navy, including any local information specific to the site of the works should be obtained. Data over a full metonic cycle of 19 years will be useful. For most ports, information on tides is provided by Indian Tide Tables published by Survey of India, Dehradun. For a new place, tide tables could be predicted and furnished by the Survey of India on request; at least two years of observed records should be studied.

For new port structures, 2 levels expected at a particular place may be required. For this purpose, tides for at least 12 months shall be observed as described and a correlation established between the observed tides and the predicted tides for the harbour in general.

For greenfield projects with navigational requirements, tidal observations shall be conducted at least for a 1-year cycle.

*Waves*: For planning and preliminary design purposes, secondary sources like wave atlases published by agencies like CSIR–National Institute of Oceanography (CSIR-NIO) National Institute of Ocean Technology (NIOT) may be used.

Wave heights can also be computed by hind casting studies using the storm data and synoptic charts from the India Meteorological Department. Wave statistics shall be derived from models that are validated with field observed wave information for a short time (typically 1 month each in monsoon and non-monsoon periods).

Separate wave observations shall be carried out for long period and short period waves.

Sample of single parameter and joint distribution of wave statistics are given in Tables 2.2 and 2.3. Long period waves affect the harbour seiches/resonance, long period drift and mooring loads, wave set-up near shore, etc. Figure 2.1 shows the significant wave height and mean wave direction.

*Storm surges:* Storm surges may be inferred from tidal gauges if the gauges had functioned through the period of the storm. Alternatively, a numerical simulation of storm surge shall be carried out for a nearby cyclone event and calibrated with known/observed tide gauge data.

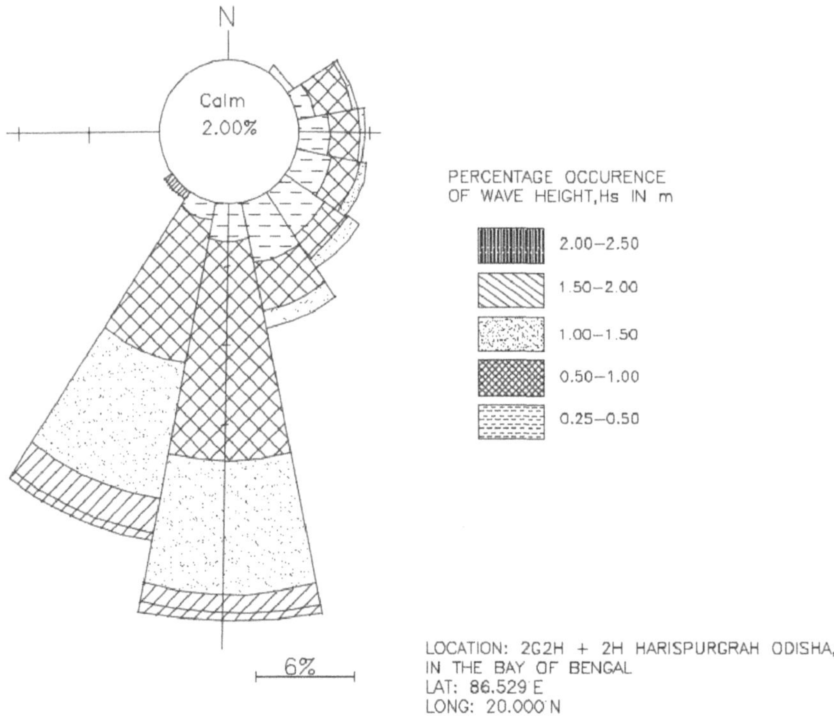


**FIGURE 2.1** Significant wave height and mean wave direction.

## TABLE 2.2
## Single Parameter Distribution

| | Significant Wave Height (percent) | | | | | | | | | | | |
|---|---|---|---|---|---|---|---|---|---|---|---|---|
| Hs (m) | Jan. | Feb. | Mar. | Apr. | May | June | July | Aug. | Sep. | Oct. | Nov. | Dec. |
| 0.0–0.50 | 61.50 | 51.43 | 21.40 | 2.31 | | | | | 1.92 | 15.88 | 47.82 | 58.25 |
| 0.50–1.00 | 38.46 | 47.00 | 65.69 | 56.67 | 37.41 | 23.85 | 32.26 | 44.29 | 60.90 | 70.84 | 48.08 | 38.77 |
| 1.00–1.50 | | 1.57 | 12.28 | 33.72 | 51.05 | 57.24 | 49.69 | 43.98 | 31.09 | 10.05 | 2.95 | 2.30 |
| 1.50–2.00 | | | 0.62 | 6.92 | 10.61 | 15.32 | 15.63 | 9.06 | 4.49 | 2.48 | 0.90 | 0.68 |
| 2.00–2.50 | | | | 0.38 | 0.93 | 2.82 | 1.99 | 2.23 | 1.28 | 0.68 | 0.26 | |
| 2.50–3.00 | | | | | | 0.45 | 0.43 | 0.25 | 0.32 | 0.06 | | |
| 3.00–3.50 | | | | | | 0.32 | | 0.19 | | | | |
| 3.50–4.00 | | | | | | | | | | | | |
| 4.00–4.50 | | | | | | | | | | | | |
| 4.50–5.00 | | | | | | | | | | | | |
| >5.00 | | | | | | | | | | | | |

**TABLE 2.3**
**Multi-Parameter Distribution (Hs and Tz)**

| Sl. No. | Tz (s) | Significant Wave Height Versus Wave Period (percent) Significant Wave Height, Hs (m) | | | | | | | | | |
|---|---|---|---|---|---|---|---|---|---|---|---|
| | | <0.50 | 0.50–1.00 | 1.00–1.50 | 1.50–2.00 | 2.00–2.50 | 2.50–3.00 | 3.00–3.50 | 3.50–4.00 | >4 | Percent |
| i) | < 02 | 0.03 | | | | | | | | | 0.03 |
| ii) | 02–03 | 4.60 | 4.42 | | | | | | | | 9.02 |
| iii) | 03–04 | 10.11 | 25.09 | 7.59 | 0.03 | | | | | | 42.82 |
| iv) | 04–05 | 5.54 | 10.90 | 13.57 | 4.31 | 0.23 | | | | | 34.55 |
| v) | 05–06 | 1.10 | 5.15 | 2.33 | 1.01 | 0.61 | 0.12 | | | | 10.31 |
| vi) | 06–07 | 0.17 | 1.26 | 1.08 | 0.14 | 0.04 | 0.01 | 0.04 | | | 2.74 |
| vii) | 07–08 | 0.02 | 0.19 | 0.19 | 0.09 | 0.01 | | | | | 0.49 |
| viii) | 08–09 | 0.01 | | | 0.01 | | | | | | 0.02 |
| ix) | >09 | | | | | | | | | | |
| **Percentage** | | 21.58 | 47.02 | 24.76 | 5.59 | 0.89 | 0.13 | 0.04 | | | 100.00 |

*Currents*: The direction, magnitude and duration of current during complete tidal cycles at maximum spring and neap tide over a year should be recorded. In riverine ports where there is fresh water discharge, the current pattern at the highest expected flood should be assessed and recorded for a period of at least 1 year for the purposes of alignment of berth, dock entrances, moorings, etc. Current readings should be taken at a minimum of three points, preferably at depths of 0.10 d, 0.50 d and 0.90 d, where d is the depth of water recorded.

*Sea bed*: Classification of sea bed material in the vicinity of structures and approaches end up to an area in the sea where the depth is more than the maximum depth for which the harbour is being designed, should be ascertained and recorded. Sea bed slope, which is also a design parameter, may be ascertained from hydrographic survey charts.

*Suspended load, salinity and temperature*: These observations shall be carried out both during the dry and wet seasons at different locations in the harbour and channels over the full tidal cycle during neap and spring tides. The suspended load and salinity shall be measured at depths of 0.10 d, 0.50 d and 0.90 d below the water level (where d is the depth of water) at every hour during the tide cycle. Salinity and suspended load should be assessed from the water samples collected at these depths. Salinity may also be measured at the site by using direct meters which are available for this purpose. Sea water temperature shall also be measured during these observations. The direction of drifts at the site may be ascertained from a validated numerical wave model and observations of the coast line in the vicinity of the structure, such as dredging records, shoreline changes, accretion erosion, wave data and the orientation of river mouths. These studies may be conducted up to a location where the depth of water is 12 m, which is more than the design depth.

Estimates of littoral drift and morphological changes in the project vicinity shall be assessed using relevant numerical modelling techniques.

#### 2.1.1.5 Geological Data

Any detailed published information on the geological condition of the area of the project, if available, should be carefully studied. A combination of shallow seismic surveys and borehole investigations shall be conducted to ascertain the nature of sub-sea strata.

The geophysical survey method of locating base rock aims at giving a continuous record of strata. For general purposes (for example, port elements, breakwaters, dredge channel), information on the nature of the material for at least 30 m below sea bed shall be acquired. Depending on site conditions and specific structural requirements (for example, pile foundation), greater depths may have to be probed. The geophysical survey by seismic refraction method may also be resorted to find out more reliable information about the strata.

Stratification survey by high-resolution seismic profiling (known commonly as a sparker survey) may be suitably used to determine the thickness of the various sediment layers and disposition of underlying bedrock. Other surveys such as side scan survey and magnetic survey are also used for the collection of geological data.

The following data about the geology of the area should be compiled:

1. Type of bedrock, including information on its origin and method of formation
2. Any faults, fissures, folds and other nonconformities in the area of the project
3. Crushing strength and other properties of the rock in the project area and its suitability for use in marine works

#### 2.1.1.6 Soil Investigation

In a site which has been partially developed, enquiries should be made regarding the past structures' layout of pipes and obstructions likely to be met in the area for new works. Enquiries should also be made regarding old creek lets, excavated pits, etc., which might have either silted up or reclaimed. This information will be particularly useful in deciding the number and location of trial pits and borings.

The spacing of borings required for subsurface investigation depends upon the type of structure and the nature of the site. The number of bore holes should be sufficient to give the probable variation in the subsurface strata over the site, and their depth should be such as to include all strata likely to affect the stability of the structure.

Recommended maximum separation between boreholes for different purposes are as follows:

Navigation channels: three rows, one at the centre of the channel and one on either edge of the channel and every 500 m along a channel.
Breakwater: 500 m (subject to a minimum of three).
Harbour basin: 500 m (or four numbers per $km^2$).

Tests like the standard penetration test and the cone penetration test may be conducted. The subsoil investigation should be carried out generally according to IS 1892 (2021) Subsurface Investigation for Foundations. The main borings may be along the top edge of the shore and are spaced 50 m apart and taken up to a depth of 3 m into hard strata or a depth equal to twice the difference in the elevation of the ground surface on either side of the structure. In a few cases, the borings may be taken deeper to investigate the nature of the underlying strata. It is desirable to use large diameter bores in reclaimed areas and where embedded boulder layers are encountered, depending on the size of the project.

### 2.1.2 IS 4651 (Part 2) (2020) – Code of Practice for Planning and Design of Port and Harbours (Geotechnical Engineering)

#### 2.1.2.1 Definitions

##### Angle of Wall Friction

The roughness of retaining walls causes the active earth pressure and passive earth resistance to be inclined. The angle of this inclination to the normal to the wall is known as the angle of wall friction.

**Arching earth pressure**

When the top of the wall is restricted, and there is freedom of movement towards the base, the earth pressure exerted by the backfill is known as arching earth pressure. For movement away the backfill, it is termed arching active earth pressure. For movement toward the backfill, it is termed arching passive earth resistance.

**Earth pressure**

The pressure is exerted by the backfill on the wall. The nature and extent of the wall movement alter the lateral pressure of the backfill exerted on it. If there is no wall movement, the pressure exerted by the backfill is termed as earth pressure at rest. If the wall moves sufficiently away from the backfill by translatory motion or rotation about the base of their combination, the lateral pressure of the backfill is reduced and is termed as active earth pressure. The movements required are small, and about 0.001 H for sand and 0.05 H (H is the height of the wall) for clay are sufficient to mobilize the active pressure. If the wall moves sufficiently towards the backfill, by translatory motion or rotation about the base of their combination, the backfill offers resistance which is termed passive earth resistance or passive earth pressure.

**Retaining wall**

A wall which holds back a soil mass is a retaining wall. The soil mass is referred to as backfill.

### 2.1.2.2 Wall Movements

In waterfront structures, wharves constructed as gravity retaining walls have sufficient outward movement to mobilize active pressure, and Coulomb's pressure distribution obtains/develops.

Wharves constructed as sheet-pile retaining walls, either dredged in the front or backfilled, sufficiently yield to justify the adoption of Coulomb's active pressure distribution. In highly plastic clays, pressures approaching at rest condition may develop unless wall movement can constitute with time.

Walls in the locks and in the dry docks, being monolithically constructed with the floors, may not yield, and earth pressure at rest in these cases should be taken.

Gravity wharf walls funded on cohesionless subgrade may not yield forward sufficiently to mobilize full passive resistance. Corresponding to the yield for active pressure, not more than half of the passive resistance may be available.

### 2.1.2.3 Earth Pressure at Rest

The intensity of earth pressure at rest for a horizontal backfill at a depth z is given by:

$$p_0 = K_0 \gamma Z$$

The field values of $K_0$ are given in Table 2.4.

**TABLE 2.4**
**Field Values of Coefficient of Earth Pressure at Rest, $K_o$**

| Sl. No. | Soil | $K_o$ |
|---|---|---|
| 1 | Loose sand | 0.40 |
| 2 | Dense sand | 0.50–0.60 |
| 3 | Sand, well tamped | 0.80 |
| 4 | Soft clay | 0.60 |
| 5 | Hard clay | 0.50 |

### 2.1.2.4 Active Earth Pressure

In waterfront structures, the backfill is horizontal. With uniform surcharge intensity and uniform backfill, the active earth pressure varies for the type of soil.

The formulae and procedure for calculating active earth pressure are described in Chapter 3.

### 2.1.2.5 Passive Resistance

When a retaining wall is pushed out due to the active thrust of the backfill and the surcharge thereon, partial or full passive resistance will be mobilized in the front soil.

Waterfront structures are usually vertical in the front, and the passive resistance varies for different soil types. The formulae and procedure for calculating passive earth pressure are described in Chapter 3.

### 2.1.2.6 Stability Factor for Soil in the Backfill

The stability factor for the retaining structure to retain backfill soil shall be calculated using the expression:

$$Stability\ factor = \frac{Total\ passive\ force}{Total\ active\ force + Differential\ water\ force}$$

This stability factor should be greater than 2.5 for the retaining structure to be stable.

## 2.1.3 IS 4651 (Part 3)-1 (2020) – Code of Practice for Planning and Design of Port and Harbours (Loading)

This standard (Part 3) deals with the loading on waterfront structures. It covers vertical imposed loads, horizontal forces due to berthing, bollard pulls, wave forces, currents and winds reference is given to earthquake forces.

### 2.1.3.1 Dead Load

All dead loads of and on structures relating to docks and harbours should be assessed and included in the design.

### 2.1.3.2 Imposed Load

**Vertical imposed loads:**

Surcharges due to stored and stacked material, such as general cargo, bulk cargo, containers and loads from vehicular traffic of all kinds, including trucks, trailers, railway, cranes, containers handling equipment and construction plant constitute vertical imposed loads.

*Truck loading and uniform loading*: the berths shall be designed for truck loading and uniform loading as given in Table 2.5.

*Crane loads*: concentrated loads from crane wheels should be considered. An impact of 25% shall be added to wheel loads in the normal design of deck-hand stringers, 15% where two or more cranes act together, and 15% in the design of pile caps and secondary framing members. For other specialized mechanical handling equipment like reach stacker, fork lift and mobile harbour crane, etc., impact effect shall be considered as per manufacturer's specification.

For the geotechnical and structural design of piles, sheet piles and diaphragm walls, no impact shall be considered.

*Railway loads*: concentrated wheel loads due to locomotive wheels and wagon wheels in accordance with the specification of the Indian Railways for the type of gauge and service at the locality in consideration.

For impact due to trucks and railways, one-third of the impact factors specified in the relevant standards may be adopted.

*Special loads*: special loads like pipeline loads or conveyor loads, or exceptional loads, such as surcharge due to ore stacks, transfer towers, heavy machinery or any other type of heavy lifts, should be individually considered.

When the imposed loads act on the fill behind the structure, such as in a sheet pile wharf so that the loads are transmitted to the structure through increased earth pressure, the retaining structure may be designed for a uniformly distributed equivalent surcharge of half the value given in column 3 of Table 2.5. In cases of higher load intensity, the actual value of the surcharge may be taken.

**TABLE 2.5**
**Truck Loading and Uniform Loading**

| Function of Berth | Truck Loading (IRC Class) | Uniform Vertical Live Loading ($t/m^2$) |
|---|---|---|
| Passenger berth | B | 1.00 |
| Bulk unloading and loading berth | A | 1.00–1.50 |
| Container berth | A or AA or 70 R | 3.00–5.00 |
| Cargo berth | A or AA or 70 R | 2.50–3.50 |
| Heavy cargo berth | A or AA or 70 R | 5.00 or more |
| Small boat berth | B | 0.50 |
| Fishing berth | B | 1.00 |

**Container Crane:**

The container terminal shall be designed to accommodate the largest vessel that can possibly arrive at the berth, considering the future growth of the port and access facilities unless specifically restricted by the owner of the facility. The container crane size and wheel spacing shall also be determined based on the size of vessels arrived at. The wheel loads shall be based on the data of the manufacturer of such cranes. Normally, a set of manufacturer's data would have been available by the time the structural design of the wharf begins, and the information shall be based on this data. Crane operating conditions shall consider load due to wind of 20 m/s in one of the four grid directions. When there are multiple cranes operating in a wharf, they should be considered to operate with a minimum spacing of 5 m between extreme wheels. Separate loading combinations shall be considered for parking the crane during extreme wind conditions.

The crane, when loaded from the barge, shall be on a properly designed skid area on the jetty. The centre line of the seaside crane line shall be more than 3 m from the face of the berth to avoid damage to the crane due to the rolling of vessels.

**Mobile harbour crane:**

If it is decided to have mobile cranes on the specific wharf, the numbers, type and capacity shall be finalized in the planning stage. The load data shall be considered from manufacturer's information. The design shall consider adequate load conditions to ensure the safety of the structure as per this code of practice for all scenarios of loading.

The pad loading from mobile harbour cranes shall be considered for the design of the berth, especially for punching shear and bending, since the intensity due to the pad loading can be about 100 to 200 kN/m$^2$.

For detailed engineering load data and wheel mobile harbour crane, specification of crane manufacturers shall be sought.

Berthing force, mooring force and differential water pressure are described in Chapter 3.

### 2.1.3.3 Earthquake Forces

In areas susceptible to seismic disturbance, a horizontal force equal to a fraction of the acceleration of gravity times the weight applied as its centre of gravity should be taken. The fraction will depend upon the likely seismic intensity of the area and shall be taken in accordance with IS 1893 (Part 1) (2016). The weight to be used is the total dead load plus one-half of the imposed load.

### 2.1.3.4 Forces due to Current

Pressure due to current will be applied to the area of the vessel below the water line when fully loaded. It is approximately equal to $wv^2/2g$ per square metre of area, where v is the velocity in m/s and w is the unit weight of water in tonnes/m$^3$. The ship is generally berthed parallel to the current. With strong currents and where berth alignment materially deviates from the direction of the current, the likely force should be calculated by any recognized method and considered.

#### 2.1.3.5 Wave Forces

It is well recognized that similar wave condition gives rise to different wave forces depending on the form of wave breaking and distance from the structure. The theory used depends on whether the structure is subjected to non-breaking (pulsating), impulse breaking (impact) or broken wave condition. The method of estimation of wave forces on structure depends on wave theories and the type of the structure.

The significant wave height shall be taken from the following relationship:

$$\frac{H_m}{H_s} = 1.7 \text{ to } 1.9$$

where

$H_m$ = maximum wave height
$H_s$ = significant wave height

The three types of waves are as follows:

*Non-breaking Waves*

When the depth of the water against the structure is greater than 1.5 times the maximum expected wave height, non-breaking wave conditions occur. Forces due to non-breaking waves are essentially hydrostatic. 'Sainflou method' may be used for the determination of pressure due to non-breaking waves. Goda's method is widely used for shallow or intermediate water for non-breaking and breaking waves (average force) on monolithic and composite structures.

*Breaking Waves*

Breaking waves cause both static and dynamic pressures. Determination of the design wave for breaking wave conditions may be based on the depth of water about seven breaker heights, seaward of the structure, instead of the water depth at which the structure is located. The actual pressure caused by a breaking wave is obtained by following the method suggested by Minikin.

*Broken Waves*

Locations of certain structures like protective structures will be such that waves will break before striking them.

Wave forces on vertical cylindrical structures such as piles are described in Chapter 3.

#### 2.1.3.6 Wave Condition for Design of Breakwater

The depth limited wave breaking shall be considered for shallow water to estimate significant wave height *H*s (0.5 times the design water depth at the seaside toe of the breakwater), whereas the significant wave height with a return period of 100 years is to be considered for intermediate and deep water. A minimum height of 0.5 m shall be provided above wave run-up + storm surge + high tide level for fixing the crest elevation.

The significant wave height shall be multiplied by the following factor to obtain the design wave height for a different types of structures:

The operating wave condition during which the construction will be carried out shall be specified, and suitable protection under armour shall be part of the design.

#### 2.1.3.7 Open Sea Jetty

The wave force shall be calculated based on the maximum wave height with an appropriate return period, whereas significant wave height shall be considered along with run-up for large diameter piles to fix the top level. High tide level along with storm surge/tsunami shall be accounted using independent or associated method with about 1.5 m air gap.

The recommended return period is 100 years for wave force estimation and 1000 years for fixing the top level.

#### 2.1.3.8 Wind Forces

Wind forces on structures shall be taken in accordance with IS 875 (Part 3) (2015) as applicable.

#### 2.1.3.9 Combined Loads

The combination of loadings for design is dead load, vertical imposed loads, plus either berthing load, or line pull, or earthquake or wave pressure. If the current and alignment of the berth are likely to give rise to line pull in excess of that given in Table 2.6, provisions for such extra pull in a combination of likely wind should be made. The worst combination should be taken for design.

The mooring line force due to oscillations of vessels either due to wave and current in an indirect direction or bypassing vessel nearby shall be considered. The design of bollards shall be safe till the mooring line breaks, and the structure shall be safe till the bollard fails.

### 2.1.4 IS 4651 (Part 4) (2023) – Code of Practice for Planning and Design of Port and Harbours (General Design Considerations)

This standard (Part 4) lays down the general design considerations for dock and harbour structures. The information required is grouped under the following headings as per IS 4651 (Part 4) (2014) (Fourth Revision).

#### 2.1.4.1 Loads, Forces and Stresses

The loads, forces and stresses to be considered in designing the structures are the following:

1. Dead load (DL)
2. Vertical live load
3. Impact or dynamic effect of live load
4. Forces caused by the tractive effort or breaking of vehicles, cranes, ship loader/unloader and material handling equipment
5. Centrifugal forces of vehicles moving on curve

6. Earth pressure
7. Hydrostatic and hydrodynamic forces
8. Berthing forces from vessels
9. Mooring forces
10. Forces due to wind
11. Secondary stresses (stresses due to shrinkage, creep, temperature, etc. as applicable)
12. Erection stage stresses
13. Live load (LL)
14. Seismic forces or earthquake loads
15. Tsunami force

The above-mentioned loads, forces and stresses should be worked out based on provisions in IS 4651 (Part 3) (1974) and other relevant Indian Standards.

#### 2.1.4.2 Combination of Loads, Forces and Stresses

All members shall be designed to sustain safely the effect of the combination of various loads, forces and stresses (specified in 2.1.4.1) that can possibly coexist. All calculations shall distinctly tabulate the various combinations of the above loads and stresses covered by design. The load combinations given in Tables 2.6 and 2.7 should be considered.

Crane/machine operating loads along with dead load, combined with wind load specified by the manufacturer or the normal wheel load with maximum wind load as specified in IS 875 (Part 3) (1987), whichever is more severe, should be taken for design purposes.

As the berthing of the vessel is done in relatively calm conditions, the berthing forces may not be considered to occur simultaneously with crane/machine operating load and wind loads or seismic forces. Wind load and seismic forces need not act simultaneously. Seismic need not be combined with erection stage stresses. Tsunami forces need not be combined with erection stage stresses.

#### 2.1.4.3 Method of Design

Structures and structural elements may be designed by the following methods:

1. Limit state method
2. Working stress method

Reinforced Cement Concrete (RCC) and pre-stressed concrete members can be designed by any of the above two methods, whereas for designing structures with other materials, working stress method should be adhered to.

*Limit State Method*

In the limit state method of design, a structure is considered unfit for use when it reaches a particular state called the limit state, at which it ceases to fulfil the function or satisfy the conditions for which it is designed. The structure shall be designed to withstand safely all loads liable to act throughout its life and it shall also satisfy the serviceability requirements, such as limitations on deflection and cracking, etc. All relevant limit states shall be considered in design to ensure an adequate degree of safety and serviceability. In general, the structure shall be designed based on the most critical limit state and shall be checked for other limit states.

*Partial Safety Factors for Loads*

Under various combinations of loading for use in working stress design, the values for the partial safety factor in Table 2.6 shall normally be used.

*Partial Safety Factors for Material Strength*

When assessing the strength of a structure or structural member for the limit state of collapse, the values of partial safety factors shall be taken as 1.5 for concrete and 1.15 for steel.

*Working Stress Method*

Working stress method or permissible stress method may also be adopted in the designs till such time the complete changeover to the limit state method is made in other relevant Indian Standards on the subject. However, as the limit state method is more rational and adoptable, the designs may be carried out by the limit state method. Limit state method affords easy calculation in the case of circular columns, which involve a complicated procedure under the working stress method.

#### 2.1.4.4 Permissible Stresses

The following permissible stresses shall be adopted for the working load method of design. When the load items mentioned in 2.1.5.1(a) to 2.1.5.1(e) are considered for possible appropriate combination, the following standard shall be applicable, subject to the stipulations contained in an increase in permissible stresses below.

1. For reinforced concrete structural members, the permissible stresses shall be as specified in IS 456 (2000).
2. For pre-stressed concrete structural members, the permissible stresses shall conform to those specified in IS 1343 (2012).
3. For structural steel members, the permissible stresses shall be as specified in IS 800 (2007).
4. For timber structural members, the permissible stresses shall be as specified in IS 883 (1970).
5. For timber or concrete piles, the safe bearing capacity and/or factor of safety shall be in accordance with IS 2911 (Part 2) (2010) and IS 2911 (Part l/Secs 1–3) (2010).

**Increase in Permissible Stresses**

The permissible stresses in normal loading may be exceeded up to the limits as indicated in Table 2.7 provided in no case does the stress exceed the yield stress or 0.20% of the proof stress of the material concerned.

**TABLE 2.6**
**Partial Load Factors for Loads in Limit State Design**

| | | Partial Load Factor | | | | | |
|---|---|---|---|---|---|---|---|
| | | Limit State of Serviceability | | Limit State of Collapse | | | |
| Sl. No. | Loading | Short Term | Long Term | Normal | Extreme/ Survival | Temporary | Reversal |
| 1 | Dead load (DL) | 1.00 | 1.00 | 1.50 | 1.20 | 1.20 | 0.90 |
| 2 | Vertical live load | | | | | | |
| | a) Impact or dynamic load ($D_yL$) | 1.01 | 0.05 | 1.50 | 1.20 | 1.20 | 0.09 |
| | b) Static (LL) | 1.00 | 0.50 | 1.50 | 1.20 | 1.20 | 0.90 |
| 3 | Earth Pressure (EP) | 1.00 | 1.00 | 1.20 | 1.00 | 1.00 | 1.00 |
| 4 | Hydrostatic force ($H_yF$) | 1.00 | – | 1.00 | 1.00 | 1.00 | 1.00 |
| 4 | Wave and current force (WL and CL) | 1.00 | – | 1.20 | 1.00 | 1.00 | 1.00 |
| 6 | Berthing force (BF) | 1.00 | – | 1.50 | 1.00 | – | 1.50 |
| 7 | Mooring force (MF) | 1.00 | – | 1.50 | – | – | 1.50 |
| 8 | Working wind force ($WW_iF$) | 1.00 | – | 1.00 | – | – | – |
| 9 | Extreme wind force ($EW_iF$) | – | – | – | 1.20 | – | 1.50 |
| 10 | Shrinkage | – | 1.00 | – | – | – | – |
| 11 | Creep | – | 1.00 | – | – | – | – |
| 12 | Temperature (Temp L) | – | 1.00 | – | – | – | – |
| 13 | Seismic force (EL) | 1.00 | – | – | 1.20 | – | 1.50 |
| 14 | Tsunami force (TL) | – | – | – | 1.20 | – | – |
| 15 | Secondary stresses (SS) | 1.00 | – | – | – | – | – |
| 16 | QRMH capacity (QC) | – | – | – | 1.0 | – | – |

**NOTES:**

1. If dead load alone is only combined with wind/seismic force, the partial load factor of ±1.5 shall be considered for wind/seismic forces under normal operating conditions similar to IS 456 and IS 1893 (Part 1).
2. Minimum wind speed for normal loading shall be 60 km/h.
3. DL + 50% live load shall be considered for seismic force estimation.
4. Importance factor of 1.50 shall be taken for waterfront structures for seismic design.

5. Tsunami load shall be considered in the design of structures with additional scour due to tsunami.
6. Additional weight due to marine growth shall be considered in dead load case. The weight of marine growth shall be calculated from its density and thickness of the marine growth around the structure. As a general guideline, it is to be taken as 100 mm, if the location is known to have higher marine growth.
7. Limit state of serviceability load combinations shall be given as below:
   a. Transient load combination for crack width check only
      i. 1.00(DL) + 1.10(DyL) + 1.00(LL + EP + HyF + WL + CL + BF + WWiF + SS)
      ii. 1.00(DL) + 1.10(DyL) + 1.00(LL + EP + HyF + WL + CL + MF + WWiF + SS)
   b. Sustained load combination for crack width check
      i. 1.00(DL) + 0.5(LL) + 1.0(EP)
   c. Short-term effect for deflection check near expansion joint only
      i. 1.00(DL) + 1.10(DyL) + 1.00(LL + EP + HyF + WL + CL + EL + SS)
8. Horizontal deflection check for long term effect shall be limited to L/250, where L shall be from point of zero deflection below dredge level till top idealized level.
   a. 1.0(DL) + 0.5(DyL + LL) + 1.0(EP + Temp) + 1.0OWL
   b. Exclude CL with temp.

   For shrinkage, creep and temperature loads, IS 456 and IS 875 (Part 5) may be referred.
9. For normal loading, berthing energy shall be considered for BF calculation, whereas for extreme loading, ultimate berthing energy shall be considered for BF calculation.
10. Limit state of collapse load combinations
    a. Normal load condition
       i. 1.50(DL + DyL + LL) + 1.20(EP) + 1.00(HyF) + 1.20(WL + CL) + 1.50(BF) + 1.00(WWiF)
       ii. 1.50(DL + DyL + LL) + 1.20(EP) + 1.00(HyF) + 1.20(WL + CL) + 1.50(MF) + 1.00(WWiF)
    b. The seismic load combination in limit state design of RCC and Pre-stressed concrete structures are
       i. 1.50(DL + IL)
       ii. 1.20(DL + IL ± EL)
       iii. 1.50(DL ± EL)

       where IL is imposed load such as weight of stairways, handrails, equipment reactions and super structures etc.
    c. Extreme/survival load condition
       i. 1.20(DL + DyL +LL) + 1.00(EP) + 1.00(HyF + WL + CL) + 1.00(BF)
       ii. 1.20(DL + DyL +LL) + 1.00(EP) + 1.00(HyF + WL + CL) + 1.20(EL)
       iii. 1.20(DL + DyL +LL) + 1.00(EP) + 1.00(HyF + WL + CL) + 1.20(TL)
       iv. 1.20(DL + DyL + LL) + 1.00(EP) + 1.00(HyF + WL + CL) + 1.20(EWiF)

**TABLE 2.7**
**Increase in Permissible Stresses**

| Sl. No. | Combination of Loads | Increase in Permissible Stress: Reinforced Concrete | Increase in Permissible Stress: Other Materials Such as Steel and Timber | Increase in Allowable Bearing Pressure |
|---|---|---|---|---|
| 1 | DL + LL + impact of breaking or traction or vehicles + centrifugal forces of vehicles | Nil | Nil | Nil |
| 2 | DL + LL with impact, breaking or tractive and centrifugal forces + earth pressure, percent | 15 | 15 | 15 |
| 3 | DL with/without LL including impact, breaking or tractive and centrifugal forces + earth pressure + hydrodynamic and hydrostatic forces + berthing or mooring forces, percent | 25 | 33 1/3 | 25 |
| 4 | Wind forces on structures + load combination of (1) + (2) or (3) | see IS 875 (Part 3) | | |
| 5 | Seismic forces + load combination of (1), (2) or (3). Berthing or Mooring forces | see IS 1893 (Part 1) | | |
| 6 | Secondary stresses + load combination of (1), percent | 15 | 15 | 15 |
| 7 | Erection stage stresses with DL and appropriate LL + earth pressure + hydrodynamic forces + wind forces, percent | 15 | 33 1/3 | 25 |

d. Temporary load condition
   i. 1.20(DL + DyL + LL) + 1.00(EP) + 1.00(HyF + WL + CL)

e. Reversal load condition
   i. 0.90(DL + DyL + LL) + 1.00(EP) + 1.00(HyF + WL + CL) + 1.50(BF or MF)
   ii. 0.90(DL + DyL + LL) + 1.00(EP) + 1.00(HyF + WL + CL) + 1.50(EWiF)
   iii. 0.90(DL + DyL + LL) + 1.00(EP) + 1.00(HyF + WL + CL) ± 1.50(EL)

### 2.1.4.5 Choice of Construction Material

The basic criteria adopted in the general choice of construction material, such as easy availability, easy working, mechanical properties suited to the purpose for which it is to be used and economic considerations, hold equally good for dock and harbour structures. The durability under the environmental condition is of importance in these marine structures. The aggressive action of sea water and/or the marine environment of the principal construction materials, such as steel, concrete and timber, require special attention.

**Steel:**

Unless otherwise specified, the steel shall conform to IS 2062. The corrosion of steel varies in different conditions of sea air or sea water exposure. Severe corrosion, however, occurs in saline water and under-marine growth, especially in the splash zone and in the reaches of the tidal range with alternate wetting and drying. Steel buried in the ground is also subjected to corrosion under certain conditions. Any one or a combination of the following remedial measures may be taken against the corrosion:

1. Cathodic protection
   Corrosion of steel completely immersed under water or buried in the ground (where the possibility of electrolytic corrosion exists) can be substantially eliminated, and corrosion of steel alternatively exposed to the wet and dry conditions can be significantly protected by cathodic protection using an impressed current system or sacrificial anode system.
2. Increased section/reduced stresses
   Any mild steel used in marine structure should have a minimum thickness of 6 mm when cathodic protection is provided and a minimum thickness of 10 mm when cathodic protection is not provided. In any case, no structural steel should be used in marine conditions without protective coatings.
3. Use of special steel
   Special alloy steels, such as those with 2% copper content, can significantly arrest corrosion.
4. Jacketing with underwater micro-concrete in splash zone

Special care has to be taken in the splash zone where the protection could be given by an underwater micro-concrete by jacketing with suitable anti-washout polymer-based cementitious grout.

*Concrete*: Concrete has extensive use in harbour structures, such as dock walls and floors, piles, sheet piles, caissons and monoliths, deck structures for jetties and wharves and breakwater armour blocks apart from the use in dock buildings and in other structures above ground.

The concrete structures built in an aggressive environment are subject to attack by sea water penetrating the mass. Concrete shall be made impermeable to such a degree that it is not penetrated by the constituents of sea water. The densest concrete will give the best result. Concrete grade not less than M 30 for RCC, M 20 for Plain Cement Concrete and M 40 for pre-stressed concrete construction shall be used. Concrete grade not less than M 15 shall be used in mass concrete construction.

Blast furnace slag cement (see IS 455) should preferably be used for marine structures. As an alternative, ordinary Portland cement/Portland Pozzolana Cement [see IS 1489 (Parts 1 and 2)] may be used provided the next higher grade of concrete (see IS 456) than that required from strength consideration alone is used. The minimum cement content

**TABLE 2.8**
**Maximum Crack Width in Different Zones**

| Sl. No. | Exposure Zone | Maximum Crack Width (mm) Sustained Load | Transient Load |
|---|---|---|---|
| 1 | Atmospheric zone – above splash zone and where direct wave or spray impingement is infrequent | 0.2 | 0.3 |
| 2 | Splash zone – zone between the chart datum and the design wave height above the mean high-water springs | 0.1 | 0.2 |
| 3 | Continuous sea water immersion zone – below splash zone up to sea bed level | 0.2 | 0.3 |
| 4 | Below sea bed level | 0.3 | 0.3 |

Sustained load – Dead load plus 50% of full uniformly distributed live load + earth pressureTransient load – Dead load plus berthing load and full crane load or full live load uniformly distributed + earth pressure

and maximum water-cement ratio for all grades of concrete for RCC and prestressed concrete construction shall be in conformity with Table 5 of IS 456 (2000).

As cracking in concrete members is to be minimized, reduced stresses are recommended for concrete and steel to be used in the design of RCC members subject to marine environments unless the structure is checked against the formation of cracks. The stresses in steel may be reduced to 165 N/mm$^2$ in working stress design. As a guide, the assessed surface width of cracks at points nearest to the main reinforcement should not exceed 0.004 times the cover of the main reinforcement or maximum crack width in different zones given in Table 2.8, whichever is minimum. Limit state of serviceability load combinations shall be considered for crack width calculation

The nominal thickness of the cover is to be provided for the structures in the marine atmosphere. It is recommended for structures immersed in sea water, in splash zone nominal cover shall be 75 mm and for all other exposure conditions (the zones above deck level, immersed and below mud line), the nominal cover shall be 50 mm.

The use of precast concrete elements is preferred for marine structures as they are cast under strict quality control and, therefore, are able to withstand the destructive influence of marine environments better.

All the limit state of serviceability load combinations excepting the one with seismic force shall be used for maximum crack width check of structural members and not for checking the deflection. The permissible deflection of berthing structure need not be checked for berthing or mooring load combinations. The seismic load combination for limit state of serviceability shall be used for checking the deflection of the berthing structures near the expansion joints at the deck level. The deflection for the berthing structure need not be controlled by limiting the deflection of pile at the soil/dredge level.

*Timber*: Timber has wide use in dock and harbour structures. It can be used for sheet piles, bearing piles, structural members in jetties, fenders, transit sheds and warehouses, as structural members and/or for door and window frames.

**TABLE 2.9**
**Nominal Thickness of Cover in Different Zones**

| Sl. No. | Exposure Zone | Adequate Thickness of Cover (mm) |
|---|---|---|
| 1 | Atmospheric zone – above splash zone and where direct wave or spray impingement is infrequent | 50 |
| 2 | Splash zone – zone between the chart datum and the design wave height above the mean high-water springs | 75 |
| 3 | Continuous sea water immersion zone – below splash zone up to sea bed level | 75 |
| 4 | Below sea bed level | 50 |

The hazards which face timber are the attack by fungi and insecticides; and in sea water the attack by the marine borers. When used in dockside buildings, the design shall be primarily guided by provisions of the relevant standards and protection against the attack of fungi and insecticides. The timber used in marine structure, particularly if subjected to fluctuating tides, is prone to attacks by marine borers and require preservative treatment.

Effective preservative treatment of timber is creosoting, which is normally applied by pressure impregnation [see IS 401 (2001)].

### 2.1.4.6 Fenders

The various types of fenders absorb energy in various ways but in most of them, the kinetic energy of the ship is stored in the energy of strain. Shear permits the absorption of the larger amount of energy than tension and compression. Flexure does not permit efficient use of the material as only a small percentage of the material of the fender may reach the allowable stresses. Torsion permits a good use of all the materials of a fender, particularly if tubes with thin walls are used. The flexibility of the berthing structure together with that of the fendering system may be taken into consideration in computing the total energy absorption capacity of the whole system. Any accepted method of analysis (such as beams on elastic foundation) may be adopted for such computation.

1. Reaction force and deflection of the fender
   The maximum amount of reaction force of a fender system on a ship shall be chosen considering:
   a. The strength of hull
   b. The strength of the berthing structure
   The manufacturers of proprietary fenders usually supply the energy absorption, deflection and reaction force characteristics of the specific fenders. These should be adopted in design. It is, however, recommended that a factor of safety of 1.4 should be applied over the ultimate energy absorption capacity of such fenders.
2. Disposition of fenders

The disposition of the fenders shall depend upon the location of the berths, the type of berthing structure, conditions of berthing, etc. The longitudinal spacing for a continuous quay should not normally exceed 0.15 L, where L is the length of the smallest ship. For island berths, the spacing of primary fenders should be in the range of 0.25–0.40 L, where L is the length of the smallest ship. If the range of design vessels is a large liner, an outer set of primary fenders will become necessary. The vertical as well as horizontal disposition of fenders shall be so designed as to prevent the ship's hull and/or berthing structure being damaged under all tidal conditions. Ship's bow radius and approach, and flair angle and in specific cases, may extend from the cope level to the low water level.

#### 2.1.4.7 Expansion Joints

A sufficient number of expansion joints shall be provided depending upon the type of the structure, the sub-soil and the atmospheric conditions in order to accommodate movements arising from shrinkage, temperature changes and some yielding of the foundation. The steel reinforcement shall be so designed that it also acts to provide shrinkage and temperature cracks.

As a general rule, a length of 60–300 m between the expansion joints is recommended for structures such as solid quay walls or pile-supported deck structures. An expansion gap shall be provided; however, in general, depending on the type of structure, type of soil and deflection during the earthquake. The expansion gap shall be calculated considering the limit state of serviceability load combinations and provide suitably. The crane track shall be designed suitably at expansion joints. For suspended deck resting on flexible long piles, length of the expansion joint shall be based on the detailed analysis of the whole structure. The expansion joints in the sections shall be keyed where horizontal forces are required to be transferred to the adjoining section and shall be so designed that changes in the length of the sections are not hindered. The arrangement of keys for vertical support shall depend on the soil conditions, the construction of the structure and the type of its loading. The expansion joints shall be covered to prevent the backfill from being washed out.

### 2.1.5 IS 4651 (Part 5) (1980) – Code of Practice for Planning and Design of Port and Harbours (Layout and Functional Requirements)

This code is under revision and hence the details are not discussed.

## 2.2 IS 2911 PART I – DESIGN AND CONSTRUCTION OF PILE FOUNDATIONS – CODE OF PRACTICE

There are four types of pile depending upon the construction method: driven cast in-situ concrete piles, bored cast in-situ concrete piles, driven precast concrete piles and bored precast concrete piles. There are four sections in IS 2911 Part

I corresponding to each type of pile. Predominantly, bored cast in-situ concrete piles are used, and the corresponding code book is explained below

### 2.2.1 IS 2911 (Part 1) (2010) Design and Construction of Pile Foundation Section 2 – Bored Cast In-Situ Concrete Piles (Second Revision)

#### 2.2.1.1 Bored Cast In-Situ Pile

A pile formed by boring a hole in the ground by percussive or rotary method with the use of temporary/permanent casing or drilling mud and subsequently filling the hole with reinforced concrete.

#### 2.2.1.2 Cut-Off Level

It is the level where a pile is cut off to support the pile caps or beams or any other structural components at the level.

#### 2.2.1.3 Diameter of Piles

Piles of 600 mm in diameter are commonly known as small piles, while piles greater than 600 mm are called large diameter piles. The minimum pile diameter shall be 450 mm.

#### 2.2.1.4 Factor of Safety

It is the ratio of the ultimate load capacity of the pile to the safe load on the pile. Factor of safety should be chosen after considering:

1. The reliability of the calculated value of ultimate load capacity of a pile
2. The types of superstructure and the type of loading
3. Allowable total/differential settlement of the structure

The minimum factor of safety on static formula shall be 2.5.

The higher value of factor of safety for determining the safe load on piles may be adopted where:

1. Settlement is to be limited or unequal settlement avoided
2. Large impact or vibrating loads are expected
3. The properties of the soil may deteriorate with time

#### 2.2.1.5 Load Bearing Pile

A pile formed in the ground for transmitting the load of a structure to the soil by the resistance developed at its tip and/or along its surface. It may be either vertically or at an inclination (batter pile) and may be required to uplift forces.

If the pile supports, the load primarily by resistance developed at the pile tip or base it is called 'end-bearing pile' if primarily by friction along its surface then 'friction pile'.

### 2.2.1.6 Pile Spacing

The spacing of piles means centre to centre distance between adjacent piles. The minimum centre to centre spacing of piles is considered from three aspects, namely

1. Practical aspects of installing the piles
2. Diameter of the pile
3. Nature of the load transfer to the soil and possible reduction in the load capacity of piles group

In the case of piles founded on hard stratum and deriving their capacity mainly from the end-bearing the minimum spacing shall be 2.50 times the diameter of the circumscribing circle corresponding to the cross-section of the pile shaft. In case of piles resting on the rock, spacing of 2 times the said diameter may be adopted.

Piles deriving their load-carrying capacity mainly from friction shall be spaced sufficiently apart to ensure that the zones of soils from which the piles derive their support do not overlap to such an extent that their bearing values are reduced. Generally, the spacing in such cases shall not be less than 3 times the diameter of the shaft.

### 2.2.1.7 Design Considerations

*Pile Capacity*

The load-carrying capacity of a pile depends on the properties of the soil in which it is embedded. Axial load from a pile is normally transmitted to the soil through skin friction along the shaft and end-bearing at its tip. A horizontal load on a vertical pile is transmitted to the soil primarily by a horizontal subgrade reaction generated in the upper part of the shaft. The lateral load capacity of a single pile depends on the soil reaction developed and the structural capacity of the shaft under bending.

The ultimate load capacity of a single pile may be obtained by using static analysis, the accuracy being dependent on the reliability of the soil properties for various strata. When computing capacity by static formula, the shear strength parameters obtained from a limited number of borehole data and laboratory tests should be supplemented, wherever possible, by in-situ shear strength obtained from field tests. The two separate static formulae, commonly applicable for cohesive and non-cohesive soil, are indicated in the following sections.

*Piles in Granular Soils*

The ultimate load capacity ($Q_u$) of piles, in kN, in granular soils is given by the following formula:

$$Q_u = A_p\left(\tfrac{1}{2} D\gamma N_\gamma + P_D N_q\right) + \sum_{i=1}^{n} K_i P_{Di} \tan \delta_1 A_{si}$$

The first term gives end-bearing resistance, and the second term gives skin friction resistance.

where

$A_p$ = Cross-sectional area of pile tip, in $m^2$

D = Diameter of pile shaft, in m

$\gamma$ = Effective unit weight of the soil at pile tip, in kN/m$^3$
$N_\gamma$ = Bearing capacity of factors depending upon
$N_q$ = Angle of internal friction, $\phi$ at pile tip
$P_D$ = Effective overburden pressure at pile tip, in kN/m$^2$ (see note 5)
$\sum_{i=1}^{n}$ = Summation for layers 1 to n in which pile is installed and which contribute to positive skin friction
$K_i$ = Coefficient of earth pressure applicable for the *i*th layer (see Note 3)
$P_{Di}$ = Effective overburden pressure for the *i*th layer, in kN/m$^2$
$\delta$ = Angle of wall friction between pile and soil for the *i*th layer, in m$^2$
$A_{si}$ = Surface area of pile shaft in the *i*th layer, in m$^2$

**Notes:**

1. $N_\gamma$ factor can be taken for general shear failure according to IS 6403.
2. $N_q$ factor will depend on the nature of the soil, type of pile, the L/B ratio and its method of construction. The values applicable for bored piles are given in Figure 1 in Annexure B IS 2911 (Part 1).
3. $K_i$, the earth pressure coefficient depends on the nature of soil strata, type of pile, spacing of pile and its method of construction. For driven piles in loose to dense sand with $\phi$ varying between 30° and 40°, $K_i$ values in the range of 1.00–1.50 may be used.
4. $\delta$, the angle of wall friction may be taken equal to the friction angle of the soil around the pile stem.
5. In working out pile capacity by static formula, the maximum effective overburden at the pile tips should correspond to the critical depth, which may be taken as 15 times the diameter of the pile shaft for $\phi \le 30°$ and increasing to 20 times for $\phi \ge 40°$.
6. For piles passing through cohesive strata and terminating in a granular stratum, a penetration of at least twice the diameter of the pile shaft should be given into the granular stratum.

### *Piles in Cohesive Soils*

The ultimate load capacity ($Q_u$) of piles, in kN, in cohesive soils is given by the following formula:

$$Q_u = A_p N_c C_p + \sum_{i=1}^{n} \alpha_i C_i A_{si}$$

The first term gives the end-bearing resistance, and the second term gives the skin friction resistance.

where

$A_p$ = Cross-sectional area of pile tip, in m$^2$
$N_c$ = Bearing capacity factor, may be taken as 9
$C_p$ = Average cohesion at pile tip, in kN/m$^2$
$\sum_{i=1}^{n}$ = Summation for layers 1 to n in which the pile is installed and which contribute to positive skin friction

$\alpha_i$ = Adhesion factor for the *i*th layer depending on the consistency of the soil (see Note)
$C_i$ = Average cohesion for the *i*th layer, in kN/m²
$A_{si}$ = Surface area of the pile shaft in the *i*th layer, in m²

NOTE: The value of the adhesion factor $\alpha_i$ depends on the undrained shear strength of the clay and may be obtained from Figure 2 in ANNEX B IS 2911 (Part 1).

*Piles in Hard Rock*

When the crushing strength of the rock is more than the characteristic strength of pile concrete, the rock should be deemed as hard rock. Piles resting directly on hard rock may be loaded to their safe structural capacity.

*Piles in Weathered/Soft Rock*

For pile found in weathered/soft rock, different empirical approaches are used to arrive at the socket length necessary for utilizing the full structural capacity of the pile.

Since it is difficult to collect cores in weathered/soft rocks, the method suggested by Cole and Stroud using 'N' values is more widely used. The allowable load on the pile, $Q_a$, in kN, by this approach, is given by:

$$Q_a = c_{u1} N_c \cdot \frac{\pi B^2}{4F_s} + \alpha c_{u2} \cdot \frac{\pi BL}{F_s}$$

where

$C_{u1}$ = Shear strength of rock below the base of the pile, in kN/m² (see Figure 2.2)
$N_c$ = Bearing capacity factor taken as 9
$F_s$ = Factor of safety usually taken as 3
$\alpha$ = 0.9 (recommended value)
$C_{u2}$ = Average shear strength of rock in the socketed length of pile, in kN/m² (see Figure 2.2)
B = Minimum width of pile shaft (diameter in case of circular piles), in m
L = Socket length of pile, in m

**NOTES**

1. For N ≥ 60, the stratum is to be treated as a weathered rock than soil.
2. Standard penetration test may not be practicable for N values >200. In such cases, design may be done on the basis of the shear strength of rock.

*Negative Skin Friction or Drag Down Force*

When a soil stratum, through which a pile shaft has penetrated into an underlying hard stratum, compresses as a result of either it being unconsolidated or it being under a newly placed fill, or as a result of remoulding during installation of the pile, a drag down force is generated along the pile shaft up to a point in depth where the surrounding soil does not move downward relative to the pile shaft. The existence of

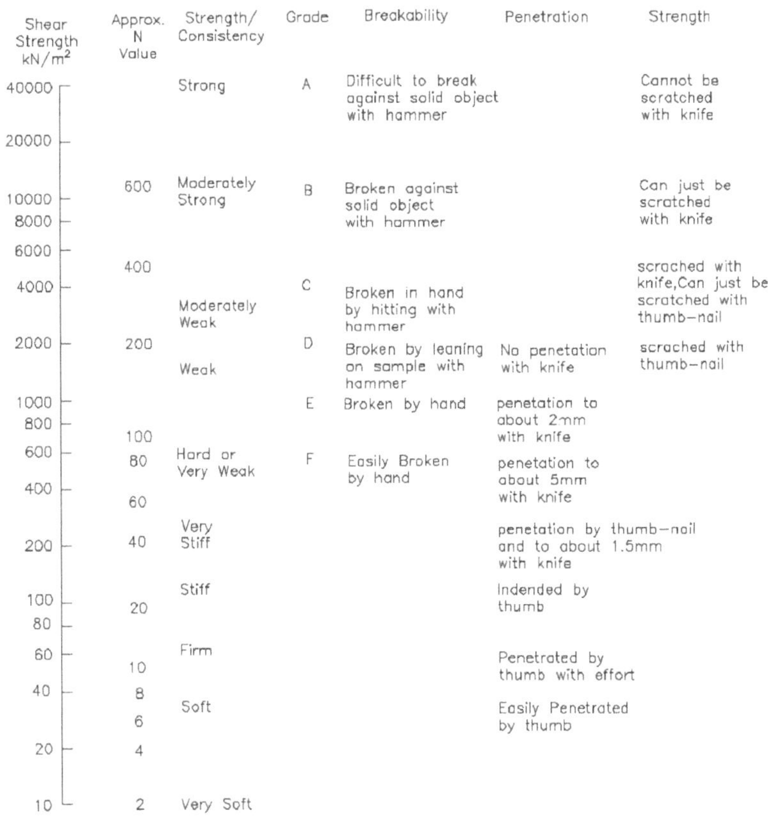


**FIGURE 2.2** Consistency and shear strength of weathered rock.

such a phenomenon shall be assessed, and suitable correction shall be made to the allowable load where appropriate.

### For piles in sand and normally loaded clays

$$\text{Stiffness factor } T\text{, in m} = \sqrt[5]{\frac{EI}{\eta_h}}$$

where

$E$ = Young's modulus of pile material, in MN/m$^2$

$I$ = Moment of inertia of the pile cross-section, in m$^4$

$\eta_h$ = Modulus of the subgrade reaction, in MN/m$^3$

### For piles in preloaded clays

$$\text{Stiffness factor } R\text{, in m} = \sqrt[4]{\frac{EI}{KB}}$$

**TABLE 2.10**
**Criteria for Behaviour of Pile Based on Its Embedded Length**

| Sl. No. | Soil Type | Relation of Embedded Length With Stiffness Factor | |
|---|---|---|---|
| | | Linearly Increasing | Constant |
| 1 | Short (rigid) pile | $L \leq 2.00T$ | $L \leq 2.00R$ |
| 2 | Long (elastic) pile | $L \geq 4.00T$ | $L \geq 3.50R$ |

The intermediate $L$ shall indicate a case between rigid pile behaviour and elastic pile behaviour.

where

$E$ = Young's modulus of pile material, in $MN/m^2$

$I$ = Moment of inertia of the pile cross-section, in $m^4$

$K = \frac{k_1}{1.5} \times \frac{0.3}{B}$ (see Table 4 of IS 2911 Design and Construction of Pile Foundation–Code of Practice (Part 1/Sec 1) (2010) for values of $k_1$, in $MN/m^3$)

$B$ = Width of pile shaft (diameter in case of circular piles), in m

*Criteria for Short Rigid Piles and Long Elastic Piles*

Having calculated the stiffness factor $T$ or $R$, the criteria for behaviour as a short rigid pile or as a long elastic pile are related to the embedded length $L$ as given in Table 2.10.

### 2.2.1.8 Reinforcement

The design of the reinforcing cage varies depending upon the installation conditions, the nature of the subsoil and the nature of load to be transmitted by the shaft-axial, or otherwise. The minimum area of longitudinal reinforcement of any type or grade within the pile shaft shall be 0.4% of the cross-sectional area of the pile shaft. The minimum reinforcement shall be provided throughout the length of the shaft.

The curtailment of reinforcement along the depth of the pile, in general, depends on the type of loading and subsoil strata. In case of piles subject to compressive load only, the designed quantity of reinforcement may be curtailed at appropriate level according to the design requirements. For piles subjected to uplift load, lateral load and moments, separately or with compressive loads, it would be necessary to provide reinforcement for the full depth of pile. In soft clays or loose sands, or where there may be danger to green concrete due to installation of adjacent piles, the reinforcement should be provided up to the full pile depth, regardless of whether or not it is required from uplift and lateral load considerations.

Piles shall always be reinforced with a minimum amount of reinforcement as dowels keeping the minimum bond length into the pile shaft below its cut-off level and with adequate projection into the pile cap, irrespective of design requirements.

## 2.3 IS 1893 (2016) – CRITERIA FOR EARTHQUAKE-RESISTANT DESIGN OF STRUCTURES

### 2.3.1 IS 1893 (2016) Criteria for Earthquake-Resistant Design of Structures (Part 1) (2016) General Provisions and Buildings

This standard primarily deals with the earthquake hazard assessment for earthquake resistant design of buildings, liquid retaining structures, bridges, embankments and retaining walls, industrial and stacklike structures, concrete masonry and earth dams.

#### 2.3.1.1 Seismic Force

For the purpose of determining the design seismic force, the country is classified into four seismic zones, as given in Figure 1 Seismic Zones of India, IS 1893 (Part 1) (2016).

The design horizontal seismic coefficient $A_h$ for a structure shall be determined by

$$A_h = \frac{\left(\frac{Z}{2}\right)\left(\frac{S_a}{g}\right)}{\left(\frac{R}{I}\right)}$$

where

$Z$ = Seismic zone factor, it is the value of peak ground acceleration considered by this standard for the design of structures located in each seismic zone. It is given in Table 2.11.

$I$ = Importance factor, it is a factor used to estimate design seismic force depending on the functional use of the structure, characterized by hazardous consequences of failure, post-earthquake functional needs, historical value or economic importance. The values are given in IS 1893 (Parts 1 to 5) for the corresponding structures; when not specified, the minimum values of I shall be:

1. 1.50 for critical and lifeline structures
2. 1.20 for business continuity structures
3. 1.00 for the rest

$R$ = Response reduction factor, it is the factor by which the base shear induced in a structure if it were to remain elastic, is reduced to obtain the design base shear. It depends on the perceived seismic damage performance of the structure, characterized by ductile or brittle deformations, redundancy in the structure or over strength inherent in the design process. The values are given in IS 1893 (Parts 1–5) for the corresponding structures.

$\frac{S_a}{g}$ = design acceleration coefficient for different soil types. It is a factor denoting the normalized design acceleration spectrum value to be considered for the design of

**TABLE 2.11**
**Seismic Zone Factor Z**

| Seismic Zone Factor | II | III | IV | V |
|---|---|---|---|---|
| Z | 0.10 | 0.16 | 0.24 | 0.36 |

**TABLE 2.12**
**Comparison of Load Combinations for Various Countries**

| Country | Load Combinations |
|---|---|
| IS 456 | 1.50 (D+L) |
| | 1.20 (D+L±W) |
| | 1.50 (D±W) |
| | 0.90D±1.5W |
| ACI318 | 1.40D |
| | 1.20D+1.60L+0.50Lr |
| | 1.20D+1.60Lr+(L or 0.80W) |
| | 1.20D+1.60W+1.00L+0.50Lr |
| | 0.90D+1.60W |
| BS800 | 1.40D+1.60L |
| | 1.40D+1.40W |
| | 1.00D+1.40W |
| | 1.20D+1.20L+1.20W |
| CSA | 1.40D |
| | 1.25D+1.50L |
| | 1.25D+1.40W |
| | 1.00D+1.00E |
| EC2 | 1.35D+1.50L |
| | 1.00D+1.50W |
| | 1.35D+1.50L+0.90W |

structures subjected to earthquake ground shaking; this value depends on the natural period of oscillation of the structure and damping to be considered in the design of the structure. It shall be given in parts 1 to 5 of IS 1893 for the corresponding structures; when not specified, it shall be taken as that corresponding 5 damping, given by expressions below and in Figure 2 in IS 1893 (Part 1) (2016).

## 2.4 COMPARISON WITH INTERNATIONAL CODES

The various international codes available are British design codes, European design codes, American design codes and Canadian Standards. The British codes used in port and harbour engineering are BS 6349 Code of Practice for Maritime Structures, Part 1 Code of Practice for General Criteria, Part 2 Design of Quay Walls, Jetties and

**TABLE 2.13**

**Comparison of Different Parameters of Slab**

| Parameters | IS 456 | BS 8110 | EC 2 | | | ACI 318 | A 23.3 |
|---|---|---|---|---|---|---|---|
| Criteria for one way or two-way slab | $\frac{ly}{lx} > 2$ | $\frac{ly}{lx} \geq 2$ | $\frac{ly}{lx} > 2$ | | | $\frac{ly}{lx} \geq 2$ | $\frac{ly}{lx} \geq 2$ |
| $\frac{span}{effective depth}$ | One-way slab | One-way slab | One-way slab | | | One-way slab | One-way slab |
| | Cantilever = 7 SS = 35 Continuous = 40 | Cantilever = 7 SS = 20 Continuous = 26 | Type | P = 1.5% | P = 0.5% | Cantilever = 10 | Cantilever = 10 |
| | For high strength deformed bars of grade 415, the values given above multiplied by 0.80 | End Span = 23 | cantilever | 6 | 8 | SS = 20 | SS = 20 |
| | | | SS | 14 | 20 | One end Cont. = 24 | One end Cont. = 24 |
| | | | one-way cont. or two-way spanning cont. over one long side | 18 | 26 | Both ends Cont. = 28 | Both ends Cont. = 28 |
| | | | one way or two-way spanning slab | 20 | 30 | | |
| Minimum reinforcement | Mild steel reinforcement in either direction should not be less than equal to 0.15%<br>For high strength deformed bars or welded wire fabric should not be less than equal to 0.12% | For mild steel in both directions. $\frac{0.24bh}{100}$<br>For high yield steel (460MPa) $\frac{0.13bh}{100}$ | $A_{smin} = \frac{0.26btdf_{ck}}{f_{yk}}$<br>But not less than 0.0013btd<br>$A_{smax} = 0.04A_c$ Secondary transverse reinforcement of not less than 20% of the principal reinforcement should be provided in one-way slab. | | | Not less than $0.0014\ A_c$ | $0.00\ 2A_g$ in each direction |

**TABLE 2.14**
**Different Parameters for Column**

| Parameters | IS 456 | BS 8110 | EC 2 | ACI 318 | A 23.3 |
|---|---|---|---|---|---|
| Slenderness ratio | $\frac{lex}{D}$ or $\frac{ley}{B}$ | $\frac{lex}{h}$ or $\frac{ley}{b}$ | $\frac{lo}{i}$ | $\frac{kl_e ff}{r}$ | $\frac{kl_u}{r}$ |
| Condition for compression member (short column) | $\frac{lex}{D}$ and $\frac{ley}{B}<12$ | $\frac{lex}{D}$ and $\frac{ley}{b}<15$ (braced)<br>$\frac{lex}{h}$ and $\frac{ley}{b}<10$<br>(unbraced) | $\frac{lo}{i}$ < limiting value of slenderness ratio | $\frac{kl_u}{r}\leq 40$<br>Short column | $\frac{kl_u}{r}\leq 34-12\left(\frac{M1}{M2}\right)$<br>(for non-sway frames)<br>$\frac{kl_u}{r}\leq 22$<br>(sway frames) |
| Ultimate axial load for short columns | $P_u = 0.4f_{ck} \cdot A_c + 0.67f_y A_{sc}$ | $P_n = 0.8\left[0.85f_c'\left(A_g - A_{st}\right) + A_{st} f_y\right]$ | $P_u = 0.57f_{ck} A_c + 0.87f_y kA_s$ | $P = 0.4f_{cu}\left(A_g - A_{sc}\right) + 0.8A_{sc} f_y$ | $P_{ro} = \alpha 1 \phi_c$<br>$f_c'\left(A_g - A_{st} A_t - A_p\right)$<br>$+ \phi_s f_y A_{st} + \phi_a f_y A_t - f_{pr} A_p$ |
| Longitudinal reinforcement<br>Minimum<br>Maximum | 0.008Ag<br>0.06Ag | 0.004Ag<br>0.06Ag | $A_{s(min)} = \frac{0.10N_{ED}}{f_{yd}}$<br>or $0.002A_c$<br>(whichever is greater)<br>$A_{s(max)} = 0.04A_c$<br>(outside lap location)<br>$= 0.008A_c$ (at lap) | $0.004\ A_g$<br>$0.01\ A_g$ | $0.01\ A_g$<br>$0.08\ A_g$ |

| | | | | | |
|---|---|---|---|---|---|
| Diameter of lateral ties | Should not be greater than<br>1) diameter of main bar/4<br>2) 6 mm (whichever is less) | 1) At least one quarter of the largest longitudinal bar size<br>2) 8 mm | Should not be less than<br>1) Diameter of main bar/4<br>2) 6 mm (whichever is less) | Should not be less than<br>1) 10 mm for longitudinal bars of 32 mm diameter or smaller<br>2) 13 mm for diameter of longitudinal bar larger than 32 mm | Calculated from Interaction diagram |
| Spacing of Lateral ties | Should not be greater than<br>1) 16xdiameter of main bar<br>2) 300 mm (whichever is less) | Maximum<br>12 x diameter of smaller longitudinal bar | Should not be greater than<br>1) 20 x diameter of ties<br>2) Least dimension of column<br>3) 400 mm (whichever is less) | Should not be greater than<br>1) 16 x diameter of main bar<br>2) 48xdiameter of ties<br>3) Least lateral dimension of column | Should not be greater than<br>1) 16 x diameter of smallest longitudinal Bar<br>2) 48 x diameter of ties<br>3) Least lateral dimension of column<br>4) 300 mm |

**TABLE 2.15**

**Beam Parameters for Codes of Different Countries**

| Parameters | IS 456 | BS 8110 | EC 2 | ACI 318 | A 23.3 |
|---|---|---|---|---|---|
| L/d ratio<br>Cantilever<br>simply supported<br>continuous | 7<br>20<br>26 | 7<br>20<br>26 | ρ = 1.5% ρ = 1.5%<br>6 8<br>14 20<br>18 26(end)<br>20 30(interior) | 8<br>16<br>18.5(end)<br>21(interior) | 8<br>16<br>18.5(end)<br>21(interior) |
| Minimum Tensile for flexure $\frac{A_{st}}{b_{wd}}$ | $\frac{0.85}{f_y}$ | 0.8( $f_y$=250 N/mm$^2$ )<br>0.45( $f_y$ = 460 N/mm$^2$ ) | $\frac{0.25 f_{ctm}}{f_y} \geq 0.0013$ | $\frac{0.25 f_{cr}}{f_y}$ (or)<br>$\frac{0.024\sqrt{f_{ck}}}{f_y} \geq \frac{1.4}{f_y}$ | $\frac{0.25 f_{c'}}{f_y}$ (or)<br>$\frac{0.18\sqrt{f_{ck}}}{f_y}$ |
| Maximum Tensile steel for flexure | 0.04bD | 0.04bD | 0.04bD | Net tensile strain in extreme tensile steel ≥ 0.005 | Tensile reinforcement limited to satisfy $\frac{c}{d} \leq \frac{700}{700+f_y}$ |
| Maximum shear reinforcement $\frac{A_s}{b_w S_v}$ | $\frac{0.4}{0.87 f_y}$ | $\frac{0.4}{0.95 f_{yv}}$<br>When 0.5$V_c$ V<$V_c$ +0.4<br>$\frac{V-Vc}{0.95 f_{yv}}$<br>When<br>$Vc + 0.4 < V < 0.8\ \sqrt{fcuor5}$ N/mm$^2$ | $\frac{0.08\sqrt{fck}}{f_y}$<br>When applied shear is less than shear strength of concrete | $\frac{0.9\sqrt{f_{ck}}}{16 f_y} \geq \frac{0.33}{f_y}$<br>When applied shear is greater than 0.58 concrete strength<br>$\frac{0.062\sqrt{f_{cr}}}{f_{yt}} \geq \frac{0.32}{f_{yt}}$ | $\frac{0.054\sqrt{f_{ck}}}{f_y}$ or<br>$\frac{0.06\sqrt{f_{cr}}}{f_y}$<br>When applied shear is greater than concrete strength |
| Spacing | 0.75d ≤ 300 mm | 0.75d should not be greater than d | 0.75d ≤ 600 mm | 0.5d ≤ 600 mm and<br>0.25 ≤ 300 mm when,<br>$Vs > \frac{bwd\sqrt{f_c}}{3}$ | 0.63d ≤ 600 mm and<br>0.32d ≤ 300 mm when,<br>$Vu > \frac{\varnothing cfcbwd}{8}$ |

**TABLE 2.16**
**Recommended Diameters of Transverse Reinforcement**

| | |
|---|---|
| Links, hoops or helical reinforcement | ≥6 mm and<br>≥0.25 of the maximum diameter of longitudinal bar |
| Wires of welded mesh transverse reinforcement | ≥5 mm |

**TABLE 2.17**
**Comparison of Mooring Point Load**

| | Mooring Point Load (tons) | |
|---|---|---|
| **Ship Displacement (tons)** | **IS 4651 (Part 3) (1974)** | **BS 6349 (Part 4) (1994)** |
| 20,000 up to including 50,000 | 80 | 80 |
| Above 50,000 and including 100,000 | 100 | 100 |
| Above 100,000 including 200,000 | 150 | 150 |
| Above 200,000 | Greater than 200 | 200 |

Dolphins, Part 3 Design of Dry-Docks, Locks, Slipways and Ship Building Berths, Ship Lifts and Dock and Lock Gates, Part 4 Design of Fendering and Mooring Systems, Part 5 Dredging and Land Reclamation, Part 6 Design of Inshore Moorings and Floating Structures and Part 7 Guide to Design and Construction of Breakwaters. BS 8004 Code of Practice for Foundations is used in the design of sub-structure.

The following table illustrates the comparison of international codes IS 456 (Indian Standards), ACI318 (American design code), BS800 (British design code), CSA (Canadian Standards) and EC2 (European code)

### Shear Reinforcement for Circular Piles

#### *As per ACI 318 (Cl.11.4.7.3)*

$$V_s = A_v f_{yt} d/S$$

Where

$A_v$ is the area of shear reinforcement within spacing S.

$A_v$ shall be taken as two times the area of bar in a circular tie, hoop or spiral at a spacing S; S is measured in a direction parallel to longitudinal reinforcement.

$f_{yt}$ is the specified yield strength of circular tie, hoop or spiral reinforcement.

#### *As per EN 1536*

Bored-cast in-situ clear distance shall not be less than the clear distance as set out of the main reinforcement.

The following illustrates the comparison of IS 4651 Code of Practice for Planning and Design of Ports and Harbours and BS 6349 Maritime Structures.

**TABLE 2.18**
**Comparison of Partial Load Factors**

| | | IS 4651 (Part 4) (2014) | | | | | | | |
|---|---|---|---|---|---|---|---|---|---|
| | | Limit State of Serviceability | | Limit State of Collapse | | | | BS 6349 (Part 2) (2010) | |
| Sl. No. | Loads | Short term | Long term | Normal | Extreme/ Survival | Temporary | Reversal | Unfavorable | Favorable |
| 1 | Dead load | 1.00 | 1.00 | 1.50 | 1.20 | 1.20 | 0.90 | | |
| 2 | Live load | | | | | | | | |
| | Impact or dynamic load | 1.10 | 0.50 | 1.50 | 1.20 | 1.20 | 0.90 | 1.35 | 0 |
| | Static load | 1.00 | 0.50 | 1.50 | 1.20 | 1.20 | 0.90 | 1.50 | 0 |
| 3 | Earth pressure | 1.00 | 1.00 | 1.20 | 1.00 | 1.00 | 1.00 | 1.50 | 0 |
| 4 | Hydrostatic force | 1.00 | – | 1.00 | 1.00 | 1.00 | 1.00 | 1.20 | 0 |
| 5 | Wave and current force | 1.00 | – | 1.20 | 1.00 | 1.00 | 1.00 | 1.40 | 0 |
| 6 | Berthing force | 1.00 | – | 1.50 | 1.00 | – | 1.50 | 1.40 | 0 |
| 7 | Mooring force | 1.00 | – | 1.50 | – | – | 1.50 | 1.40 | 0 |
| 8 | Wind force | | | | | | | 1.40 | 0 |
| | Working wind force | 1.00 | – | 1.00 | – | – | – | | |
| | Extreme wind force | – | – | – | 1.20 | – | 1.50 | | |
| 9 | Shrinkage | – | 1.00 | – | – | – | – | | |
| 10 | Creep | – | 1.00 | – | – | – | – | | |
| 11 | Temperature | – | 1.00 | – | – | – | – | 1.30 | |
| 12 | Wave loading | – | – | – | 1.20 | – | – | 1.40 | |
| 13 | Seismic force | 1.00 | – | – | 1.20 | – | 1.50 | | |
| 14 | Secondary stresses | 1.00 | – | – | – | – | – | | |
| 15 | Snow load | | | | | | | 1.40 | 0 |
| 16 | Crane loads | | | | | | | 1.35 | 0 |
| 17 | Forces from ship propulsion | | | | | | | 1.50 | 0 |

It can be compared from IS 4651 (1974) (Part 3) Code of Practice for Planning and Design of Port and Harbours – Loading and BS 6349 Part 4 (1994) Code of Practice for the Design of Fendering and Mooring Systems that the mooring points loads are the same.

Table 2.18 explains the difference in partial load factors of IS 4651 (Part 4) (2014) Code of Practice for Planning and Design of Port and Harbours – General Design Considerations and BS 6349–2 (2010) Maritime Structures (Part 2) Code of Practice for Design of Quay Walls, Jetties and Dolphins).

**General Recommendations for Design**

The following recommendations can be followed to satisfy the design considering the practical limitations:

1. It is a good practice to provide 0.8% as the minimum reinforcement and 2% as the maximum of piles. As the reinforcement will be about 4% at lap zone, it is good practice to limit the maximum reinforcement to 2%.
2. The stress levels are less to satisfy crack width of 0.1 mm at splash zone, and hence the development length shall be calculated as per the stress in the reinforcement. About 20 to 30 times diameter of reinforcement may be sufficient for lap length depending on the stress level.
3. The fixity depth calculation is given for soil in IS 2911, whereas the fixity depth for rock can also be calculated assuming the shear strength as per Figure 3 in IS 2911 (Part 1).
4. The crack width and deflection shall be checked only for serviceability condition. No need to consider extreme wind or wave. Only expansion gap shall be provided adequately for seismic case and no need to check deflection for seismic case.
5. If the cover provided is more than required as per code provisions, the reinforcement requirements are much higher and hence additional cover is not required.
6. Fusion bonded epoxy coated bars are recommended for piles as the coating will be disturbed while lowering the cage.
7. The lateral load test needs to be carried out for limiting the deflection for serviceability load combination only.
8. Low strain integrity test is recommended for all piles and high strain integrity test can be carried out in lieu of the conventional static pile load test.
9. In case the soil is soft clay with N < 4, the layer shall be completely neglected for calculating fixity depth.
10. In case of covered shed on top of the jetty, the deflection shall be checked considering the jetty up to founding level with superstructure.
11. The liners to pile are provided to facilitate bored concrete piles and no need to consider liner for structural strength. Liners shall be removed above LWL to reduce accumulation of salt in between liner and pile in splash zone.
12. The deck elevation shall be fixed considering the gap as per code provision below the bottom of the beam.

## REVIEW QUESTIONS

1. What are the Indian Standard codes and international codes for port and harbour engineering?
2. Write the Indian Standard used for the design of approach jetty and slipway.

3. Write the Indian Standard for seismic force.
4. What are the Indian Standard code books for the design of piles?
5. Describe in brief layout and functional requirement of a port as per the Indian Standard code.
6. Write the comparison between IS 456 and BS 800 for load combination.
7. Compare the load combination between IS 4651 and BS 6349.
8. Compare the parameters for a slab between the Indian code and ACI.
9. Compare the shear reinforcement calculation for piles between the Indian code and other international codes.

## REFERENCES

**BS 6349 (Part 2) (2010)** *Maritime Structures – Part 2: Code of Practice for Design of Quay Walls, Jetties and Dolphins*

**BS 6349 (Part 4) (1994)** *Maritime Structures – Part 4: Code of Practice for Design of Fendering and Mooring Systems*

**IS 401 (2001)** *Preservation of Timber – Code of Practice*

**IS 456 (2000)** *Plain and Reinforced Concrete – Code of Practice* (Fourth Revision)

**IS 1892 (2021)** Subsurface Investigation for Foundations – Code of Practice

**IS 1893 (Part 1) (2016)** *Criteria for Earthquake Resistant Design of Structures – General Provisions and Buildings* (Sixth Revision)

**IS 2911 (Part 1/Sec 2) (2010)** *Code of Practice for Design and Construction of Pile Foundations Part 1 – Concrete Piles, Section 2 – Bored Cast In-situ Concrete Piles* (Second Revision)

IS **4651 (Part I)** (2020) *Code of Practice for Planning and Design of Ports and Harbours – Site Investigation* (First Revision)

IS **4651 (Part 2)** (2020) *Code of Practice for Planning and Design of Port and Harbours – Geotechnical Engineering*

IS **4651 (Part 3)** (2020) *Code of Practice for Planning and Design of Port and Harbours – Loading*

IS **4651 (Part 4)** (2023) *Code of Practice for Planning and Design of Port and Harbours – General Design Considerations*

IS **4651 (Part 5)** (1980) *Code of Practice for Planning and Design of Port and Harbours – Layout & Functional Requirements*

**Tabish Izhar and Reena Dagar (2018)** Comparison of reinforced concrete member design methods of various countries, *International Journal of Civil Engineering and Technology (IJCIET)*, Volume 9, Issue 4, April 2018, pp. 637–646, Article ID: IJCIET_09_04_072 ISSN Print: 0976-6308 and ISSN Online: 0976-6316

# 3 Loads

Loads are the primary inputs for analysis and design of any structure. For all infrastructures, loads play a significant role in deciding the stability and life of structures. It includes all the factors of operational and extreme conditions. The classification of load for berthing structure and their detailed calculation is described in this chapter.

## Classification of loads:

Loads acting on berthing structure can be classified:

1. Based on the time period
2. Based on the function of site conditions

## Based on the occurrence:

Loads are classified based on time period as

1. Normal
   Any loads that may reasonably be expected to occur during the design life of the
   Structure under normal operating conditions with a one-year return period.
2. Extreme
   Any loads with 100 years of return period that may occur.
3. Temporary
   Any load that occurs during the construction stage.

## Based on the function of site conditions:

Loads classified based on the site conditions are

1. Loads from the seaside
   Loads from the seaside are due to waves, current, berthing and mooring of vessel.
2. Loads from the deck
   Loads from the deck are due to the self-weight of the deck and superimposed loads from handling equipment like cranes and trucks and from the stacking of cargoes.
3. Loads from the land side
   Loads from the land side are due to earth pressure with or without surcharge and differential water pressure.

Among these loads, berthing force, mooring force, loads from mechanical handling systems, earth pressure, differential water pressure, seismic force, wave force and current force are illustrated in sections.

DOI: 10.1201/9781003361275-3

**TABLE 3.1**
**Loads Classification Based on the Occurrence**

| Normal Load | Extreme Load | Temporary Load |
|---|---|---|
| • Dead load<br>• Live load<br>• Berthing force<br>• Mooring force<br>• Wave and current force (operational)<br>• Stacking cargo load<br>• Crane load<br>• Vehicular load (trucks) under wind force (operational)<br>• Earth pressure<br>• Hydrostatic pressure<br>• Wind pressure | • Seismic force<br>• Extreme wind<br>• Extreme wave and current pressure<br>• Tsunami load<br>• Storm surge | • Construction live load |

## 3.1 BERTHING FORCE

Berthing force is a horizontal force that acts on the berthing structure when an approaching vessel strikes the fender fixed to berth. The kinetic energy from this force can be absorbed by the fendering system. The fenders are designated for this relevant absorption energy. Deflection-reaction diagrams for the absorption energy can be obtained from fender manufacturers (discussed in Chapter 6). The corresponding reaction force of the absorption energy for which the berth is to be designed can be taken from the manufacturer's catalogues.

### 3.1.1 Berthing Energy

The different methods are used in determining the maximum amount of kinetic energy to be absorbed by the fender. These methods are:

#### Quinn Method

In the Quinn method, 50% of the energy of the ship is calculated based on the velocity of the ship normal to berthing structure is assumed as the energy absorbed by the fender.

$$E = \left(\frac{W}{G}\right)V^2/4 \tag{3.1}$$

Alternatively,

$$E = \frac{1}{2}\left(\frac{1}{2}mv^2\right) \tag{3.2}$$

where

m = (mass + added mass) of the vessel

v = Berthing velocity

### Woodruff method

In Woodruff method, the following empirical equation is used to calculate the berthing energy.

$$E = W\left(0.004 - W\right)\times 10^{-8} \tag{3.3}$$

### Vasco costa method

Vasco Costa has given the following analytical solution for a ship moving with translation velocity 'u' and angular velocity 'w' having no ship along the berth.

$$E = \left(\frac{WV^2}{2g}\right)\left(1+\frac{2D}{B}\right)\left(K+\frac{r^2COS^2r}{K^2}+r^2\right) \tag{3.4}$$

### Is 4651 (part 3) (2020) cl.6.2

When an approaching vessel strikes a berth, the horizontal force acts on the berth. The magnitude of this force depends on the kinetic energy that can be absorbed by the fendering system. The kinetic energy, E, imparted to a fendering system by a vessel moving with velocity V is given by

$$E = \frac{W_D \times V^2}{2g} C_m \times C_e \times C_s \tag{3.5}$$

where

E = Berthing energy in T-m

$W_D$ = Displacement tonnage in T

V = Berthing velocity in m/sec

$C_m$ = Mass coefficient (a)

$C_e$ = Eccentricity coefficient (b)

$C_s$ = Softness coefficient (c)

g = Acceleration due to gravity in m/sec$^2$

#### (a) Mass coefficient:

When a vessel approaches a berth, and as its motion is suddenly checked, the force of impact which the vessel imparts comprises the weight of the vessel and the effect of water moving along with the moving vessel. Such an effect, expressed in terms of the weight of water moving with the vessel, is called the additional weight ( $W_A$ ) of the vessel or the hydrodynamic weight of the vessel. Thus, the effective weight in berthing is the sum of the displacement tonnage of a vessel and its additional weight, which is known as the virtual weight ( $W_V$ ) of a vessel.

1. The mass coefficient $(C_m)$ is calculated using the following equation

$$C_m = 1 + \frac{2D}{B} \tag{3.6}$$

where
D = Draught of the vessel in m
B = Beam of the vessel in m

2. Alternative to (i) in the case of a vessel which has a length much greater than its beam or draught or generally for vessels with displacement tonnage greater than 20,000 the additional weight may be approximated to the weight of a cylindrical column of water of height equal to the length of vessel and diameter equal to the draught of vessel, then

$$C_m = 1 + \frac{\frac{\pi}{4} D^2 Lw}{W_D} \tag{3.7}$$

where
D = Draught of the vessel in m
L = Length of the vessel in m
w = Unit weight of water (1.03 T/m³ for sea water)
$W_D$ = Displacement tonnage of the vessel in tonnes
$W_v = W_D \times C_m$

### (B) Eccentricity coefficient:

A vessel generally approaches a berth at an angle denoted by θ and touches it at a point either near the bow or stern of the vessel. In such eccentric cases, the vessel imparts a rotational force at the moment of contact, and the kinetic energy of the vessel is partially expended in its rotational motion.

1. The eccentricity coefficient $(C_e)$ may then be derived as follows:

$$C_e = \frac{1 + (l/r)^2 \operatorname{Sin}^2 \theta}{1 + (l/r)^2} \tag{3.8}$$

where
$l$ = Distance from the centre of gravity of the vessel to the point of contact projected along the water line of the berth in m
$r$ = Radius of gyration of rotational radius on the plane of the vessel from its centre of gravity in m

2. The approach angle (θ), unless otherwise known with accuracy should be taken as 10°. For smaller vessels approaching the wharf structures, the approach angle should be taken as 20°.

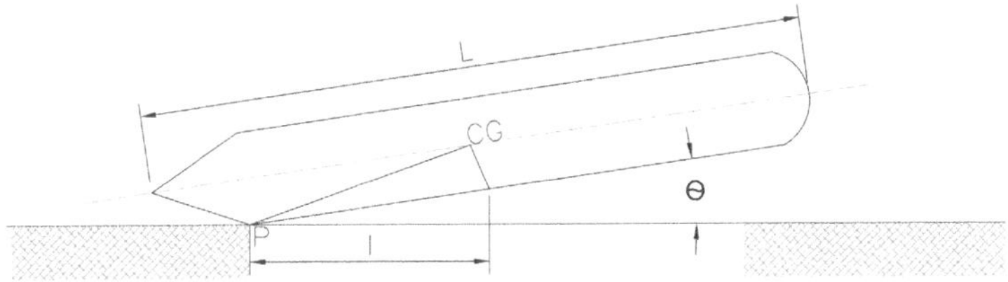


**FIGURE 3.1** Berthing force.

3. The ship is relatively rigid compared with the usually yielding fendering systems; a value of 0.90 is generally applied for this factor, or 0.95 if a higher safety margin is thought desirable.

### 3.1.2 Assigning of Berthing Force in STAAD.Pro Software

The vessel shall first contact with a fender only while berthing. The first contact may be any location of the fender. After the berthing of vessel, the berthing force exerted shall be distributed to all other fenders along berthing face. While analysing of berthing structure, two primary load cases for full berthing force shall be considered. One at end fender location and another at the centre fender location of the berth shall be applied.

The berthing force can be applied at the point of fender location as a concentrated load in the STAAD.Pro (structural analysis software) modelled structure.

## 3.2 MOORING FORCE

The mooring loads are the lateral loads caused by the mooring lines when they pull the ship into or along the dock or hold it against the forces of winds or current. The mooring force calculation as per IS 4651 (Part 3) (2020) Cl.6.3 is described below.

### 3.2.1 Forces due to Wind

The maximum mooring loads are due to the wind forces on the exposed area on the broad side of the ship in light condition.

$$F = C_{w.} A_{w.} P \tag{3.9}$$

where

$F$ = Force due to wind in kg

$C_w$ = Shape factor = 1.30–1.60 As per IS 4651 (Part 3) (2020) Cl.6.3

$A_W$ = Windage area in m$^2$

$P$ = Wind pressure in kg/m$^2$ to be taken in accordance with IS 875 (2015)

The windage area $(A_w)$ can be estimated as follows

$$A_w = 1.175\, L_v \left(D_M - D_L\right) \tag{3.10}$$

where

$L_v$ = Length between perpendicular in m
$D_M$ = Moulded depth in m
$D_L$ = Average light draft in m

When the ships are berthed on both sides of a pier, the total wind force acting on the pier should be increased by 50 to allow for wind against the second ship.

John W. Gay Thwaite (1990) Design of Marine Facilities: Engineering for Port and Harbour Structures has suggested the following formulae to calculate wind force in the longitudinal $(F_{wx})$ and lateral $(F_{wy})$ force components and a yawing moment $(M_{yw})$.

$$F_{wx} = 0.0034\, C_{Dx} V^2_w\, A_x \tag{3.11}$$

$$F_y = 0.0034\, C_{Dy}\, V_w^2\, A_y \tag{3.12}$$

$$M_{yw} = F_{wy}\, LOA\, C_{ym} \tag{3.13}$$

where

$F_{wx}$ and $F_{wy}$ = Wind force along x and y directions in pounds
$M_{yw}$ = Yawing moment in pounds-ft
$C_{Dx}$ and $C_{Dy}$ = Drag coefficients along x and y directions
$V_w$ = Wind speed in knots
$A_x$ and $A_y$ = End-on and side projected areas of vessel (including the areas of masts, stacks, rigging, deck cargoes, etc.)
LOA = Overall length of ship in feet

The hydrodynamic coefficients are the functions of the angle of wind approach (θ). The yaw moment is given in terms of the lateral force times the vessel's length overall (LOA) and $C_{ym}$. The total resultant force for wind from any direction ($F_w(\theta)$) is found in this equation:

$$F_w(\theta) = 0.0034 C_D\, (\theta)\, V_w^2\, (A_x\, \cos^2\theta + A_y\, \sin^2\theta) \tag{3.14}$$

**WIND PRESSURE (P):**

As per IS 875 (Part 3) (2015) Cl.7.2:

$$P_d = K_d K_a K_c P_z \tag{3.15}$$

As per IS 875 (Part 3) (2015):

$$P_z = 0.6\, v_z^2 \tag{3.16}$$

where

$v_z$ = Design wind speed in m/s

As per IS 875 (Part 3) (2015) Cl.6.3:

$$V_z = V_b \times K_1 \times K_2 \times K_3 \times K_4 \tag{3.17}$$

where

$V_z$ = design wind speed at height z, in m/s
$K_1$ = probability factor (risk coefficient) (from IS 875 (Part 3) (2015) Cl.6.3.1)
$K_2$ = terrain roughness and height factor (from IS 875 (Part 3) (2015) Cl.6.3.2)
$K_3$ = topography factor (from IS 875 (Part 3) (2015) Cl.6.3.3)
$K_4$ = importance factor for the cyclonic region (from IS 875 (Part 3) (2015) Cl.6.3.4)

NOTE: Wind speed may be taken as constant up to a height of 10 m. However, pressure for buildings/structures less than 10 m high may be reduced by 20% for evaluating stability and design of the framing.

Risk coefficient ($K_1$. Factor) (Refer IS 875 (Part 3) (2015) Cl.6.3.1): Figure 1 in IS 875 (Part 3) gives basic wind speed for terrain category 2 as applicable at 10 m above ground level based on a 50-year mean return period. The suggested life period to be assumed in design and the corresponding $K_1$ factors for different class of structures to design are given in Table 3.2. In the design of buildings and structures, a regional basic wind speed having a mean return period of 50 years shall be used except as specified in a note of Table 3.2.

**TABLE 3.2**
**Risk Coefficient for Different Classes of Structures in Different Wind Speed Zones (Refer to IS 875 (Part 3) (2015))**

| Sl. No. | Class of Structure | Mean Probable Design Life of Structure (Years) | $K_1$ Factor for Basic Wind Speed (m/s) | | | | | |
|---|---|---|---|---|---|---|---|---|
| | | | 33 | 39 | 44 | 47 | 50 | 55 |
| 1 | All general buildings and structures | 50 | 1.00 | 1.00 | 1.00 | 1.00 | 1.00 | 1.00 |
| 2 | Temporary sheds, structures such as those used during construction operations (for example, formwork and false work), structures during construction stages and boundary walls | 5 | 0.82 | 0.76 | 0.73 | 0.71 | 0.70 | 0.67 |
| 3 | Buildings and structures presenting a low degree of hazard to life and property in the event of failure, such as isolated towers in wooden areas, farm buildings other than residential buildings | 25 | 0.94 | 0.92 | 0.91 | 0.90 | 0.90 | 0.89 |
| 4 | Important buildings and structures such as hospitals communication buildings/towers, power plant structures | 100 | 1.05 | 1.06 | 1.07 | 1.07 | 1.08 | 1.08 |

Terrain, Height Factor ( $K_2$ Factor) (Refer IS 875 (Part 3) (2015) Cl.6.3.2.1): Selection of terrain categories shall be made with due regard to the effect of obstructions which constitute the ground surface roughness. The terrain category used in the design of a structure may vary depending on the direction of wind under consideration. Wherever sufficient meteorological information is available about the nature of wind direction, the orientation of any building or structure may be suitably planned.

Terrain in which a specific structure stands shall be assessed as being one of the following terrain categories:

1. Category 1 – Exposed open terrain with few or no obstructions and in which the average height of any object surrounding the structure is less than 1.50 m. The equivalent aerodynamic roughness height ( $z_{0.1}$ ) for this terrain is 0.002 m. Typically this category represents open sea-coasts and flat plains without trees.
2. Category 2 – Open terrain well scattered obstructions having heights generally between 1.50 m and 10.00 m. The equivalent aerodynamic roughness height ( $z_{0.2}$ ) for this terrain is 0.02 m. This is the criterion for measuring regional basic wind speeds and represents airfields, open park lands and undeveloped sparsely built-up outskirts of towns and suburbs. Open land adjacent to the sea-coast may also be classified as category 2 due to the roughness of large sea waves at high winds.
3. Category 3 – Terrain with closely spaced numerous obstructions having the size of buildings/structures up to 10 m in height with or without a few isolated tall structures. The equivalent aerodynamic roughness height ( $z_{0.3}$ ) for this terrain is 0.20 m. This category represents well wooded areas, shrubs, towns and industrial areas full or partially developed. It is likely that the next higher category than this will not exist in most design situations and that selection of a more severe category will be deliberate.
4. Category 4 – Terrain with numerous large high closely spaced obstructions. The equivalent aerodynamic roughness height ( $z_{0.4}$ ) for this terrain is 2.00 m. This category represents large city centres, generally with obstructions above 25 m high and well-developed industrial complexes.

NOTE: The factor $k_1$ *is* based on statistical concepts which consider the degree of reliability required and the period of time in years during which these will be exposed to wind, that is, the life of the structure. Whatever wind speed is adopted for design purposes, there is always a probability (however small) that it may exceed in a storm of exceptional violence; more than the period of years over which there is exposure to the wind, more is the probability. Larger return periods range from 100 to 1000 years (implying lower risk level) in association with larger periods of exposure and chosen probability of exceedance (risk level). The probability level of 0.63 is normally considered sufficient for the design of buildings and structures against wind effects, and the values of $k_1$ corresponding to this risk level are given above. Variation of wind speed with height in different terrains $(k_2)$ (Refer IS 875 (Part 3) 2015 Cl.6.3.2.2) by which the basic wind speed shall be multiplied to obtain the wind speed at different heights, in each terrain category.

**TABLE 3.3**
**Factors to Obtain Design Wind Speed Variation With Height in Different Terrains (Refer to IS 875 (Part 3) (2015))**

| | | Terrain and Height Multiplier ($k_2$) | | | |
|---|---|---|---|---|---|
| Sl. No. | Height Z (m) | Terrain Category 1 | Terrain Category 2 | Terrain Category 3 | Terrain Category 4 |
| 1 | 10 | 1.05 | 1.00 | 0.91 | 0.80 |
| 2 | 15 | 1.09 | 1.05 | 0.97 | 0.80 |
| 3 | 20 | 1.12 | 1.07 | 1.01 | 0.80 |
| 4 | 30 | 1.15 | 1.12 | 1.06 | 0.97 |
| 5 | 50 | 1.20 | 1.17 | 1.12 | 1.10 |
| 6 | 100 | 1.26 | 1.24 | 1.20 | 1.20 |
| 7 | 150 | 1.30 | 1.28 | 1.24 | 1.24 |
| 8 | 200 | 1.32 | 1.30 | 1.27 | 1.27 |
| 9 | 250 | 1.34 | 1.32 | 1.29 | 1.28 |
| 10 | 300 | 1.35 | 1.34 | 1.31 | 1.30 |
| 11 | 350 | 1.35 | 1.35 | 1.32 | 1.31 |
| 12 | 400 | 1.35 | 1.35 | 1.34 | 1.32 |
| 13 | 450 | 1.35 | 1.35 | 1.35 | 1.33 |
| 14 | 500 | 1.35 | 1.35 | 1.35 | 1.34 |

NOTE: For intermediate values of height z in a given terrain category, use linear interpolation.

Topography ($k_3$ Factor) (Refer IS 875 (Part 3) (2015) Cl.6.3.3: The basic wind speed $V_b$ considers the general level of the site above sea level. This does not allow for local topographic features such as hills, valleys, cliffs, escarpments, or ridges which can significantly affect wind speed in their vicinity. The effect of topography is to accelerate wind near the summits of hills or crests of cliffs, escarpments or ridges and decelerate the wind in valleys or near the foot of cliffs, steep escarpments, or ridges. The effect of topography shall be significant at a site when the upwind slope ($\theta$) is more than about 3°, and below that, the value of $k_3$ is confined in the range of 1.00 to 1.36 for slopes more than 3°. A method of evaluating the value of $k_3$ for values more than 1.00 is given in Annexure C. It may be noted that the value of $k_3$ varies with height above ground level, at a maximum near the ground, and reducing to 1.00 at higher levels.

Importance Factor for Cyclonic Region $(k_4)$ (Refer IS 875 (Part 3) (2015) Cl.6.3.4: The east coast of India is relatively more vulnerable to occurrences of severe cyclones. On the west coast, Gujarat is vulnerable to severe cyclones. Studies of wind speed and damage to buildings and structures point to the fact that the speeds given in the basic wind speed map are often exceeded during cyclones. The effect of cyclonic storms is largely felt in a belt of approximately 60 km width at the coast. To ensure better safety of structures in the region (60 km wide on the east coast as well as on the Gujarat coast), the following values of $k_4$ (as recommended in IS 15498) are stipulated as applicable according to the importance of the structure:

| | $k_4$ |
|---|---|
| Structures of post-cyclone importance for emergency services (such as cyclone shelters, hospitals, schools, communication towers, etc.) | 1.30 |
| Industrial structures | 1.15 |
| All other structures | 1.00 |

**Design wind pressure, from IS 875 (Part 3) (2015) Cl.7.2**

$$p_d = K_d\ K_a\ K_c\ p_z$$

where,

$K_d$ = wind directionality factor,
$K_a$ = area averaging factor, and
$K_c$ = combination factor (see 7.3.3.13)

The value of $p_d$, however shall not be taken as less than $0.70p_z$.

1. The coefficient 0.60 (in SI units) in the above formula depends on some factors and mainly on the atmospheric pressure and air temperature. The value chosen corresponds to the average Indian atmospheric conditions.
2. $K_d$ should be taken as 1.00 when considering local pressure coefficients.

Wind Directionality Factor, $K_d$ (Refer IS 875 (Part 3) (2015) Cl.7.21): Considering the randomness in the directionality of wind and recognizing the fact that pressure or force coefficients are determined for specific wind directions, it is specified that for buildings, solid signs, open signs, lattice frameworks, and trussed towers (triangular, square, rectangular), a factor of 0.90 may be used on the design wind pressure. For circular or near-circular forms this factor may be taken as 1.00. For the cyclone-affected regions also, the factor $K_d$ shall be taken as 1.00.

Area Averaging Factor, $K_a$ (Refer IS 875 (Part 3) (2015) Cl.7.2.2): Pressure coefficients given in Cl.7.3 are a result of the average of the measured pressure values over the given area. As the area becomes larger, the correlation of measured values decreases and vice versa. The decrease in pressures due to larger areas may be considered as given in Table 3.4.

When taking wind loads on frames of clad buildings, it is reasonable to assume that the pressures or suctions inside and outside the structure shall not be fully correlated. Therefore, when taking the combined effect of wind loads on the frame, a reduction factor of $K_c = 0.90$ may be used over the building envelope when the roof is subjected to pressure and internal pressure is suction, or vice versa.

### 3.2.2 Forces due to Current

Pressure due to current will be applied to the area of the vessel below the water line when fully loaded. It is approximately equal to w $v^2/2g$ per square metre of area,

**TABLE 3.4**
**Area Averaging Factor ($K_a$) (Cl.7.2.2)**

| Sl. No. | Tributary Area (A) $m^2$ | Area Averaging Factor ($K_a$)* |
|---|---|---|
| 1 | ≤10 | 1.00 |
| 2 | 25 | 0.90 |
| 3 | ≥100 | 0.80 |

* Linear interpolation for intermediate values of a is permitted.

where v is the velocity in m/s and w is the unit weight of water in $T/m^3$. The ship is generally berthed parallel to the current. With strong currents and where berth alignment materially deviates from the direction of the current, the likely force should be calculated by any recognized method and considered. A ship is aligned predominantly in head sea condition with the current direction.

### 3.2.3 Bollard Pull as per IS 4651 (Part 3) (2020) Cl.6.3.4

The mooring loads are the lateral loads caused by the mooring lines when they pull the ship into or along the dock or hold it against the forces of wind or current. Mooring loads are to be calculated only for a wind speed of about 30 m/s for mooring analysis and not for the storm wind as detailed in IS 875 (Part 3).

The maximum mooring loads are due to the wind forces on the exposed area on the broad side of the ship in light conditions.

$$F = Cw.\ Aw.\ P$$

where

F = Force due to wind in kg
Cw = Shape factor = 1.30 to 1.60
Aw = Windage area in $m^2$ and
P = Wind pressure in $kg/m^2$ to be taken in accordance with IS 875 (Part 3) (2015)

The windage area (Aw) can be estimated as

$$Aw = 1.175\ Lp\ (DM - DL)$$

where

Lp = Length between perpendiculars in m
DM = Moulded depth in m
DL = Average light draft in m

When the ships are berthed on both sides of a pier, the total wind force acting on the pier should be increased by 50% to allow for wind against the second ship.

The appropriate load on the bollard shall then be calculated, which depends upon the layout of the harbour, and the position of the bow line, stern line, spring line and

**TABLE 3.5**
**Bollard Pulls**

| Sl. No. | Displacement (Tons) | Line Pull (Tons) |
|---|---|---|
| 1 | 2000 | 10 |
| 2 | 10,000 | 30 |
| 3 | 20,000 | 60 |
| 4 | 50,000 | 80 |
| 5 | 100,000 | 100 |
| 6 | 200,000 | 150 |
| 7 | >250,000 | 200 |

breasting lines for guidance the bollard pulls independent of the number of laid-on hawsers, may be taken as given in Table 3.5 since the hawsers are not fully stressed simultaneously.

### 3.2.4 Assigning of Mooring Force in STAAD.Pro Software

The vessel shall be moored in 8 or 6 lines of bollard of the berth during load/unloading of cargoes. Due to the wind/wave, the mooring line will resist the movement towards the head sea/beam sea/following sea. The governing one is head sea/following sea, in which half of the mooring lines are active.

While analysing of berthing structure, two primary load cases for mooring force shall be considered: one at end bollard location at 45 degrees, and another at the centre bollard location of the berth at 90 degrees.

### 3.2.5 OPTIMOOR Analysis Software

OPTIMOOR is a mooring analysis software for ships and barges. The static or dynamic response of any type of inshore or offshore mooring shall be checked in this OPTIMOOR software. The vessel movements and mooring forces shall be obtained by inputting data of wind, current, waves, tide and draft changes.

## 3.3 LOADS FROM MECHANICAL HANDLING SYSTEMS

### 3.3.1 Mechanical Handling System

A mechanical handling system is for transferring loads from one place to another. Automated rail-mounted quay cranes (RMQC) with conveyor systems are also used for transferring coal (dry bulk cargoes).

Cranes to be used for berthing structures will differ accordingly with the nature of the deck and handling materials of the structure.

RMQC can also be used for handling container loads on the deck structure. The loads from the wheels railing, tie down forces during storm shall be derived for design considerations.

Rubber-tyred gantry cranes (RTGC) and reach stackers (RST) shall be used for handling all types of cargoes for stacking purpose in stack yards.

Harbour mobile cranes (HMC) shall be used for handling all types of cargoes in decks as well as for stacking. The loads from the pads of the HMC shall be distributed in the deck for the design of all components in the structure.

Automated RMQC with a conveyor system is also used for transferring coal (dry bulk cargoes).

A loading example of a typical RMQC wheel load, RMQC ship unloader with conveyor and HMC has been provided here.

### 3.3.2 Loads From RMQC

The crane loads to be considered for the berthing structure is discussed below for an overview.

Typical loading details of RMQC with gauge 20 m c/c

Sea side = 14 wheels

Land side = 14 wheels

The maximum load shall be considered as follows:

**TABLE 3.6**
**Maximum Load From Wheel**

| Description | | Specifications |
|---|---|---|
| No of wheels/bogie | | 14nos |
| Wheel spacing | | 1 m c/c |
| Rail c/c distance | | 20 m |
| Buffer to buffer distance | | 37 m |
| Buffer load | | 70 T/corner |
| C/C distance between bogies | | 19 m |
| Max wheel loading for operating condition | Sea side | 36 T/wheels |
| | Land side | 24 T/wheels |
| Max wheel loading for | Sea side | 35 T/wheels |
| Stowed case condition | Land side | 40 T/wheels |
| Tie down load | Sea side | 110 T/Corner |
| | Land side | 20 T/Corner |

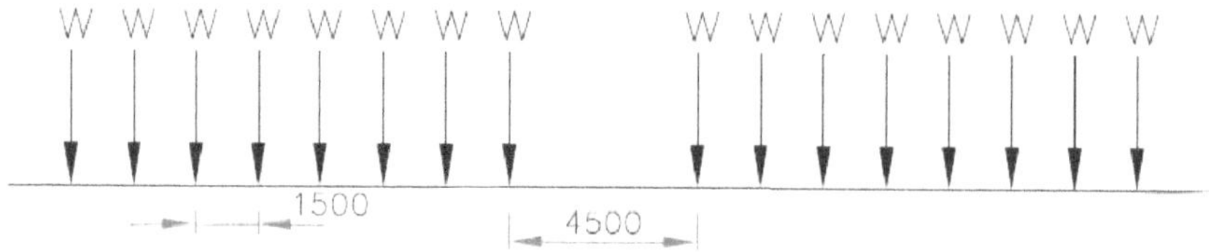


**FIGURE 3.2** Wheel loads of rail-mounted quay crane.

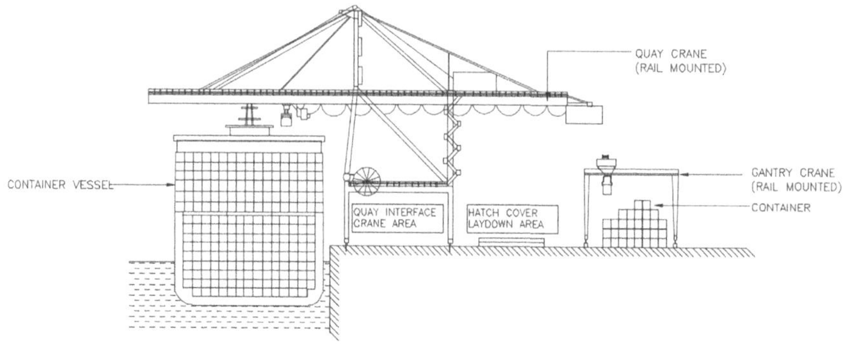


**FIGURE 3.3** Sectional view of rail-mounted quay crane.

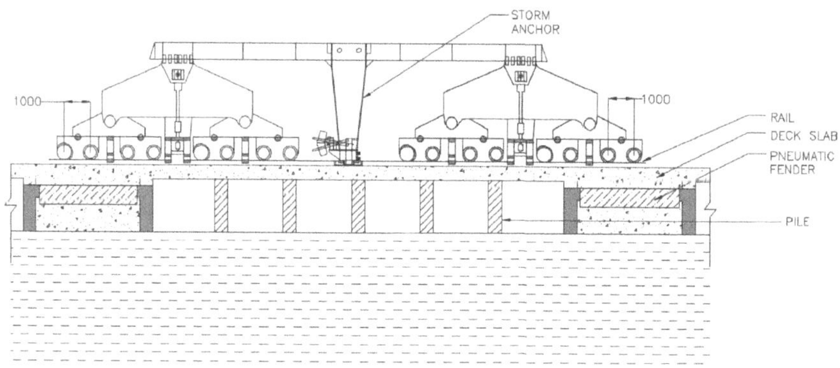


**FIGURE 3.4** Detailed view of wheel of rail-mounted quay crane.

### 3.3.3 Loads From RMQC Ship Unloader

Typical ship unloader (dry bulk cargoes) by RMQC is given below:

Two shore unloaders of 2000 TPH capacity shall be considered over 20 m rail gauge.

Sea side = 24 wheels
Land side = 20 wheels

The load points P, Q, R and S are shown in Figure 3.6. P and Q are located on the sea side, R and S are located on the lee side.

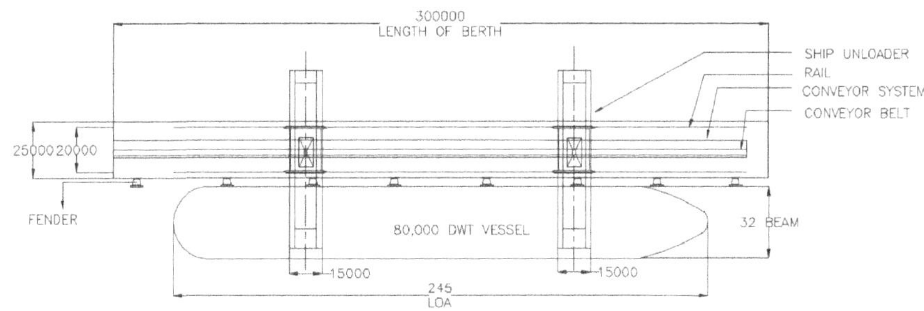


**FIGURE 3.5** Plan of ship unloader with berth.

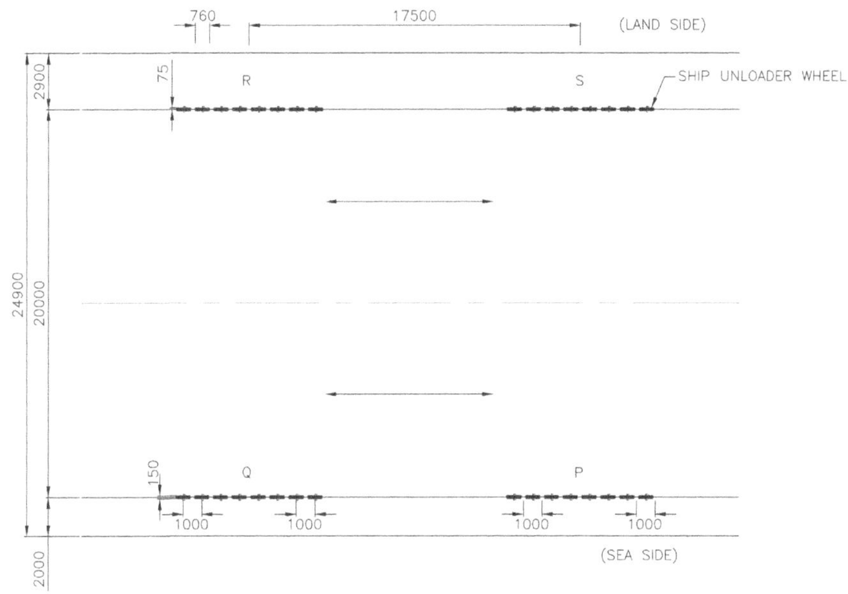


**FIGURE 3.6** Plan view showing wheel plan of ship unloader.

| | Wheel loads for shore unloaders | | | Number of wheels |
|---|---|---|---|---|
| **Load points** | **$H_x$(kN)** | **$V_y$ (kN)** | **$H_z$ (kN)** | |
| P | 122 | 643 | 157 | 12 |
| Q | 120 | 549 | 137 | 12 |
| R | 108 | 443 | 83 | 10 |
| S | 254 | 565 | 123 | 10 |

Horizontal force shall be considered on either direction

Hx – Horizontal force along longitudinal direction due to braking forces, inertia and crane motion.

Hz – Horizontal force along transverse direction due to wind load, skewing, trolley swing.

| | |
|---|---|
| Maximum load per wheel | 70T |
| Maximum load per wheel (At storm anchor points) | 80T |
| Maximum wind speed | 50 m/s |

### 3.3.4 Typical Loading Details of Harbour Mobile Crane

Maximum lifting capacity of crane = 124 t with 11 m out reach

Pad contact width = 1.80 m

Pad contact length = 5.50 m

Centre to centre distance between pads = 13.50 m

$$\text{Maximum corner load } \left(\text{Normal } 45^\circ\right) = 2433 \text{ kN}$$

Load considering impact factor = 1.25 x 2433 = 3041.25 kN

Considering dispersion angle as 45°

$$\begin{aligned}\text{Dispersed width of the track} &= \left(2\times\text{Thickness of slab and wearing coat}\right)+\text{B of track}\\ &= \left(2\times(0.50+0.10)\right)+1.80\\ &= 3.00\text{ m}\end{aligned}$$

$$\begin{aligned}\text{Dispersed length of the track} &= \left(2\times\text{Thickness of slab and wearing coat}\right)+\text{L of track}\\ &= \left(2\times(0.50+0.10)\right)+5.50\\ &= 6.70\text{ m}\end{aligned}$$

$$\begin{aligned}\text{Load intensity} &= 3041/\left(3.00*6.70\right)\\ &= 151.30\text{ kN/m}^2\end{aligned}$$

The three-dimensional view of a mobile harbour crane is shown in Figure 3.7, and the pad dimensions of a mobile harbour crane are shown in Figure 3.8. The load capacity of a mobile harbour crane (LHM 550) for bulk operation is shown in Figure 3.9.

For analysis, all wheels of two crane loads (LHM or RMQC) shall be assumed as acting on one module at the same time, positioning the cranes at equidistance from the middle of the module, keeping a gap between the cranes as 15 m. Live loads shall be considered on the remaining area leaving 5 m distance from the cranes.

## 3.4 ACTIVE EARTH PRESSURE

If the wall moves sufficiently away from the backfill by translation motion or rotation about the base of their combination, the lateral pressure of the backfill is reduced,

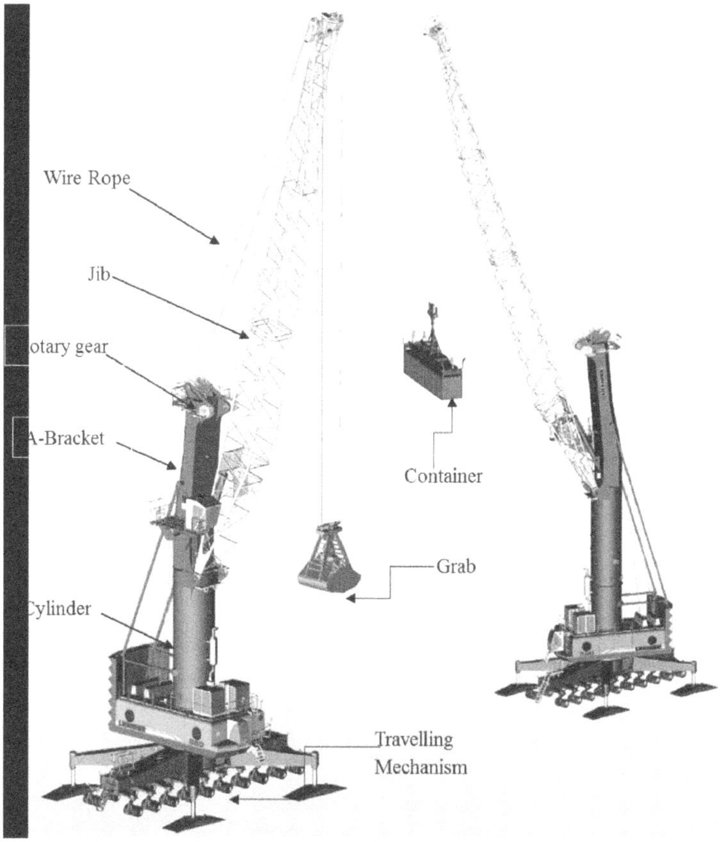


**FIGURE 3.7** Three-dimensional view of mobile harbour crane.

and it is termed as active earth pressure. The pressure (active, at rest and passive versus displacement as shown in Figure 3.10.)

The deflection corresponding to active earth pressure varies from 0.001 H to 0.004 H for sand and 0.01 H for clay. The deflection corresponding to passive earth pressure varies from 0.01 H to 0.04 H for sand and 0.05 H for clay. H is defined as a height from dredge level to top level of the structure.

### 3.4.1 Cohesionless Soils

The active earth pressure at H (see Figure 3.11) for Cohesionless soil layer is calculated by the following formula:

$$P_A = K_A \gamma H \tag{3.18}$$

$$K_A = \frac{1 - sin\phi}{1 + sin\phi} \tag{3.19}$$

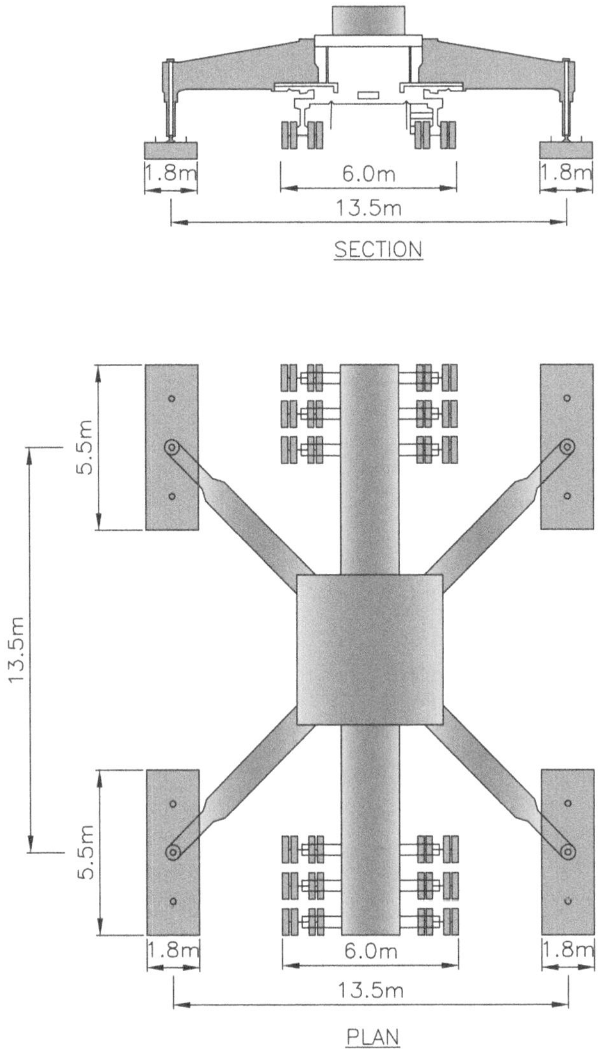


**FIGURE 3.8** Pad dimensions of mobile harbour crane.

where

$P_A$ = Active earth pressure in kN/m$^2$
$K_A$ = Coefficient of active earth pressure
$\Phi$ = Angle of internal friction
$Y$ = Unit weight of the soil layer in kN/m$^3$
$Z$ = Depth of the soil layer in m.

The effect of wall friction on the values of $K_A$ is not very large and may be neglected for sheet pile structures; that is, take $\delta = 0$. The error resulting from this is on the safer side.

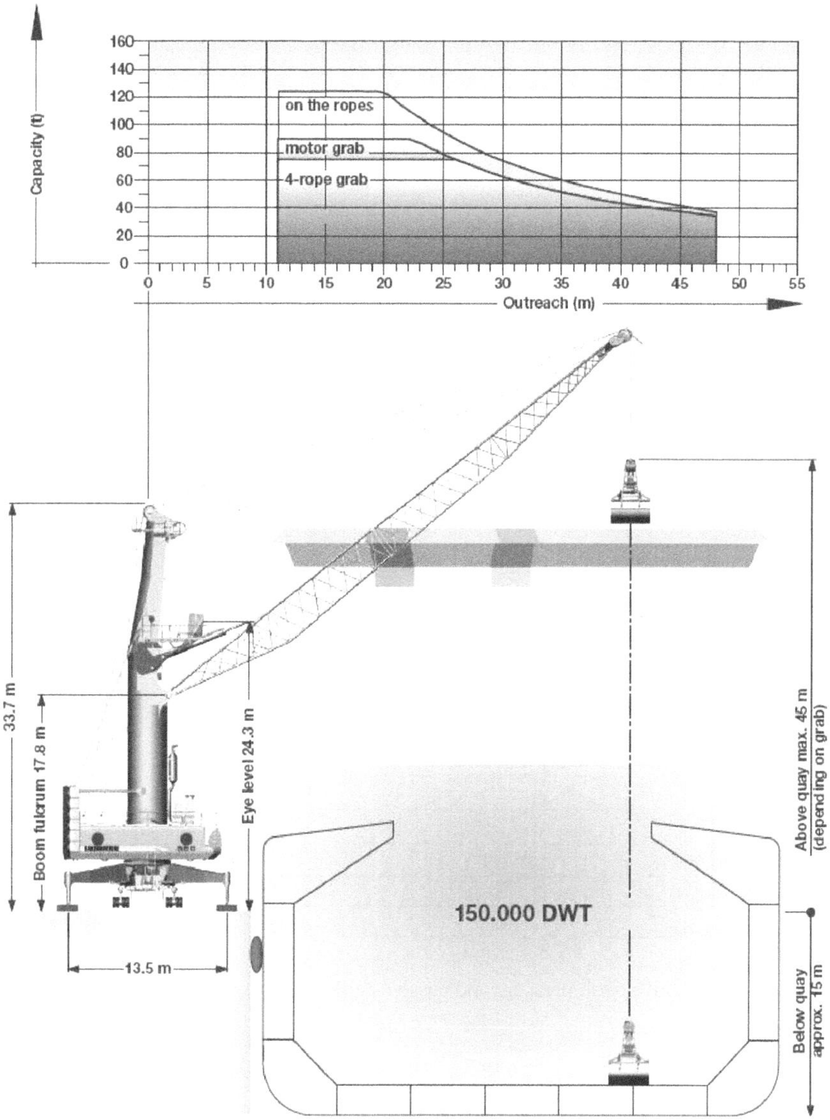


FIGURE 3.9 Load capacity of mobile harbour crane (LHM 550) for bulk operation.

### 3.4.2 Cohesive (φ = 0) and Mixed (c − φ) Soils

The active earth pressure at any depth $H$ for the cohesive soil layer is calculated by the following formula:

$$P_A = K_A \gamma H - 2C\sqrt{K_A} \tag{3.20}$$

$$K_A = \frac{1 - sin\phi}{1 + sin\phi}$$

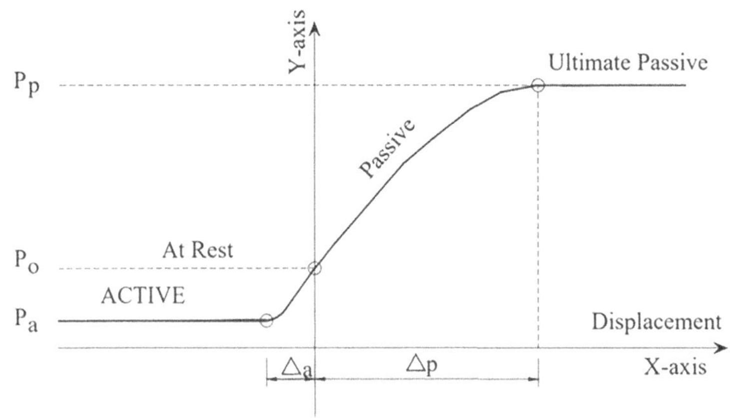


**FIGURE 3.10** Active earth pressure.

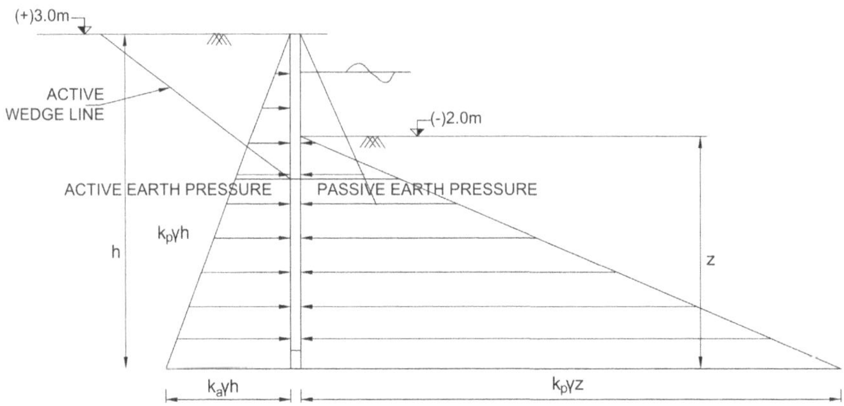


**FIGURE 3.11** Active earth pressure and passive earth pressure.

where

$P_A$ = Active earth pressure in kN/m$^2$
$K_A$ = Coefficient of active earth pressure
$\Phi$ = Angle of internal friction
$Y$ = Unit weight of the soil layer in kN/m$^3$
Z = Depth of the soil in m
C = Cohesion of the soil layer in kN/m$^2$

The effect of wall friction on the values of $K_A$ is small for sheet pile structures; that is, take $\delta = 0$. The error is on the safer side. The wall adhesion $C_w$ for calculation of active pressure should be taken as being equal to C up to a maximum of 50 kN/m$^2$.

**TABLE 3.7**
**Values of $K_A$ for Cohesionless Soils – Vertical Walls With Horizontal Ground Surface**

| | Values of φ | | | | |
|---|---|---|---|---|---|
| **Value of δ** | **25°** | **30°** | **35°** | **40°** | **45°** |
| 0° | 0.41 | 0.33 | 0.27 | 0.22 | 0.17 |
| 10° | 0.37 | 0.31 | 0.25 | 0.20 | 0.16 |
| 20° | 0.34 | 0.28 | 0.23 | 0.19 | 0.15 |
| 30° | – | 0.26 | 0.21 | 0.17 | 0.14 |

**TABLE 3.8**
**Values of $K_A$ for Cohesive and Mixed Soils**

| | | | Values of φ | | | | | |
|---|---|---|---|---|---|---|---|---|
| **Coefficient** | **Values of δ** | **Values of $C_w/C$** | **0°** | **5°** | **10°** | **15°** | **20°** | **25°** |
| $K_A$ | 0 | All | 1.00 | 0.85 | 0.70 | 0.59 | 0.48 | 0.40 |
| | φ | values | 1.00 | 0.78 | 0.64 | 0.50 | 0.40 | 0.32 |
| | 0 | 0 | 2.00 | 1.83 | 1.68 | 1.54 | 1.40 | 1.29 |

### 3.4.3 Lateral Earth Pressure due to Surcharge Loads

If the backfill carries a uniformly distributed superimposed load $q$, the lateral active earth pressure at any depth $Z$ is given by:

$$P_A = K_A(q + \gamma z), \text{ for granular soil and}$$

$$K_A(q + \gamma z) - 2c\sqrt{K_A}, \text{ for cohesive and mixed soils.}$$

## 3.5 PASSIVE EARTH PRESSURE

Passive earth pressure occurs when the wall moves towards the backfill and provides fixity to the base of the wall. In this case, the backfill offers internal shearing resistance.

### 3.5.1 Passive Resistance for Cohesionless Soils

The passive earth pressure at any depth $Z$ for the cohesionless soil layer is calculated by the following formula:

$$P_p = K_p \gamma z, \tag{3.21}$$

**TABLE 3.9**
**Values of $K_p$ for Cohesionless Soils**

| | Values of φ | | | |
|---|---|---|---|---|
| **Values of δ** | **25°** | **30°** | **35°** | **40°** |
| 0° | 2.50 | 3.00 | 3.70 | 4.60 |
| 10° | 3.10 | 4.00 | 4.80 | 6.50 |
| 20° | 3.70 | 4.90 | 6.00 | 8.80 |
| 30° | – | 5.80 | 7.30 | 11.40 |

$$K_p = \frac{1+sin\phi}{1-sin\phi}$$

where

$P_p$ = Passive earth pressure in kN/m$^2$
$K_p$ = Coefficient of passive earth pressure
Φ = Angle of internal friction
Y = Unit weight of the soil layer in kN/m$^3$
Z = Depth of the soil layer in m

For calculating passive earth resistance in the design of sheet pile walls, it is usual to take δ = 2/3 ϕ, except in the case of certain silty sands where a reduced value of δ should be taken (δ = 1/3 ϕ to 2/3 ϕ). The values of the coefficient $K_p$ for different values of angle of shearing resistance ϕ, angle of wall friction δ are given in Table 3.9.

### 3.5.2 Passive Earth Pressure for Cohesive (Φ = 0) and Mixed (c − Φ) Soils

The passive earth pressure at any depth Z for cohesive soil is calculated by the following formula:

$$P_p = K_p \gamma Z + 2C\sqrt{K_p} + K_p q \tag{3.22}$$

where

$P_p$ = Passive earth pressure in kN/m$^2$

$K_p$ = Coefficient of passive earth pressure, $K_p = \frac{1+sin\phi}{1-sin\phi}$

Φ = Angle of internal friction

Y = Unit weight of the soil layer in kN/m$^3$

Z = Depth of the soil in m

C = Cohesion of the soil layer in kN/m$^2$

**TABLE 3.10**
**Values of $K_p$**

| Coefficient | Values of δ | Values of φ 0° | 5° | 10° | 15° | 20° | 25° |
|---|---|---|---|---|---|---|---|
| $K_p$ | 0 | 1.00 | 1.20 | 1.40 | 1.70 | 2.20 | 2.50 |
| | φ | 1.00 | 1.30 | 1.60 | 2.20 | 2.90 | 3.90 |

**TABLE 3.11**
**N Versus C for Clay With References**

| N | Consistency | C (KPa) |
|---|---|---|
| <2 | Very soft | <12 |
| 2–4 | Soft | 12–25 |
| 4–8 | Medium | 25–50 |
| 8–16 | Stiff | 50–100 |
| 16–32 | Very stiff | 100–200 |
| >32 | Hard | >200 |

**TABLE 3.12**
**Relationship Between 'N' Values and Sand Properties**

| Type of Sand | SPT – N Value | Dry Unit Weight of Soil (kN/m³) | Friction Angle φ |
|---|---|---|---|
| Very loose | <4 | <14 | <30 |
| Loose | 4–10 | 14–16 | 30–32 |
| Medium dense | 10–30 | 16–18 | 32–35 |
| Dense | 30–50 | 18–20 | 35–38 |
| Very dense | >50 | >20 | >38 |

The values of the coefficient $K_p$ for different values of angle of shearing resistance φ, angle of wall friction δ and the ratio of adhesion to cohesion are given in Table 3.10.

Tables 3.11 through 3.15 are used to obtain φ and C based on SPT-N value.

**TABLE 3.13**
**Typical Friction Angle of Granular Soils**

| Type | Description | Friction Angle (degrees) |
|---|---|---|
| Cohesionless | Soft sedimentary (chalk, shale, siltstone, coal) | 30–40 |
| Compacted broken rock | Hard sedimentary (conglomerate, sandstone) | 35–45 |
| | Metamorphic | 35–45 |
| | Igneous | 40–50 |
| Cohesionless gravels | Very loose/loose | 30–34 |
| | Medium dense | 34–39 |
| | Dense | 39–44 |
| | Very dense | 44–49 |
| Cohesionless sands | Very loose/loose | 27–32 |
| | Medium dense | 32–37 |
| | Dense | 37–42 |
| | Very dense | 42–47 |
| Cohesionless sands | Loose | 25–30 |
| | Uniformly graded | 27–30 |
| | Well graded | 30–32 |
| | Dense | 36 |
| | Uniformly graded | 37–40 |
| | Well graded | 40–42 |

**TABLE 3.14**
**Typical Shear Strength of Intact Rock**

| | | Shear Strength | |
|---|---|---|---|
| **Origin** | **Rock Type** | **Cohesion (MPa)** | **Friction angle (°)** |
| Sedimentary – soft | Sandstone (TriassicTriassic), coal, chalk, shale, limestone (Triassic) | 1–20 | 25–35 |
| Sedimentary – hard | Limestone, dolomite, greywacke sandstone (carboniferous), limestone (carboniferous) | 10–30 | 35–45 |
| Metamorphic – non-foliated | Quartzite, marble, gneiss | 20–40 | 30–40 |
| Metamorphic – foliated | Schist, slate, phyllite | 10–30 | 25–35 |
| Igneous – acid | Granite | 20–30 | 45–55 |
| Igneous – basic | Basalt | 30–50 | 30–40 |

**TABLE 3.15**
**Representative Range of Unit Weight**

| Type | Soil Description | Unit Weight Range (kN/m³) Dry | Saturated |
|---|---|---|---|
| Cohesionless | Soft sedimentary (chalk, shale, siltstone, coal) | 12 | 18 |
| Compacted broken rock | Hard sedimentary (conglomerate, sandstone) | 14 | 19 |
| | Metamorphic | 18 | 20 |
| | Igneous | 17 | 21 |
| Cohesionless | Very loose | 14 | 17 |
| | Loose | 15 | 18 |
| Sands and gravels | Medium dense | 17 | 20 |
| | Dense | 19 | 21 |
| | Very dense | 21 | 22 |
| Cohesionless | Loose | | |
| | Uniformly graded | 14 | 17 |
| | Well graded | 16 | 19 |
| Sands | Dense | | |
| | Uniformly graded | 18 | 20 |
| | Well graded | 19 | 21 |
| Cohesive | Soft – organic | 8 | 14 |
| | Soft – non-organic | 12 | 16 |
| | Stiff | 16 | 18 |
| | Hard | 18 | 20 |

*NOTES:*

- Use saturated unit weight for soils below the water table and within the capillary fringe above the water table.
- Buoyant unit weight = Saturated unit weight – unit weight of water.
- The compacted rock unit weight shown is lower than the in-situ unit weight.

## 3.6 DIFFERENTIAL WATER PRESSURE

### 3.6.1 Determination of Differential Water Pressure as per IS 4651 (Part 3) (2020) Cl.6.4

In the case of water-front structures with backfill, the pressure caused by the difference in water levels at the fill side and the waterside has to be considered in the design. The magnitude of this hydrostatic pressure is influenced by the tidal range, free water fluctuations, the ground water influx, the permeability of the foundation soil and the structure, as well as the efficiency of available backfill drainage.

In the case of good and poor drainage conditions of the backfill the differential water pressure may be calculated on the guidelines given in Figure 2 of IS 4651 (Part 3) (2020). Please refer example 4 for typical calculation.

## 3.7 SEISMIC FORCE

The horizontal force caused by the earthquakes called seismic force shall be calculated using the seismic coefficient method given in IS 1893 (Part 1) (2016).

### 3.7.1 Seismic Force Terminology

- Design horizontal acceleration coefficient ( $A_h$ )
  It is a horizontal acceleration coefficient that shall be used for the design of structures.
- Seismic zone factor
  It is the value of peak ground acceleration considered for the design of structures located in each seismic zone.
- Importance factor (I)
  It is a factor used to estimate design seismic force depending on the functional use of the structure, characterized by hazardous consequences of its failure, post-earthquake functional needs, historical value or economic importance
- Response reduction factor (R)
  It is the factor by which the base shear induced in a structure, if it were remain elastic, is reduced to obtain the design base shear.
- Response acceleration coefficient of a structure ( $S_a/g$ )
  It is a factor denoting the normalized design acceleration spectrum value to be considered for the design of structures subjected to earthquake ground shaking; this depends on the natural period of oscillation of the structure and damping to be considered in the design of the structure.

### 3.7.2 Seismic Force as per IS 1893 (Part 1) (2016)

$$\text{Seismic force} = A_{h\,X}\left(\text{Dead load} + 50\ \text{Live load}\right)$$

Horizontal Seismic coefficient

$$A_h = \frac{ZI}{2R}\left(\frac{S_a}{g}\right)$$

where

Z = Zone factor

I = Importance factor

R = Response reduction factor

$S_a/g$ = Average response acceleration coefficient (as given in Figure 2 in IS 1893 (Part 1) (2016))

**Equivalent static method**

- For rocky or hard soil sites

$$2.50,\ 0 < T < 0.40\ \text{s}$$

**TABLE 3.16**
**Importance Factor for Seismic Load**

| Sl. No. | Structure | I |
|---|---|---|
| 1 | Important service and community buildings or structures (for example, critical governance buildings, schools), signature structures, monument structures, lifeline and emergency structures and large community halls | 1.50 |
| 2 | Residential or commercial buildings or structures, with occupancy more than 200 persons | 1.20 |
| 3 | All other buildings or structures | 1.00 |

$$Sa/g = 1.00 / T, \; 0.40 < T < 4.00 \text{ s}$$
$$0.25, \; T > 4.00 \text{ s}$$

- For medium soil sites

$$2.50, \; 0 < T < 0.55 \text{ s}$$
$$Sa/g = 1.36 / T, \; 0.55 < T < 4.00 \text{ s}$$
$$0.34, \; T > 4.00 \text{ s}$$

- For soft soil sites

$$2.50, \; 0 < T < 0.67 \text{ s}$$
$$Sa/g = 1.67 / T, \; 0.67 < T < 4.00 \text{ s}$$
$$0.42, \; T > 4.00 \text{ s}$$

**Response spectrum method**

- For rocky or hard soil sites

$$1+15T, \; T<0.1 \text{ s}$$
$$Sa/g = 2.50, \; 0.1 < T < 0.40 \text{ s}$$
$$1.00/ T, \; 0.40 < T < 4.00 \text{ s}$$
$$0.25, \; T > 4.00 \text{ s}$$

- For medium soil sites

$$1+15T, \; T<0.1 \text{ s}$$
$$Sa/g = 2.50, \; 0 < T < 0.55 \text{ s}$$
$$1.36 / T, \; 0.55 < T < 4.00 \text{ s}$$
$$0.34, \; T > 4.00 \text{ s}$$

- For soft soil sites

$$Sa/g = \begin{cases} 1+15T, & T<0.1 \text{ s} \\ 2.50, & 0<T<0.67 \text{ s} \\ 1.67/T, & 0.67<T<4.00 \text{ s} \\ 0.42, & T>4.00 \text{ s} \end{cases}$$

### 3.7.3 Assigning of Seismic Force in STAAD.Pro

1. Joint weight method.
   A separate STAAD model for self-weight X, Y, Z directions shall be created, and frequency analysis shall be carried out for DL+0.50 LL to get fundamental natural period(s).
   Step 1: Berth is modelled assigning all properties, supports and specifications separately for this method
   Step 2: DL+ 0.50 LL is added in load definitions
   Step 3: Self-weight load is assigned in one direction (X) in DL+ 0.50 LL
   Step 3: 50 of Live load is applied in the same DL+ 0.50 LL load case
   Step 4: Modal and Rayleigh frequency in load definitions is added
   Step 5: P-delta analysis definition with 10 iterations is assigned for the DL+ 0.50 LL load case, and the model is run for analysis.
   Step 6: Corresponding time (Fundamental natural period) with respect to frequency is taken from the output of each direction STAAD.
   Step 7: This obtained time (t in 's') is used for deriving Sa/g in $A_h$ as per IS 1893 in respective X and Z direction
   Step 8: $A_h$ value is derived, and the same is multiplied with weight (DL + 0.50 LL) to get the seismic force in the X and Z direction
   Step 9: The force is applied in the Beam column junction in X, -X, Z, -Z directions separately in main STAAD.Pro.
2. Assigning of all properties as per code in seismic load definition itself in the main STAAD.Pro model with dead load and 50% live load applied and adding seismic force in load definition separately for X, -X, Z, -Z directions. After running the model, it will show the force applied at beam column joint.

## 3.8 WAVE FORCE ON PILE AND DIAPHRAGM WALLS

Three distinct types of waves

- Non-breaking waves
- Breaking waves
- Broken waves

#### Non-breaking waves

- Generally, when the depth of water against the structure is greater than about 1.5 times the maximum expected wave height, non-breaking wave conditions occur.

- Forces due to non-breaking waves are essentially hydrostatic. The 'Sainflou method' may be used for the determination of pressure due to non-breaking waves for simple walls in deep water. The method of computation using the Sainflou Method is outlined in sections 3.8.1 and 3.8.2.
- Goda's method is widely used for shallow or intermediate water for non-breaking and breaking waves (average force) on monolithic and composite structures.

**Breaking waves**

- Breaking waves cause both static and dynamic pressures.
- Determination of the design wave for breaking wave conditions may be based on the depth of water about seven breaker heights, seaward of the structure, instead of the water depth at which the structure is located.
- The actual pressures caused by a breaking wave is obtained by following the method suggested by Minikin.

**Broken waves**

Locations of certain structures like protective will be such that waves will break before striking them.

### 3.8.1 Wave Forces on Piles as per IS 4651 (Part 3) (2020)

The total force 'F' exerted by non-breaking waves on a cylindrical pile can be divided into two components:

1. Force due to inertia
2. Force due to drag

The total force acting on the vertical pile is the summation of inertial force and drag force.

$$F_T = \int_{-d}^{0} (f_1 + f_d)\,dz \tag{3.23}$$

$$F_1 = \int_{-d}^{0} f_1 dz = \frac{w\rho g C_D H^2 D}{8} tanh(kd)\, sin(\theta) \tag{3.24}$$

$$F_D = \int_{-d}^{0} f_D dz = \rho g C_D H^2 D \left[\frac{2kd + sinh(2kd)}{16 sinh(2kd)}\right] cos(\theta)\left|cos(\theta)\right| \tag{3.25}$$

The maximum total force acting on the vertical pile is given by:

$$F_M = F_{1M} \sin(\theta_{max}) + F_{DM} \cos(\theta_{max})\left|cos(\theta_{max})\right| \tag{3.26}$$

The maximum force occurs at a phase,

$$\sin(\theta_{max}) = \frac{F_{1M}}{2F_{DM}}$$

$$\frac{F_{1M}}{\rho g C_D H^2 D} = \frac{w}{8} tanh(kd)$$

$$\frac{F_{DM}}{\rho g C_D H^2 D} = \left[\frac{2kd + \sinh(2kd)}{16\sinh(2kd)}\right]$$

$$M_M = F_{1M} \sin(\theta_{max}) S_I + F_{DM} \cos(\theta_{max}) |\cos(\theta_{max})| S_D \tag{3.27}$$

$$S_I = d\left[1 - \frac{\cosh(kd) - 1}{kdsinh(kd)}\right] \tag{3.28}$$

$$S_D = d\left[1 - \frac{1}{2n}\left(\frac{\cosh(2kd) - 1 + 2(kd)^2}{2sinh(2kd)kd}\right)\right] \tag{3.29}$$

If $\frac{F_{1M}}{2F_{DM}} > 1$ the force is predominantly inertial. The maximum force is taken as 1.25 $F_{IM}$.

The effective lever arm from the bottom of the pile is,

$$S = \frac{M_M}{F_M}$$

The drag coefficients and inertia coefficients depend upon the value of Keulegan–Carpenter number (KC),

$$KC = \frac{\pi H}{D} \coth(kd)$$

where

$F_T$ = total force on the vertical pile from the sea bottom to the surface crest elevation, in N

$F_T$ = inertia force on the vertical pile at any section of the pile, in N/m

$f_D$ = drag force on the vertical pile at any section of the pile, in N/m

$F_I$ = total inertial force on the vertical pile from the sea bottom to the surface crest elevation, in N

$F_D$ = total drag force on the vertical pile from the sea bottom to the surface crest elevation, in N

$\theta$ = wave phase, in radians.

$K = wave\ number = \frac{2\pi}{L}, in\ m^{-1}$

$L$ = wavelength $= \frac{gT^2}{2\pi} \tanh(kd)$ in m; the wave length is obtained by iterative process.

$$w = \frac{C_D D}{C_M H}$$

$C_D$ = drag force coefficient
$C_M$ = inertial force coefficient
$D$ = diameter of pile, in m
$H$ = wave height, in m
$T$ = time period of the wave, in s
$d$ = water depth, in m
$z$ = water depth at any level from still water line, in m
$\rho$ = density of sea water = 1025 kg/m$^3$
$g$ = acceleration due to gravity = 9.81 kg/m$^2$
$C_m$ = 2.0 = maximum total force on the vertical pile from the sea bottom to the surface crest elevation, in N
$F_{IM}$ = total maximum inertia force on the vertical pile from the sea bottom to the surface crest elevation, in N
$F_{DM}$ = total maximum drag force on the vertical pile from the sea bottom to the surface crest elevation, in N
$\theta_{max}$ = wave phase at which the total force is maximum, in radians
$M_M$ = maximum total moment, in Nm
$S_D$ = effective lever arm for $F_{DM}$ from the bottom of pile, in m
N = ratio of group celerity to celerity of wave

$$= \frac{1}{2}\left(1 + \frac{2kd}{\sinh(2kd)}\right)$$

$S_I$= effective lever arm for $F_{IM}$ from the bottom of pile, in m and
S = effective lever arm for $F_M$ from the bottom of pile, in m.

The above equation is applicable only when $\frac{D}{L} < 0.2$ and $\frac{H}{D} > 1$.

The wave force can also be computed using Morison equation considering the following:

1. The operating (1-year return period) and survival sea state (100 year or 50-year return period depending on life of structure to estimate the wave force and 1000-year return period to fix the top level) shall be considered. The wave slam and slap on structures shall also be considered if the air gap provided is less than 1.5 m.
2. The wave force reduction factor to account for three dimensional effects of wave spreading in reality versus 2D wave theory can be considered along with:
   a. Wave load reduction due to array of piles spaced at less than 3 times diameter of piles (shielding effect).
   b. Wave load on rectangular structures especially partly submerged, such as breasting and mooring dolphins.
3. In lieu of $C_d$ and $C_m$ variation along the depth as a function of KC, wave load on piles (assuming KC values may range from 20 to 30 for most practical

conditions across the depth), it is recommended to use the following coefficients in Morison equation:

$$C_d = 0.7 \text{ and } C_m = 2.0\text{, for piles without marine growth}$$

$$C_d = 1.05 \text{ and } C_m = 2.0\text{, for piles with marine growth}$$

The Morison equation can also be used for estimating force on piles due to breaking wave with the assumption that the wave acts on the pile as a water mass with high velocity with zero acceleration. Hence $C_m$ is set to zero and $C_d$ is increased to 1.75.

The wave and current shall be assumed to act collinear and the combined force shall be calculated using the vectoral sum velocity due to wave and current.

NOTE – Addition of wave and current force calculated separately and added is not correct due to non-linear term in the drag part of the Morison equation.

The Wave force on piles due to breaking waves shall be calculated using the following equation

$$F_b = C_D \frac{1}{2} \rho g D H_b K_{Dm} \tag{3.29A}$$

$F_b$ = Force due to breaking wave
$C_D$ shall be used as 2.5 times 0.7 (1.75)
$K_{Dm}$ = 1 (for shallow water depth)
D = diameter of the pile
$H_b$ = Breaking wave height

### 3.8.2 Calculation of Wave Force – Sainflou Method as per IS 4651 (Part 3) (2020) for Sheet Piles or Sea Wall or Breakwaters Diaphragm Wall with Both Sides Exposed to Water

Sainflou's method states that when a wave of length L and height H strikes the vertical member (wall) a standing wave or Clapotis wave pattern is formed. The pressure shall be calculated as follows:

$$h_0 = \frac{\pi H^2}{L} coth \frac{2\pi d}{L} \tag{3.30}$$

$$P_1 = \frac{wH}{\cos h \frac{2\pi d}{L}} \tag{3.31}$$

where

d = Depth from still water level
H = Height of original free wave
L = Length of wave
w = Weight per $m^3$ of water

$h_0$ = Height of orbit centre (on mean level) above still water level
$P_1$ = Pressure the Clapotis adds to or subtracts from still water pressure

The symbols are indicated in Figure 4 ANNEX B, IS 4651 (Part 3) (2020).

Assuming the same still water level on both sides of the wall, the pressure diagram will be as given in Figure 5, ANNEX B, IS 4651 (Part 3) (2020) in which

$$R_e = \frac{(d+H+h_o)(wd+p_1)}{2} - \frac{wd^2}{2} \tag{3.32}$$

$$M_e = \frac{(d+H+h_o)^2(wd+p_1)}{6} - \frac{wd^3}{6} \tag{3.33}$$

$$R_i = \frac{wd^2}{2} - \frac{(d+h_o-H)(wd-p_1)}{2} \tag{3.34}$$

$$M_i = \frac{wd^3}{6} - \frac{(d+h_o-H)(wd-p_1)}{6} \tag{3.35}$$

where

$R_e$ = resultant pressure with maximum crest level;
$M_e$ = moment due to $R_e$ about the base;
$R_i$ = resultant pressure with minimum through level; and
$M_i$ = moment due to $R_i$ about the base

### 3.8.3 Calculation of Wave Force – Goda's Method as per IS 4651 (Part 3) (2020) for Sheet Piles or Sea Wall or Breakwaters or Diaphragm Wall With Both Sides Exposed to Water

Goda's method of wave force calculation is for non-breaking wave forces acting on the wall. A factor of safety of 1.2 shall be considered for structure design against sliding and overturning. The pressure shall be calculated as follows with the reference to the Figure 6, ANNEX B, IS 4651 (Part 3) (2020).

**Procedure to calculate wave pressure:**

- Significant wave height $H_s$ and significant wave period $T_s$ shall be calculated for the design sea state.
- $h$ shall be obtained by $h = D + 5 \times mH_s$, where $m$ is the bottom slope.
- Non-breaking wave height, $H_b$ at $h$ shall be calculated. It is to note that $H_b$ should be greater than maximum wave height. The maximum wave height shall be estimated as 1.8 times the $H_s$.
- Wave length 'L' for depth 'D' shall be determined using the significant wave period of the design sea state.
- The wave forces and moments shall be now calculated using Goda's equation 3.23 to 3.29A.

The following notes shall have to be taken into point while deriving the wave pressure as per IS 4651 (Part 3) (2020) Annex B:

- When there is no water on the landward side of the wall, then the total pressure on the wall will be represented by the triangle ACB (Figure 4, ANNEX B, IS 4651 (Part 3) (2020)) when the Clapotis crest is at A.
- If there is wave action on the landward side also, then the condition of the crest of Clapotis on the seaside and trough of the wave on the harbour side will produce maximum pressure from the seaside.
- The maximum pressure from the harbour side will be produced when the trough of the Clapotis on the seaside and the crest of a wave on the land-side are at the structure.
- If the height of the wall is less than the predicted wave-height at the wall, forces may be approximated by drawing the force polygon as if the wall were higher than the impinging waves then analysing only that portion below the wall crest. Forces due to a wave crest at the wall are computed from the area *AFBSC*, as shown in Figure 7, ANNEX B, IS 4651 (Part 3) (2020).

## 3.9 CURRENT ON PILES

Current is a continuous, directed movement of sea water generated by several forces acting upon the water, including wind, breaking waves etc.

As per IS 4651 (Part 3) (2020):

- Pressure due to current will be applied to the area of the vessel below the water line when fully loaded.
- It is approximately equal to $wv^2/2g$ per square meter of area, where v is the velocity in m/s and w is the unit weight of water in tonnes/m$^3$.

### EXAMPLES

1a. Calculate the berthing energy for 80,000 DWT vessel in moderate wind and swells condition.

### Given Data

Dead weight tonnage (DWT) = 80,000

Site condition = Moderate wind and swells

From IS 4651 (Part 3) (2020) Annexure A, Dimensions of vessels will be taken.

Overall length of the vessel L = 259.00 m

Width of the vessel B = 39.20 m

Draught of the vessel D = 12.60 m

Unit weight of sea water $w = 1.03\ \text{t/m}^3$

### Solution

Berthing energy, $E = W_D \times V^2 \times C_m \times C_e \times C_s / 2g$ (Refer 3.1.1 IS code 4651 method)

To convert dead weight tonnage (DWT) into displacement tonnage (DT), IS 4651 (Part 3) (2020) Cl.3.1.2 is referred.

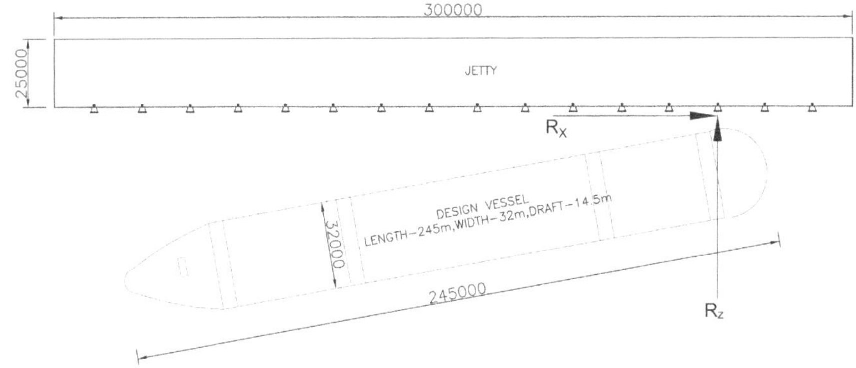


**FIGURE 3.12** Diagram representing berthing force.

Displacement tonnage of the vessel $W_D$ = DWT x 1.25
= 80,000 x 1.25
= 100,000 t.
To obtain approaching velocity of vessel (m/s), IS 4651(Part 3) (2020) Table 2 is referred.
Berthing condition = Moderate
Approaching velocity of the vessel (V) = 0.2 m/s
Approach angle (θ) = $10^o$
To get value of mass coefficient, the following formula as per IS 4651 (Part 3) (2020) p.no.8 is referred.

i. When DT < 20,000 T, Mass coefficient $C_m = 1 + (2D/B)$

ii. When DT > 20,000 T, Mass Coefficient $C_m = 1 + \frac{\frac{\pi}{4} D^2\ Lw}{W_D}$

(As per IS 4651 (Part 3) (2020) Cl.6.2.1.2)
Since DT = 100,000 T > 20,000 T, Formula (b) is used.

$$\text{Mass Coefficient, } C_m = 1 + \left(((\pi/4) \times (12.60)^2 \times 259 \times 1.03)/(100000)\right)$$
$$= 1 + 0.33$$
$$\text{Mass Coefficient, } C_m = 1.33.$$

To get value of eccentricity coefficient, the following formula as per IS 4651 (Part 3) (2020)

$$\text{Eccentricity Coefficient, } C_e = 1 + \frac{(l/r)^2 \sin^2 q}{1 + (l/r)^2}$$

Where,

l = distance from the centre of gravity of the vessel to the point of contact projected along the water line of berth in m r = radius of gyration of rotational radius on the plane of the vessel from its centre of gravity in m

$$r = L/4$$
$$r = 259/4$$
$$r = 64.75 \text{ m}$$
$$l = r \times \cos ,$$
$$l = 64.75 \times \cos 10^{\circ}$$
$$l = 64.75 \times \cos ((10 \times \text{pi}\ ())/180)$$
$$l = 64.749 \text{ m}$$
$$l/r = 1.00$$
$$C_c = 0.51 \text{ (From IS 4654 (Part-III)} - 2020 \text{ Table 3)}$$

To get value of softness coefficient, the following formula as per IS 4651(Part-III) – 2020

$$C_s = 0.90 \text{ (From IS 4654 (Part-III)} - 2020 \text{ C1.5.2.1.4)}$$

Therefore,

Berthing Energy E = (100000 X $0.2^2$ x 1.33 x 0.51 x 0.90)/(2 x 9.81)
Berthing Energy E = 124.46 T m.
Ultimate Berthing Energy Ult = E x FOS
Factor of Safety (FOS) = 2 as per IS 4651 (Part 4) (2014)
Ultimate Berthing Energy Ult = 124.46 x 2
Ultimate Berthing Energy Ult = 248.92 T m.

A suitable fender for 248.92 T.m berthing energy is chosen from fender catalogue based on the project requirement. For academic purpose, berthing energy is multiplied with 1.50 (up to 1.70) to obtain berthing force empirically.

Here a cone fender 1600H is proposed to absorb energy 272 kNm. The reaction of fender is 3278 kN.

Therefore, berthing force along long direction = 3278 kN
Berthing force along transverse direction = 30% (long berthing force)
=30 (3278)
=983.40 kN

1b. Derive the berthing force of the fender whose energy absorption given in per metre length with the following data:
Ultimate berthing energy = 67.893 T m
Top level of fender =+7.20 m CD
Bottom level of fender = +0.70 m CD
Number of fenders required at one location = 2

## Solution

Design berthing energy as per catalogue of manufacturer =71.29 Tm
ARCH fender shall be adopted
Total length = 6.50 m
Assume clearance between two fenders = 0.50 m
Total length of fender required = 6.50- (1 x 0.50) m
Total length of fender required = 6 m
Length of fender each = 3 m
Design berthing energy required for 1 m length of fender = 71.29/3 Tm = 23.77 T.m
From fender catalogue, about 23.77 T.m absorption energy for the fender of 1 m length shall be chosen.
Type of fender chosen: Arch fender (V type) – 1000H Rubber Grade R2
Each 3 m length
Absorption energy of fender/meter for Arch fender (V type) – 1000H rubber grade
R2= 25 Tm
Reaction force of fender = 64.10 t
Reaction for arch fender (V type) – 1000H rubber grade R2 23.77 tm energy
= 64.10/25x23.77 t/m
Design berthing force = 60.95 t/m
Design berthing force = 609.50 kN/m
2. Calculate the mooring force for 75,000 DWT vessel

## Given data

Dead weight tonnage (DWT) = 75,000
From IS 4651 (Part 3) (2020) Annexure A, Dimensions of vessels will be taken.

| | |
|---|---|
| Overall length of the vessel | L = 260.00 m |
| Width of the vessel | B = 30.00 m |
| Draught of the vessel | D = 12.60 m |
| Unit weight of sea water | w = 10.25 kN/m |
| Length between perpendiculars | $L_p$= 0.90 x overall length (empirically used) = 234 m |
| Moulded depth | $D_M$ = Draft + free board |
| | $D_M = 18$ m |

Average light draft $D_L = 0.45 \times$ Moulded depth (empirically used)
$= 0.45 \times 18$
$= 8.1$ m

No. of mooring lines $n = 8$
No. Mooring lines resisting the system = 4
Acceleration due to gravity $g = 9.81$ m/s$^2$
Current velocity $v = 1$ m/s

**Solution**

*1a. Force due to wind (F)*

$$F = C_w A_w P$$

Shape factor $(C_w)$

As per IS 4651 (Part 3) (2020)* the shape factor will be taken as 1.3 to 1.6
Hence shape factor taken as

$$C_w = 1.3$$

The windage area $(A_w)$

$$A_w = 1.175\, L_v \left(D_M - D_L\right)$$
$$A_w = 1.175 \times 234\,(18 - 8.1)$$
$$A_w = 1.175 \times 234(9.9) = 2722.01\ m^2$$

Design Wind Pressure (P)

$$p_d = K_d\ K_a\ K_c\ p_z$$

Where, (according to IS 875 Part – 3 (2015)

$$p_z = 0.60 \times V_z^2$$

Where,

$V_z = V_b \times K_1 \times K_2 \times K_3 \times K_4$
$V_b$ = the basic wind speed in m/s
= 39 m/s
$K_1 = 1.0$

According to IS 875 (Part 3) (2015) Cl.6.3.2.1 – Category 1, the $K_2$ is taken from Table 2

$$K_2 = 1.05$$

$K_3$, topography factor taken from IS 875 (Part 3) (2015) Cl.6.3.3.1
The slope is less than 3°
$K_3 = 1$
$K_4$ =Importance factor for the cyclonic region taken from IS 875 (Part 3) (2015) Cl.6.3.4
$K_4 = 1$

Hence, the wind speed

$$V_z = V_b \times K_1 \times K_2 \times K_3 \times K_4$$
$$= 39 \times 1.0 \times 1.05 \times 1 \times 1$$

Design Wind speed, $V_z = 40.95$ m/s

Wind Pressure $p_z = 0.6 \times V_z^2$

$$= 0.6 \times 40.95^2$$

$$p_z = 1006.14 \text{ N/m}^2$$

Design Wind Pressure, $p_d = K_d\ K_a\ K_c\ p_z$

$K_d$ wind directionality factor, from IS 875 (Part 3) (2015) Cl.7.2.1

$$K_d = 1$$

$K_a$ Area Averaging factor, from IS 875 (Part 3) (2015) Cl.7.2.2

$$K_a = 0.80$$

$K_c$ frames, from IS 875 (Part 3) (2015) Cl.7.3.3.13

$$K_c = 0.90$$

Hence,

$$p_d = K_d K_a\ K_c\ p_z$$
$$= 1 \times 0.8 \times 0.9 \times 1006.14$$
$$p_d = 724.14 \text{ N/m}^2$$

As per IS 875 (Part 3) (2015) Cl.7.2, the value of $p_d$ however shall not be taken as less than $0.70p_z$

Therefore,

$$= 0.70 \times p_z$$
$$= 0.70 \times 1006.14$$
$$= 704.3 < 724.14 \text{ N/m}^2$$

Wind pressure to be considered,

$$p_d = 724.14 \text{ N/m}^2$$

$$p = 0.72 \text{ kN/m}^2$$

Hence,

*1b. Force due to wind* $(F_w)$

$$\begin{aligned} F_w &= C_w A_w P \\ &= 1.3 \times 2722.005 \times 0.72 \\ &= 2547.80\ kN \end{aligned}$$

*1c. Force due to current* $(F_c)$

$$F_c = \left(wv^2/2g\right) \times A_p$$

$A_p$ = Area of ship under water= 30 x 8.1= 243 $m^2$

$$\begin{aligned} F_c &= (wv^2/2g) \times A_p \\ &= (10.25 \times 1.00^2/2.00 \times 9.81) \times 243 \\ &= (10.25/19.62) \times 243 \\ &= 0.522 \times 243 \end{aligned}$$

$$F_c = 126.95\ \text{kN}$$

*Resultant Mooring Force (F)*

$$\begin{aligned} F &= \sqrt{F_w^2 + F_c^2} \\ &= \sqrt{2457.8^2 + 126.95^2} \\ F &= 2550.96\ \text{kN} \end{aligned}$$

Assuming vessel to be moored by 8 mooring lines and considering only 50 of the lines resist the force.

Mooring force on each mooring line = Resultant Mooring force/Number of mooring lines

$$\begin{aligned} &= 2550.96/8 \\ &= 318.87\ \text{kN} \end{aligned}$$

3. Calculate the Active earth pressure of Pile with the following inputs:
   Top level is +3.90 m CD, founding level is –30.00 m CD.

*Note: If the properties of soil are not provided in soil investigation report, shall be assumed as in 'Foundation Analysis and design book by Joseph E. Bowles'.

Active earth pressure Pressure at -30.60 m

Diameter of Pile= 1.00 m

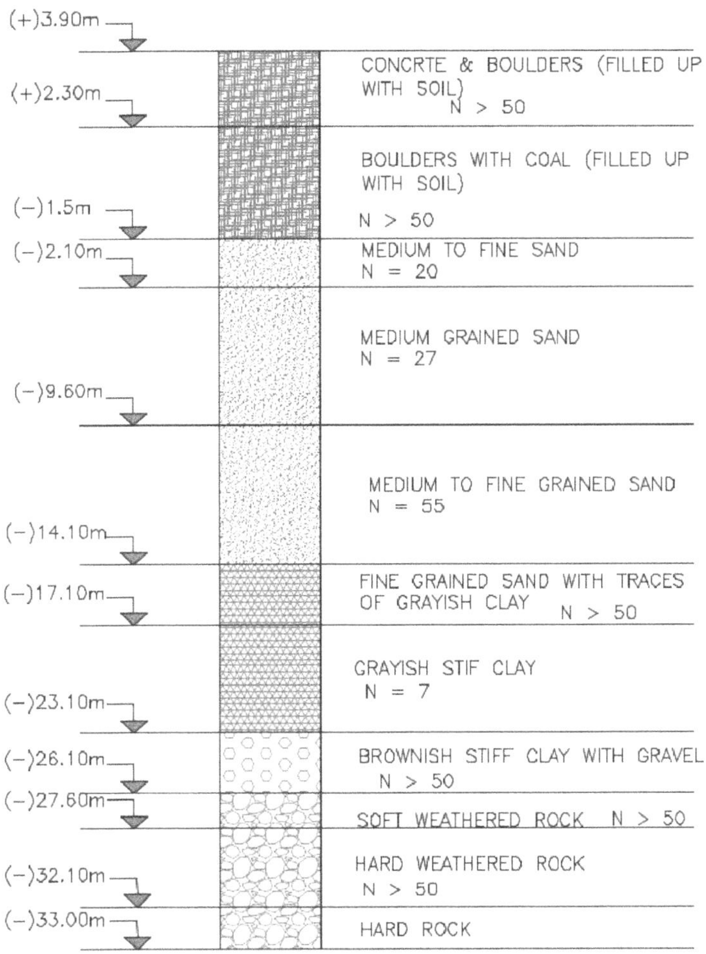


**FIGURE 3.13** Bore log for active earth pressure calculation.

Length of sheet pile = 0.30 m

Surcharge, q = 5 kN/m$^2$

Top level = 3.90 m

Assumed ground water table as per DWP calculation = -1.50 m

Density of Sea water = 10.25 kN/m$^3$

Bed/dredge level = -9.50 m

## Layer 13.90 to 2.30 m

Soil = Concrete & boulders filled up soil

$$\gamma_{sat} = 22.00 \text{ kN/m}^3$$

$$\gamma_{sub} = 22.00 \text{ kN/m}^3$$

$$\varphi = 38 \text{ degree}$$

$$N = 50$$

$$c = 300 \text{ kN/m}^2$$

$$K_a = 1 - \sin\varphi/1 + \sin\varphi = 0.24$$

Pressure at 3.90 m

$$P1 = K_a \gamma_1 0 - 2C(Ka)\text{^}1/2$$
$$= -292.64 \text{ kN/m}^2$$

$$\text{Surcharge, } P = k_a \cdot q = 1.19 \text{ kN/m}^2$$

$$AEP = -291.45 \text{ kN/m}^2$$

Force to be applied in STAAD = – 378.89 kN/m

Pressure at 2.30 m

$$h_1 = 1.60 \text{ m}$$

$$P2 = K_a \gamma h1 - 2C(Ka)\text{^}1/2$$
$$= -284.27 \text{ kN/m}^2$$

$$\text{Surcharge, } P = k_a . q = 1.19 \text{ kN/m}^2$$

$$AEP = -283.00 \text{ kN/m}^2$$

Force to be applied in STAAD = – 367.90kN/m

Force acting in Layer 1= $0.50 \times (-378.88 + 367.9) \times 1.60 = -597.43$ kN

### Layer 2 2.30 to –1.50 m

Soil = Boulders with coal filled up soil

$\gamma_{sat} = 22.00 \text{ kN/m}^3$

$\gamma_{sub} = 22.00 \text{ kN/m}^3$

$\varphi = 38$ degree

$N = 50$

$c = 300 \text{ kN/m}^2$

$K_a = 1 - \sin\varphi/1 + \sin\varphi = 0.24$

Pressure at 2.30 m

$$P3 = K_a\gamma_2 h_2 - 2C(Ka)^\wedge 1/2 + K_a\gamma_1 h_1$$
$$= -283.08 \text{ kN/m2}$$

$$\text{Surcharge, } P = k_a . q = 1.19 \text{ kN/m}^2$$

$$AEP = -283.08 \text{ kN/m}^2$$

Force to be applied in STAAD = -368.00 kN/m

Pressure at -1.50 m

$$h_2 = 3.80 \text{ m}$$

$$P4 = K_a\gamma_2 h_2 - 2C(Ka)^\wedge 1/2 + Ka\gamma_1 h_1$$
$$= -264.38 \text{ kN/m}^2$$

$$\text{Surcharge, } P = k_a.q = 1.19 \text{ kN/m}^2$$

$$AEP = -263.00 \text{ kN/m}^2$$

Force to be applied in STAAD = – 341.90 kN/m

$$\text{Force acting in Layer 2} = 0.50 \times \left(-368.00 + \left(-341.90\right)\right) \times 3.80 = -1348.82 \text{ kN}$$

From -1.50 m CD soil becomes submerged; unit weight of sea water is deducted from unit weight of soil.

### Layer 3–1.50 to –2.10 m

Soil = Medium to fine sand

$\gamma_{sat} = 17.80 \text{ kN/m}^3$

$\gamma_{sub} = 7.55 \text{ kN/m}^3$

$\varphi = 29$ degree

$N = 28$

$c = 0 \text{ kN/m}^2$

$K_a = 1 - \sin\varphi/1 + \sin\varphi = 0.35$

Pressure at -1.50 m

$$P5 = K_a\gamma_3 0 - 2C(Ka)^\wedge 1/2 + K_a\gamma_1 h_1 K_a\gamma_2 h_2$$
$$= 41.22 \text{ kN/m}^2$$

$$\text{Surcharge, } P = k_a . q = 1.73 \text{ kN/m}^2$$

$$AEP = 42.90 \text{ kN/m}^2$$

Force to be applied in STAAD =55.74 kN/m

Pressure at -2.10 m

$$h_3 = 0.60 \text{ m}$$

$$P6 = K_a\gamma_3 h_3 - 2C(Ka)^\wedge 1/2 + K_a\gamma_1 h_1 K_a\gamma_2 h_2$$
$$= 125.51 \text{ kN/m}^2$$
$$\text{Surcharge, } P = k_a . q = 1.73 \text{ kN/m}^2$$

$$AEP = 45.00 \text{ kN/m}^2$$

Force to be applied in STAAD = 58.50kN/m

Force acting in Layer 3 = 0.50 x $(55.74 + 58.50)$ x 0.60= 34.27 kN

### Layer 4 –2.10 to –9.60 m

Soil = Medium grained sand

$\gamma_{sat} = 17.80 \text{ kN/m}^3$

$\gamma_{sub} = 7.55 \text{ kN/m}^3$

$\varphi = 32$ degree

$N = 69$

$c = 0 \text{ kN/m}^2$

$K_a = 1 - \sin\varphi/1 + \sin\varphi = 0.31$

Pressure at -2.10 m

$$P7 = K_a\gamma_4 0 - 2C(K_a)^\wedge 1/2 + K_a\gamma_1 h_1 + K_a\gamma_2 h_2 + K_a\gamma_3 h_3$$
$$= 37.89 \text{ kN/m}^2$$

$$\text{Surcharge, } P = k_a . q = 1.54 \text{ kN/m}^2$$

$$AEP = 39.00 \text{ kN/m}^2$$

Force to be applied in STAAD = 50.70kN/m

Pressure at -9.60 m

$$h_4 = 7.50 \text{ m}$$

$$AEP = 57.0 \text{ kN/m}^2$$

$$P8 = K_a\gamma_4 h_4 - 2C(K_a)^\wedge 1/2 + K_a\gamma_1 h_1 + K_a\gamma_2 h_2 + K_a\gamma_3 h_3$$
$$= 55.29 \text{ kN/m}^2$$

$$\text{Surcharge, } P = k_a . q = 1.54 \text{ kN/m}^2$$

$$AEP = 57.0 \text{ kN/m}^2$$

Force to be applied in STAAD = 74.10kN/m

Force acting in Layer 4 = 0.50 x $(50.70 + 74.10)$ x 7.50
= 468.00 kN

### Layer 5 −9.60 to −14.10 m

Soil = Medium to fine grained sand

$\gamma_{sat} = 20.00\ \text{kN/m}^3$

$\gamma_{sub} = 9.75\ \text{kN/m}^3$

$\varphi = 38$ degree

$N = 85$

$c = 0\ \text{kN/m}^2$

$K_a = 1 - \sin\varphi/1+\sin\varphi = 0.24$

Pressure at -9.60 m

$\text{AEP} = 55.0\ \text{kN/m}^2$

Surcharge, $P = k_a.q = 1.19\ \text{kN/m}^2$

$\text{AEP} = 44.00\ \text{kN/m}^2$

Force to be applied in STAAD = 57.20 kN/m

Pressure at -14.10 m

$h_5 = 4.50\ \text{m}$

$$P10 = K_a\gamma_5 h_5 - 2C(Ka)^\wedge 1/2 + K_a\gamma_1 h_1 + K_a\gamma_2 h_2 + K_a\gamma_3 h_3 + K_a\gamma_4 h_4$$
$$= 53.25\ \text{kN/m}^2$$

Surcharge, $P = k_a.q = 1.19\ \text{kN/m}^2$

$\text{AEP} = 55.0\ \text{kN/m}^2$

Force to be applied in STAAD = 71.50 kN/m

$$\text{Force acting in Layer 5} = 0.50 \times (57.20 + 71.50) \times 4.50$$
$$= 289.58\ \text{kN}$$

### Layer 6 −14.10 to −17.10 m

Soil = Fine grained sand with traces of clay

$\gamma_{sat} = 17.10\ \text{kN/m}^3$

$\gamma_{sub} = 6.85\ \text{kN/m}^3$

$\varphi = 29$ degree

$N = 50$

$c = 0\ \text{kN/m}^2$

$K_a = 1 - \sin\varphi/1+\sin\varphi = 0.35$

Pressure at -14.10 m

$$P11 = K_a\gamma_6 0 - 2C(Ka)^{\wedge}1/2 + K_a\gamma 1h1 + K_a\gamma_2 h_2 + K_a\gamma_3 h_3 + K_a\gamma_4 h_4 + K_a\gamma_5 h_5$$
$$= 77.66 \text{ kN/m}^2$$

Surcharge, $P = k_a . q = 1.73 \text{ kN/m}^2$

$AEP = 79.0 \text{ kN/m}^2$

Force to be applied = 102.70kN/m

Pressure at -17.10 m

$h_6 = 3$ m

$$P12 = K_a\gamma_6 h_6 - 2C(Ka)^{\wedge}1/2 + K_a\gamma_1 h_1 + K_a\gamma_2 h_2 + K_a\gamma_3 h_3 + K_a\gamma_4 h_4 + K_a\gamma_5 h_5$$
$$= 84.79 \text{ kN/m}^2$$

Surcharge, $P = k_a . q = 1.73 \text{ kN/m}^2$

$AEP = 87.0 \text{ kN/m}^2$

Force to be applied in STAAD = 113.10kN/m

Force acting in Layer 6 = 0.50 x $(102.7+113.10)$ x 3.00
= 323.70 kN

### Layer 7 –17.10 to –23.10 m

Soil = Greyish stiff clay

$\gamma_{sat} = 17.5 \text{ kN/m}^3$

$\gamma_{sub} = 7.25 \text{ kN/m}^3$

$\varphi = 10$ degree

$N = 9$

$c = 54 \text{ kN/m}^2$

$K_a = 1 - \sin\varphi/1{+}\sin\varphi = 0.70$

Pressure at -17.10 m

$$P13 = K_a\gamma_7 0 - 2C(Ka)^{\wedge}1/2 + K_a\gamma_1 h_1 + K_a\gamma_2 h_2 + K_a\gamma_3 h_3 + K_a\gamma_4 h_4 + K_a\gamma_5 h_5 + K_a\gamma_6 h_6$$
$$= 81.44 \text{ kN/m}^2$$

Surcharge, $P = k_a . q = 3.52 \text{ kN/m}^2$

$AEP = 85.00 \text{ kN/m}^2$

Force to be applied in STAAD= 110.50kN/m

Pressure at -23.10 m

$h_7 = 6.00$ m

$$P14 = K_a\gamma_7 h_7 - 2C(Ka)^{\wedge}1/2 + K_a\gamma_1 h_1 + K_a\gamma_2 h_2 + K_a\gamma_3 h_3 + K_a\gamma_4 h_4 + K_a\gamma_5 h_5 + K_a\gamma_6 h_6$$
$$= 112.07 \text{ kN/m}^2$$

$$\text{Surcharge, } P = k_a . q = 3.52 \text{ kN/m}^2$$

$$\text{AEP} = 116.00 \text{ kN/m}^2$$

Force to be applied in STAAD = 150.80kN/m

$$\text{Force acting in Layer 7} = 0.5 \times (110.5 + 150.8) \times 6.00$$

$$= 783.90 \text{ kN}$$

## Layer 8 −23.10 to −26.10 m

Soil = Brownish stiff clay with gravel

$$\gamma_{sat} = 18.60 \text{ kN/m}^3$$

$$\gamma_{sub} = 8.35 \text{ kN/m}^3$$

$$\varphi = 35 \text{ degree}$$

$$N = 70$$

$$c = 420 \text{ kN/m}^2$$

$$K_a = 1 - \sin\breve{} / 1 + \sin\varphi = 0.27$$

Pressure at -23.10 m

$$P15 = K_a \gamma_8 0 - 2C(Ka)^\wedge 1/2 + K_a \gamma_1 h_1 + K_a \gamma_2 h_2 + K_a \gamma_3 h_3 + K_a \gamma_4 h_4 + K_a \gamma_5 h_5 + K_a \gamma_6 h_6 + K_a \gamma_7 h_7 = -359.26 \text{ kN/m}^2$$

$$\text{Surcharge, } P = k_a . q = 1.35 \text{ kN/m}^2$$

$$\text{AEP} = -358.00 \text{ kN/m}^2$$

Force to be applied in STAAD = -465.40kN/m

Pressure at -26.10 m

$$h_8 = 3 \text{ m}$$

$$P16 = K_a \gamma_8 h_8 - 2C(Ka)^\wedge 1/2 + K_a \gamma_1 h_1 + K_a \gamma_2 h_2 + K_a \gamma_3 h_3 + K_a \gamma_4 h_4 + K_a \gamma_5 h_5 + K_a \gamma_6 h_6 + K_a \gamma_7 h_7 = -352.48 \text{ kN/m}^2$$

$$\text{Surcharge, } P = k_a . q = 1.35$$

$$\text{AEP} = -351.00 \text{ kN/m}^2$$

Force to be applied in STAAD = -456.30kN/m

$$\text{Force acting in Layer 8} = 0.50 \text{ x } (-465.40 + 456.30) \text{ x } 3.00$$

$$= -1382.55 \text{ kN}$$

## Layer 9 −26.10 to −27.60 m

Soil = soft disintegrated rock

$$\gamma_{sat} = 22.0 \text{ kN/m}^3$$
$$\gamma_{sub} = 11.75 \text{ kN/m}^3$$
$$\varphi = 35 \text{ degree}$$
$$N = 50$$
$$c = 300 \text{ kN/m}^2$$
$$K_a = 1 - \sin\varphi / 1 + \sin\varphi = 0.27$$

Pressure at −26.1 m

$$P17 = K_a\gamma_9 0 - 2C(K_a)^{\wedge}1/2 + K_a\gamma_1 h_1 + K_a\gamma_2 h_2 + K_a\gamma_3 h_3 + K_a\gamma_4 h_4 + K_a\gamma_5 h_5 + K_a\gamma_6 h_6 + K_a\gamma_7 h_7 + K_a\gamma_8 h_8 = -242.68 \text{ kN/m}^2$$

Surcharge, $P = k_a . q = 1.35 \text{ kN/m}^2$

$$AEP = -242.00 \text{ kN/m}^2$$

Force to be applied = − 314.60kN/m

Pressure at − 27.60 m

$$h_9 = 1.50 \text{ m}$$

$$P18 = K_a\gamma_9 h_9 - 2C(Ka)^{\wedge}1/2 + K_a\gamma_1 h_1 + K_a\gamma_2 h_2 + K_a\gamma_3 h_3 + K_a\gamma_4 h_4 + K_a\gamma_5 h_5 + K_a\gamma_6 h_6 + K_a\gamma_7 h_7 + K_a\gamma_8 h_8 = -244.90 \text{ kN/m}^2$$

Surcharge, $P = k_a . q = 1.35 \text{ kN/m}^2$

$$AEP = -243.00 \text{ kN/m}^2$$

Force to be applied in STAAD = − 315.90kN/m

$$\text{Force acting in Layer 9} = 0.50 \times \left(-314.60 + \left(-315.90\right)\right) \times 1.50$$

$$= -472.88 \text{ kN}$$

### Layer 10–27.60 to −30.60 m

Soil = Hard weathered rock

$$\gamma_{sat} = 28.00 \text{ kN/m}^3$$
$$\gamma_{sub} = 17.75 \text{ kN/m}^3$$
$$\varphi = 38 \text{ degree}$$
$$N = 100$$
$$c = 600 \text{ kN/m}^2$$
$$K_a = 1 - \sin\check{}\ / 1 + \sin\varphi = 0.24$$

Pressure at -27.60 m

$$P19 = K_a\gamma_{10} 0 - 2C(Ka)^{\wedge}1/2 + K_a\gamma_1 h_1 + K_a\gamma_2 h_2 + K_a\gamma_3 h_3 + K_a\gamma_4 h_4 + K_a\gamma_5 h_5 + K_a\gamma_6 h_6 + K_a\gamma_7 h_7 + K_a\gamma_8 h_8 + K_a\gamma_9 h_9 = -506.65 \text{ kN/m}^2$$

$$\text{Surcharge},\ P = k_a . q = 1.19$$

$$AEP = -505.00\ \text{kN/m}^2$$

$$\text{Force to be applied in STAAD} = -656.50\text{kN/m}$$

$$\text{Pressure at} - 30.60\ \text{m}$$

$$h_{10} = 3.00\ \text{m}$$

$$P20 = K_a\gamma_{10}h_{10} - 2C(Ka)^\wedge 1/2 + K_a\gamma_1h_1 + K_a\gamma_2h_2 + K_a\gamma_3h_3 + K_a\gamma_4h_4 + K_a\gamma_5h_5 + K_a\gamma_6h_6 + K_a\gamma_7h_7 + K_a\gamma_8h_8 + K_a\gamma_9h_9 = -493.98\ \text{kN/m}^2$$

$$\text{Surcharge},\ P = k_a . q = 1.19\ \text{kN/m}^2$$

$$AEP = -492.00\ \text{kN/m}^2$$

$$\text{Force to be applied in STAAD} = -639.60\text{kN/m}^2$$

$$\text{Force acting in Layer } 10 = 0.50 \times (-656.50 + -639.60) \times 3.00 = -1944.15\ \text{KN}$$

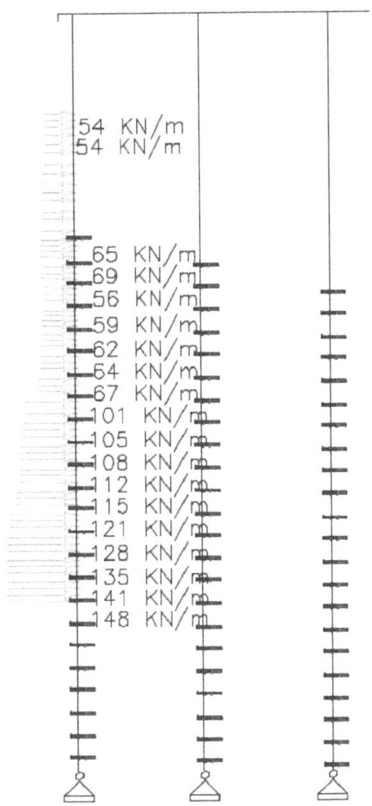


**FIGURE 3.14** Active earth pressure application in STAAD.Pro.

* Negative values of Active earth pressure are neglected while applying in STAAD.Pro model for analysing the structure

4. Calculate the differential water pressure for the following data:

**Given Data**

| | | | | | |
|---|---|---|---|---|---|
| Highest High-Water Level | HHWL | = | (+) | 2.06 | m |
| Mean High Water Level Springs | MHWS | = | (+) | 2.06 | m |
| Mean High Water Level Neaps | MHWN | = | (+) | 1.50 | m |
| Mean Sea Level | MSL | = | (+) | 0.80 | m |
| Mean Low Water Springs | MLWS | = | (–) | 0.16 | m |
| Mean Low Water Neaps | MLWN | = | (+) | 0.50 | m |
| Chart datum | CD | = | (+) | 0.00 | m |
| Lowest Low Water Level | LLW | = | (–) | 0.55 | m |

**Solution**

Differential water pressure calculation – Poor drainage condition as per IS 4651 (Part 3) (1974)

| | | | | | |
|---|---|---|---|---|---|
| Mean High Water | MHW | = | | (MHWS+MHWN)/2 | |
| | | = | (+) | 1.78 | m |
| Mean Low Water | MLW | = | | (MLWS+MLWN)/2 | |
| | | = | (+) | 0.33 | m |
| Difference in Water level on sea side | | = | | (MLWS+LLW)/2 | |
| | | = | (–) | 0.36 | m |

Assumed Ground water for bad drainage condition,

| | | | | |
|---|---|---|---|---|
| Assumed Ground Water Level | = | (MHW+MLW)/2 | | |
| | = | (+) | 1.06 | m |

Differential water pressure +1.06 m to –0.36 m

| | | | | |
|---|---|---|---|---|
| Height | h | = | 1.42 | m |
| Unit weight of water | $\gamma$ | = | 10.25 | kN/m$^3$ |
| Water pressure | P | = | $\gamma \times h$ | |
| Water pressure | P | = | 14.56 | kN/m$^2$ |

Water pressure at +1.06 m 0 kN/m$^2$ water pressure at –0.36 m 14.56 kN/m$^2$

Water pressure from –0.36 m to founding level 14.56 kN/m$^2$

Differential water pressure calculation – Good drainage condition as per IS 4651 (Part 3) (2020)

| | | | |
|---|---|---|---|
| Mean High Water | MHW | = | (MHWS+M-HWN)/2 |

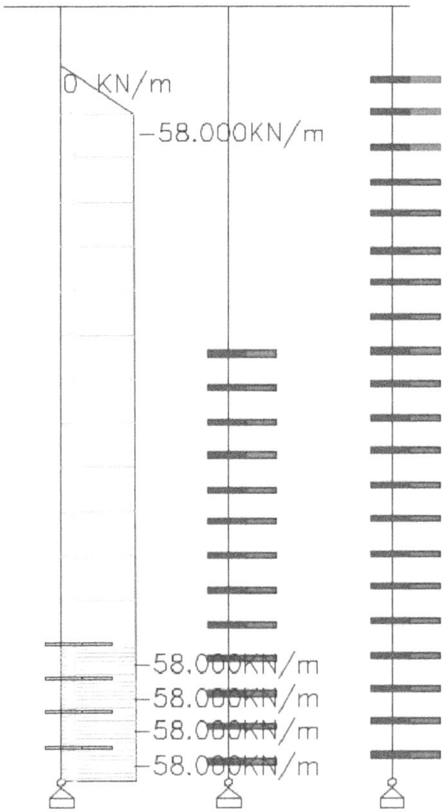


**FIGURE 3.15** Differential water pressure application in STAAD.

| | | | | | |
|---|---|---|---|---|---|
| | | = | (+) | 1.78 | m |
| Mean Low Water | MLW | = | | (M L W S + M - LWN)/2 | |
| | | = | (+) | 0.33 | m |
| Difference in Water level on sea side | | = | | (MLWS+LLW)/2 | |
| | | = | (–) | 0.36 | m |

Assumed Ground water for good drainage condition,

| | | | | | |
|---|---|---|---|---|---|
| Assumed Ground Water Level | =0.3 m above mean low water level | | | | |
| | | = | (+) | 0.63 | m |

Differential water pressure +0.63 m to –0.36 m

| | | | | |
|---|---|---|---|---|
| Height | h | = | 0.99 | m |
| Unit weight of water | $\gamma$ | = | 10.25 | $kN/m^3$ |
| Water pressure | P | = | $\gamma \times h$ | |
| Water pressure | P | = | 10.15 | $kN/m^2$ |

Water pressure at +0.63 m 0 kN/m² water pressure at −0.36 m 10.15 kN/m²
Water pressure from −0.36 m to founding level 10.15 kN/m² (14.56 × 4.00 = 58.24 kN/m)

5. Calculation of stability factor with AEP, PEP and DWP of examples 2, 3 and 4 respectively:
   Stability factor for soil in the backfill:
   The stability factor for the retaining structure to retain backfill soil shall be calculated by

$$\text{Stability factor} = \frac{\text{Total passive force}}{\text{Total active force} + \text{Differential Water force}}$$

   This stability factor should be greater than 2.50 for the retaining structure to be stable.

| Level | Active force | |
|---|---|---|
| 2.04 | 0.54 | 15.00 |
| 0.54 | −0.96 | 84.54 |
| −0.96 | −3.46 | 268.35 |
| −3.46 | −6.96 | −882.00 |
| −6.96 | −16.46 | −17138.00 |
| −16.46 | −24.06 | −30400.00 |
| −24.06 | −26.56 | −9410.00 |

| Level | | Passive force |
|---|---|---|
| −16.10 | −20.00 | 19078.80 |

| Level | | Differential water pressure |
|---|---|---|
| 1.06 | −0.36 | 7.28 |
| −0.36 | −16.10 | 161.34 |
| | | |

Stability factor = Passive force/(Active force + Differential water pressure force) = 36.55 > 2.5 as per IS 4651 (Part 2) (2020) described in Chapter 3.4.4

6. Calculate the wave force for sheet pile by the Sainflou method.

### Given data

High Tide Level HTL = 1.50 m
Bed level = 10.00 m
Sheet bottom pile level = 6.00 m
Wave height $H_s$ = 2.40 m
Wave period T = 6.32 s

Centre to centre distance between sheet pile = 1.64 m

**Solution**

| | | | | |
|---|---|---|---|---|
| Unit weight of water, w | | = | 10.25 | kN |
| Water Depth, $d_w$ | | = | HTL + bed level | |
| | | = | 11.50 | m |
| Sheet pile bottom level | | = | 6.00 | m |
| Depth of sheet pile below HTL water, d | | = | HTL + sheet pile bottom level | |
| | | = | 7.50 | m |
| Wave length,(Deep waters), $L_0$ = | $(1.56 \times T^2)$ | = | 62.31 | m |
| $d_w/L_0$ | 7.5/62.31 | = | 0.18 | |
| $d_w/L$ | | = | 0.2083 | |
| L | $0.2083 \times 7.5$ | = | 55.21 | m |

Height of orbit centre above still water, $h_o$

$$h_0 = \frac{\pi H^2}{L} coth \frac{2\pi d}{L}$$

$$= \frac{\pi\left(2.40^2\right)}{55.21} coth \frac{2\pi\left(7.50\right)}{55.21}$$

$$= 0.47 \text{ m}$$

*Case (i): Crest of the Clapotis on the sea side of pile & sheet pile wall and still water on the other side*

Increase in water level due to Clapotis formation from $\text{HTL} = H_s + h_o$

$$= 2.4 + 0.47 = 2.87 \text{ m}$$

Total pressure at the bottom $(\text{wd}) = \text{w}\left(\text{HTL} + \text{Depth of sheet pile} + \left(H_s + h_o\right)\right)$

$$= 10.25\left(1.50 + 6.00 + 2.87\right)$$

$$= 106.29 \text{ kN/m}^2$$

Force on sea side $= 106.29\left(0.50 \times 1.64 \times \left(2.87 + 1.50 + 6.00\right)\right)$

$$= 903.83 \text{ kN}$$

*Case (ii) Static water pressure on the rear side of pile & sheet pile wall due to HTL*

Static water pressure from HTL $\left(\text{w} \times \text{d}\right) = \text{w}\left(\text{Depth of sheet pile} + \text{HTL}\right)$

$$= 10.25\left(6.00 + 1.50\right)$$

Total pressure at the bottom $(wd)$ in rear side = 76.875 kN/m$^2$

Force on rear side =76.875$(0.50 \text{ x } 1.64 \text{ x } (6.00 + 1.50))$

= 472.78 kN

7. Calculate the current force for pile of 1.20 m diameter and length of 17 m.

**GIVEN DATA**

Diameter = 1.20 m
Length = 17 m
Velocity = 1.0 knots
Unit weight of water = 1.03 t/m$^3$

**SOLUTION**

*Pressure due to Current (As per IS 4651 (Part 3) (2020) Cl.6.6)*

Pressure due to current= $\gamma v^2/2g$

Velocity $(V) = 1.0$ knots

= 0.5 m/s

$P_c = (1.03 \times 0.5^2)/(2 \times 9.81)$

= 0.0131 T/m$^2$

Force due to current = $P_c \times D \times L = 0.0131 \times 1.20 \times 17 = 0.27$ T

$F_c = 2.69$ kN

8. Calculate the seismic force as per IS 1893: 2016 considering following data total dead load = 2500.0 kN, total live load = 1750 kN. The structure being in Zone II with natural time period of structure 0.7 s and the soil type is rock.

**GIVEN DATA**

| | | |
|---|---|---|
| Total Dead Load | = | 2500.0 kN |
| Total Live load | = | 1750 kN |
| Zone | = | II |
| Natural Time period of the structure | = | 0.7 s |
| Soil type | = | Rock |

**Solution**

As per IS 1893 (2016)

Horizontal seismic coefficient

$$A_h = \frac{ZI}{2R}\left(\frac{S_a}{g}\right)$$

Zone II as per IS 1893 (2016) Table 3

Zone Factor, Z = 0.16

As per IS 1893 (2016) Cl.6.4.2 (Table 8) and as per IS 4651 (Part 4) (2014)

| | | |
|---|---|---|
| Importance Factor, I | = | 1.5 |
| As per IS 1893 (2016) Table 9 | | |
| Response reduction factor, R | = | 3 |
| Average response acceleration co-efficient( $S_a/g$ ): | | |
| As per IS 1893 (Part 1) (2016) Cl.6.4.2 | | |
| Time period | = | 0.7 |
| Soil Type | | Rock |
| | = | 1/t = 1/0.7 |
| $S_a/g$ | | |
| | = | 1.43 |
| $A_h$ | = | 0.16 × 1.5 × 1.43 / (2 X 3) |
| | = | 0.0571 |
| Seismic Force | = | $A_h$(Dead Load + 50% live load) |
| | = | 0.0571 (2500 + (50/100) × 1750) |
| | = | 192.9 kN |

## REVIEW QUESTIONS

1. How are loads classified acting on the berthing structure?
2. Name the loads acting on the berthing structure.
3. What is berthing force? Write the formula for determination of berthing energy as per Indian Standard.
4. What are coefficients used in the determination of berthing energy?
5. Describe the approach angle with a diagram.
6. Define mooring force and how it is determined in brief.
7. Name the loads from the mechanical handling system acting on the berth.
8. What are the types of crane used in ports?
9. Explain the difference between passive earth pressure and active earth pressure.
10. How is the stability factor for the soil in backfill calculated?
11. Describe in brief three types of waves acting on piles and diaphragm walls.
12. What is meant by Clapotis or a standing wave?
13. Calculate the horizontal seismic coefficient for the berthing structure located in Visakhapatnam Port as per the Indian Standard.

## REFERENCES

**IS 875 (Part 3) (2015)** *Design Loads (Other Than Earth Quake) for Buildings and Structures – Code of Practice Part 3 – Wind Loads* (Third Revision)

**IS 1893 (Part 1) (2016)** *Criteria for Earthquake Resistant Design of Structures – General Provisions and Buildings* (Sixth Revision)

**IS 4651 (Part 2) (2020)** *Code of Practice for Planning and Design of Port and Harbours - Earth Pressures* (First Revision)

**IS 4651 (Part 3) (2020)** *Code of Practice for Planning and Design of Port and Harbours – Loading*

**IS 4651 (Part 4) (2023)** *Code of Practice for Planning and Design of Port and Harbours – General Design Considerations*

**Shore Protection Manual: Volume II (1984)** *Coastal Engineering Research Centre Department of the Army Waterways Experiment Station*, Corps of Engineers

# 4 Breakwaters

Breakwaters are constructed to reduce wave action in the lee side of the structure. Wave action is reduced through a combination of reflection and dissipation of incoming wave energy. When used for harbours, breakwaters are constructed to create sufficiently calm waters for safe berthing, mooring and loading operations of ships, and protection of harbour facilities.

Breakwaters are also built to improve manoeuvring conditions at harbour entrances and by creating areas with differing levels of tranquility. Protection of water intakes for power stations and protection of coastlines against tsunami waves are other applications of breakwaters.

When used for shore protection, breakwaters are built in nearshore waters and usually oriented parallel to the shore, like detached breakwaters. The layout of breakwaters used to protect harbours is determined by the size and shape of the area to be protected as well as by the prevailing directions of storm waves, net direction of currents and littoral drift, and the manoeuvrability of the vessels using the harbour. Breakwaters protecting harbours and channel entrances can be either detached or shore-connected.

The cost of breakwaters increases drastically with water depth and wave climate severity. Also, poor foundation conditions significantly increase costs. These three environmental factors heavily influence the design and positioning of the breakwaters and the harbour layout.

## 4.1 TYPES

Generally, breakwaters are shore-connected or offshore type and floating;

Three major categories according to the functional requirements:

- Rubble mound breakwaters (mound type)
- Vertical breakwaters (wall type)
- Vertical wall on rubble base (composite type)

### 4.1.1 Rubble Mound Breakwater

The rubble mound breakwater is constructed by a heterogeneous assemblage of rubble stone. The filter layer (1–50 kg) consists of well graded stones. The centre core of the breakwater is made up of quarry run rock (10–500 kg) and shall be with very wide gradation, like Rosin Rammler particle size distribution ( $D_{85}/D_{15}$=2.5 to 5 ). The outside layer consists of large armour units that can be either rock or specially designed concrete units. This primary armour layer is intended to be statically stable with respect to the environmental conditions imposed on it (the waves and currents do not move the armour stones under design conditions). It is usual to build

DOI: 10.1201/9781003361275-4

the primary armour layer roughly two-unit diameters thick or one layer (Coreloc, Accropode, etc.) and to place the units randomly. If the armour units were placed directly over the core, the finer core material would be removed by the waves through the openings of the armour layer. The under layer below the main armour shall be 1/10th to 1/5th of main armour weight. 75% of the secondary armour shall be more than 1/5th of the main armour weight. The toe armour shall be placed above the core or filter depending of the water depth and is about twice the size of secondary armour. Types of rubble mound breakwater include:

- Berm breakwaters armoured with natural rock
- Conventional multi-layer breakwater with armour over crest (no crown wall)
- Conventional multi-layer breakwater with crown wall

**Advantages:**

1. Use of natural material is a big advantage, especially when large quantities of rubble stones are available near the site. It reduces the material cost.
2. Use of smaller construction equipment.
3. Less environmental impact due to smaller reflected waves.
4. Creation of a natural reef and the slope of the rubble mound breakwater provides a suitable place for sea life.

### 4.1.2 Vertical Breakwater

A breakwater is formed by the construction in a regular and systematic manner of a vertical wall of masonry concrete blocks or mass concrete, with vertical face on the seaward side. The vertical face reflects the incident waves without dissipating much wave energy. Normally it is constructed in locations where the depth of the sea is greater than twice the design wave height and the sea bed is rocky or dense sand. The main component of a vertical breakwater cross-section is normally a concrete caisson; a large, hollow concrete box that is floated to the site and filled with granular material to sink it to form a stable structure. The caissons are placed on a prepared bed of quarry stone. In some locations, it has become practice to decrease wave reflection and prevent waves from breaking directly onto the caisson by covering its seaward face with an armour protection (horizontally composite breakwater).

### 4.1.3 Composite Breakwater

A breakwater is constructed by the combination of a rubble mound base and a vertical wall or other forms of the superstructure (vertically composite breakwater). It is adopted where the depth of water is much higher than twice the design wave height.

**Advantages:**

1. Smaller body width/quantity of material.
    - This is one of the biggest advantages of using a composite breakwater. Construction is more economical, especially in deep water.

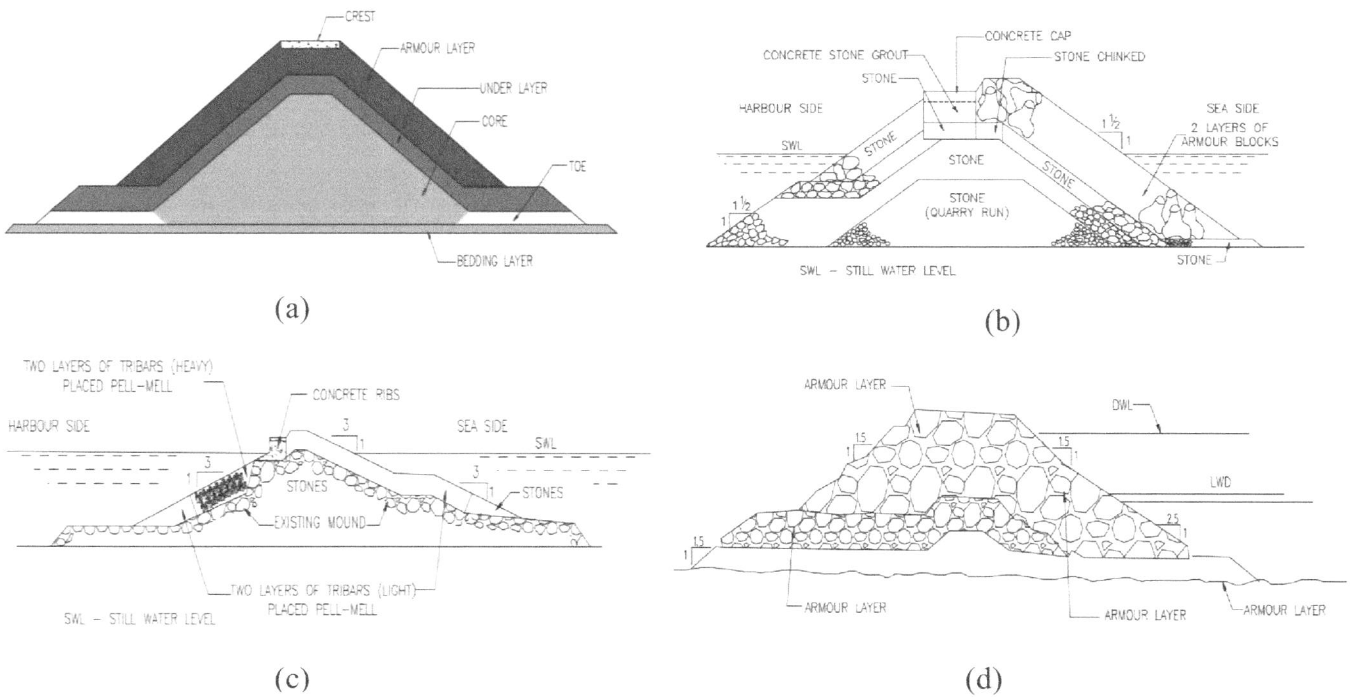


**FIGURE 4.1** Rubble mound breakwater: (a) structural components; (b) armoured with tetrapod; (c) armoured with tri-bar; (d) as offshore breakwater.

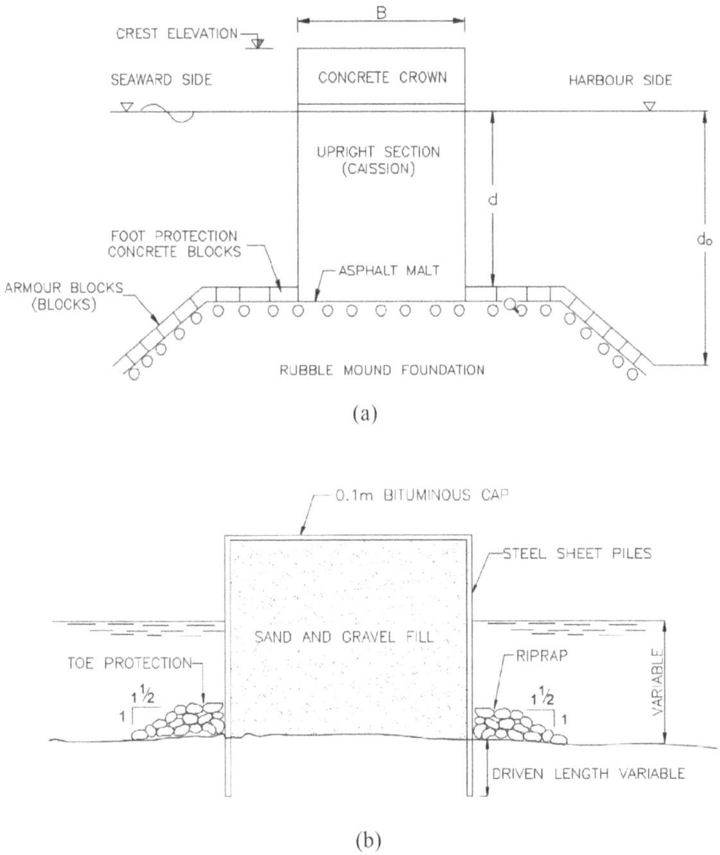


**FIGURE 4.2** Vertical breakwater: (a) definition sketch; (b) vertical wall – caisson breakwater.

- In addition, a small breakwater width, limits the impact on sea bed life and increases the usable water area. Lee side can be used for berthing vessels.

2. Composite breakwater requires relatively less maintenance.
3. It can be constructed rapidly and is fully stabilized once its caissons are filled with sand.
4. In addition, not much quarry work is required, and the environmental damage is also less.

### 4.1.4 Floating Breakwater

A removable breakwater constructed by caissons or pontoons with valves for trimming whilst afloat and for sinking when in position. It is ineffective for long waves. It is inexpensive, easily moved from site to site, and often very quickly fabricated. Some of the typical floating breakwaters are shown in below Figure 4.3.

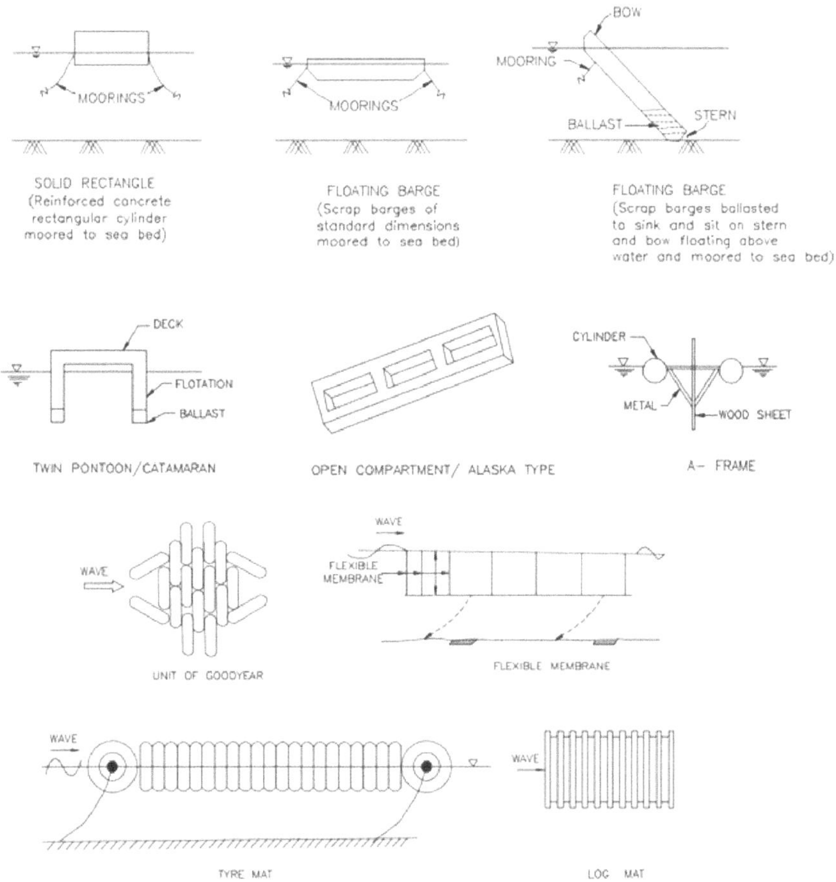


**FIGURE 4.3** Typical floating breakwater.

### 4.1.5 Detached Breakwater

Detached breakwaters are built parallel to the shore just seaward of the shoreline in shallow water depths. They are relatively short (100–200 m) with a gap in between them with the principal function of reducing beach erosion and creating tombolo formation. Multiple detached breakwaters (more than three) spaced along the shoreline can provide protection to substantial shoreline frontages. In most cases, the gaps between the breakwaters are about 0.5–1 times the length of one individual structure.

Each breakwater reflects and dissipates some of the incoming wave energy, thus reducing wave heights in the lee of the structure and reducing shore erosion. Beach material transported along the beach moves into the sheltered area behind the breakwater where it is deposited in the lower wave energy region. The nearshore wave pattern, which is strongly influenced by diffraction at the heads of the structures, will cause salients, and sometimes tombolos are formed, thus making the coastline

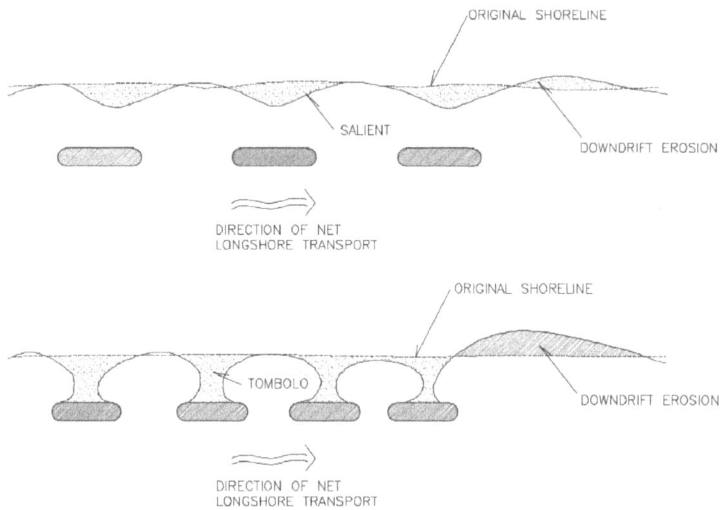


**FIGURE 4.4** Typical beach configurations with detached nearshore breakwaters.

similar to a series of pocket beaches. Once formed, the pockets will cause wave refraction, which helps to stabilize the pocket-shaped coastline.

Detached breakwaters are normally built as rubble-mound structures with fairly low crest levels that allow significant overtopping during storms at high water. The low-crested structures are less visible and help promote a more even distribution of littoral material along the coastline. In some cases, submerged detached breakwaters are used because they do not spoil the view, but they do represent a severe nonvisible hazard to boats and swimmers.

Properly designed detached breakwaters effectively reduce erosion and build up beaches using natural littoral drift. Moreover, they are effective in holding artificially nourished beach material.

Optimizing detached breakwater designs is difficult when large water level variations are present, as is the case on coastlines with a large tidal range.

### 4.1.6 Berm Breakwater

Berm breakwater is a special type used in an artificial harbour. It is also known as a reshaping breakwater or 'S' type breakwater. The horizontal berm is constructed above or just below the design water level. The incident waves will propagate over the horizontal berm, resulting in more dissipation of wave energy. The stones move along with the action of waves. So, the structure will reshape and attain an equilibrium 'S' profile. The structure will become stable after reshaping.

#### Advantages

- Smaller stones or locally available stones are utilized for construction.
- Nearly 100% quarry yield can be achieved.

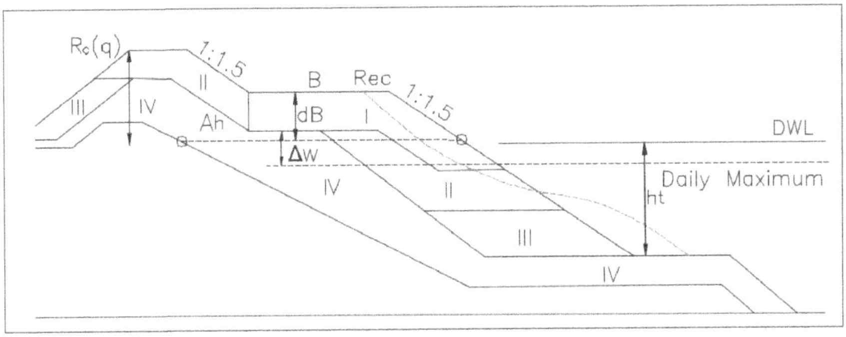


FIGURE 4.5 Berm breakwater – 'S' profile.

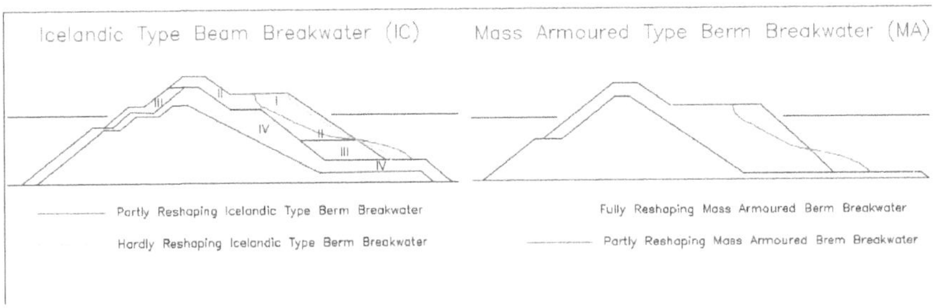


FIGURE 4.6 Icelandic type and mass armoured berm breakwater.

**TABLE 4.1**
**Classification of Berm Breakwater**

| Classification | Abbreviation | $H_s/D_{n50}$ | $S_d$ | $R_{ec}/D_{n50}$ |
|---|---|---|---|---|
| Hardly reshaping berm breakwater (Icelandic type) | HR-IC | 1.7–2.0 | 2–8 | 0.5–2 |
| Partly reshaping Icelandic type berm breakwater | PR-IC | 2.0–2.5 | 10–20 | 1–5 |
| Partly reshaping mass armoured berm breakwater | PR- MA | 2.0–2.5 | 10–20 | 1–5 |
| Fully reshaping berm breakwater (mass armoured) | FR-MA | 2.5–3.0 | – | 3–10 |

- The construction is labour-intensive.
- Sophisticated equipment is eliminated.
- Thus, berm breakwater is economical and being adopted in developing countries and suitable for the east coast of India, where cyclone risk is high.

**Classification of berm breakwater**

According to the movement of stones (as per PIANC – WG 40)

- Statically stable non-reshaped
- Statically stable reshaped
- Dynamically stable reshaped

According to stability number (*Sigurdharsan and Van der Meer (2013)*)
Icelandic type berm breakwater (IC)
Mass armoured type breakwater (MA)

## 4.2 MATERIALS USED IN CORE AND ARMOUR

The principal function of materials used in the core of a hydraulic structure is volume filling. It is usually acceptable to use lower-quality materials with widely varying mean particle sizes compared to materials serving a filtering or armouring function.

Core materials may consist of primary rock fill, such as quarry run, or alternative granular materials. In the core of a structure – especially for underwater placement – the placed bulk density is the most important indicative material parameter, as it closely relates to shear strength, potential settlement, and permeability. Bulk density is influenced by grading width and particle shape.

### 4.2.1 Grading Width and Common Terminology

The overall steepness of the grading curve indicates the uniformity in mass, generally referred to as grading width or gradation. A quantitative indication of the uniformity is the ratio $M_{85}/M_{15}$ or its cube root (size) which reduces to $D_{n85}/D_{n15}$ or $D_{85}/D_{15}$. Table 4.2 gives the description of the various grading widths, expressed in above mass and size ratios.

For wide gradations, the term *well-graded* simply means that there are no significant gaps in material sizes across the total grading range.

In most cases, armourstone is narrowly graded. The term *riprap* usually refers to armourstone with a combination of the following characteristics: wide gradation, bulk placement, and application as a cover layer. The gradations required by designers and controlled by producers generally become wider as the material becomes lighter.

**TABLE 4.2**
**Armourstone Grading Width Related to the Uniformity**

| Grading Width | *D85/D15* | *M85/M15* |
|---|---|---|
| Narrow or single-sized gradation | <1.5 | 1.7–2.7 |
| Wide gradation | 1.5–2.5 | 2.7–16 |
| Very wide or quarry run gradation | 2.5–5.0 | 16–125+ |

Masses may be converted into sizes in terms of nominal diameter (Dn) or sieve diameter (D), which should not be confused. Conversion of mass to size is achieved by dividing the mass by density to obtain volume, and then taking the cube root to get Dn. To express particle size in terms of D using Dn, divide Dn by 0.84.

### 4.2.2 Rosin-Rammler Grading Curves for Mass and Size Distributions

The Rosin-Rammler grading curves can be fitted to represent most grading data sets obtained during quarry production and processing, including core materials.

The typical curve follows the Rosin-Rammler (Ros-Ram) equation:

$$y = 1 - exp\left\{ ln\left(\frac{1}{2}\right)\left(\frac{M_y}{M_{50}}\right)^{n_{RRM}} \right\} \cong 1 - exp\left\{ -0.693\left(\frac{M_y}{M_{50}}\right)^{n_{RRM}} \right\} \tag{4.1}$$

Or its inverse

$$M_y = M_{50}\left\{ \frac{\ln(1-y)}{\ln\left(\frac{1}{2}\right)} \right\}^{1/n_{RRM}} \cong M_{50}\left\{ \frac{-\ln(1-y)}{0.693} \right\}^{1/n_{RRM}} \tag{4.2}$$

where

$y$ is the fraction passing value;

$M_y$, is the mass corresponding to that value using a percentage subscript to express that fraction,

$n_{RRM}$ the uniformity index, being a measure for the steepness of the grading curve.

Given any two fixed points on the Rosin-Rammler curve, $M_{50}$ and $n_{RRM}$ can be determined. If the nominal lower limit mass of a grading is NLL and the fraction passing at that value is $y_{NLL}$, and similarly the nominal upper limit mass is NUL and the fraction passing at that value is $y_{NUL}$, then by solving the following two equations:

$$M_{50} \cong NLL\left(\frac{\ln(1-y_{NLL}}{-0.693}\right)^{-1/n_{RRM}} \tag{4.3}$$

$$M_{50} \cong NUL\left(\frac{\ln(1-y_{NUL}}{-0.693}\right)^{-1/n_{RRM}} \tag{4.4}$$

To give

$$n_{RRM} = log\left(\frac{\ln(1-y_{NUL})}{\ln(1-y_{NLL})}\right) / \log(NUL / NLL) \tag{4.5}$$

Therefore, besides providing idealized standard gradings, the Ros-Ram curve shape can be fitted to give a useful representation of most sets of grading data measured during production and processing in the quarry, including core materials.

## 4.3 TRANQUILITY

The word *tranquil* means free from disturbance from heavy waves. In the design of the breakwater layout of the harbour, priority must be given to maintaining tranquility. Harbour tranquility analysis is carried out based on the hindcast waves as the incident waves.

## 4.4 COASTAL EROSION AND ACCRETION

### 4.4.1 Coastal Erosion

A coastline is a complex series of interlinked physical systems in which both offshore and onshore processes are involved. Erosion is defined as the encroachment of land by the sea after over an average period which is sufficiently long to eliminate the impacts of weather, storm events and local sediment dynamics (such as sand waves). Coastal erosion is the loss or displacement of land or the long-term removal of sediment and rocks along the coastline due to the action of waves, currents, tides, wind-driven water, waterborne ice, or other impacts of storms. The landward retreat of the shoreline can be measured and described over a temporal scale of tides, seasons, and other short-term cyclic processes. Coastal erosion may be caused by hydraulic action, abrasion, impact and corrosion by wind and water, and other forces, natural or unnatural.

On non-rocky coasts, coastal erosion results in rock formations in areas where the coastline contains rock layers or fracture zones with varying resistance to erosion. Softer areas become eroded much faster than harder ones, which typically results in landforms such as tunnels, bridges, columns and pillars. Over time the coast generally evens out. The softer areas fill up with sediment eroded from hard areas, and rock formations are eroded away. Also, abrasion commonly happens in areas where there are strong winds, loose sand, and soft rocks. The blowing of millions of sharp sand grains creates a sandblasting effect. This effect helps to erode, smooth and polish rocks. The definition of abrasion is the grinding and wearing away of rock surfaces through the mechanical action of other rock or sand particles.

The accretion and erosion may happen for different coastal types resulting from a coastal structure. The coastal structure in this example is a large port with an extension greater than the width of the surf zone, but the structure could also be a set of tidal inlet jetties or a long groyne.

Coastal erosion results in three different types of impact: loss of land with economical value (such as the beaches) or with ecological value; a specific mechanism is the collapse of properties located on the top of cliffs and dunes.

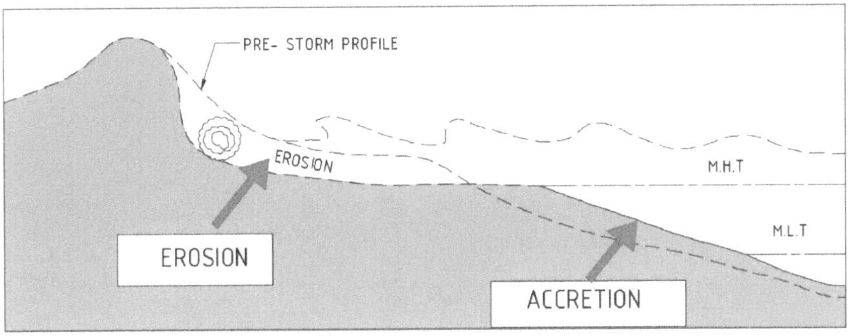


**FIGURE 4.7** Coastal erosion and accretion.

### 4.4.2 Coastal Accretion

Coastal accretion can be defined as the accumulation of (beach) sediment deposited by natural fluid flow processes.

### 4.4.3 Causes of Erosion

- Most severe incidents of coastal erosion occur during storms; there are many other causes, both natural and man-induced, which need to be examined.
- Natural causes of erosion are those which occur because of the response of the beach to the effects of nature.
- Man-induced erosion occurs when human endeavours impact the natural system.

#### Natural causes

- Sea level rise
- Variability in sediment supply to the littoral zone
- Storm waves
- Wave and surge overwash
- Deflation
- Longshore sediment transport
- Sorting of beach sediment

#### Man-induced causes

- Land subsidence from removal of subsurface resources
- Interruption of material in transport
- Reduction of sediment supply to the littoral zone
- Concentration of wave energy on beaches
- Increased water level variation
- Change natural coastal protection
- Removal of material from the beach

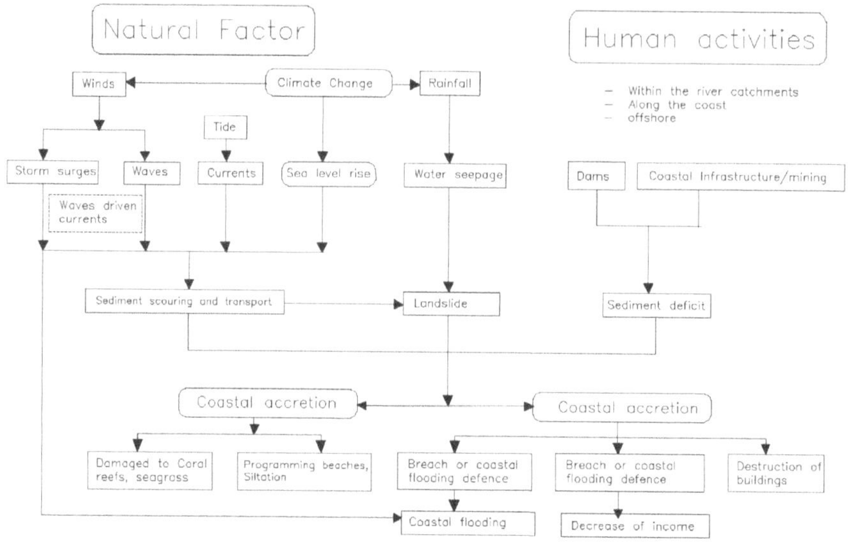


**FIGURE 4.8** Process of coastal erosion and accretion.

### 4.4.4 Coastal Protection Methods

#### 4.4.4.1 Bulkheads

Soil retaining structures are designed to also resist wave attacks. Bulkheads are generally anchored vertical pile walls or gravity walls, i.e., cribs or cellular steel pile structures. Walls of soldier beams and lagging have also been used at some sites. Three structural types of bulkheads are available, namely concrete, steel and timber. Cellular steel sheet pile bulkheads are used where rock is near the surface, and adequate penetration is impossible for the anchored sheet-pile bulkhead. When vertical or nearly vertical bulkheads are constructed and the water depth at the wall is less than twice the anticipated maximum wave height, the design should provide for riprap armouring at the base to prevent scouring.

#### 4.4.4.2 Seawalls

Seawalls are massive structures built to resist high wave action and reduce scour. There are two types of seawall, namely a curved-face seawall and a combination of stepped and curved face seawall. Both seawalls have sheet-pile cutoff walls to prevent loss of foundation material by wave scour and leaching from overtopping water or storm drainage beneath the wall. The curved-face seawall also has an armouring of large rocks at the toe to reduce scouring by wave action. The stepped-face seawall is designed for stability against moderate waves.

Rubble-mound seawalls are built to withstand severe wave action. Although scour of the fronting beach may occur, the quarry stone comprising the seawall can readjust and settle without causing structural failure.

**TABLE 4.3**
**Possible Natural Factors and Human Activities That Affect Shoreline Change**

| Factor | Effects | Time Scale | | | | | | | | | |
|---|---|---|---|---|---|---|---|---|---|---|---|
| | | secs | hours | days | months | years | 10 years | 50 years | 100 yrs | 1000 yrs | 10,000 yrs |
| **Natural Factors** | | | | | | | | | | | |
| Short wave period | Erosion | | | | | | | | | | |
| Waves of small steepness | Accretion | | | | | | | | | | |
| Large wave height | Erosion | | | | | | | | | | |
| Storm surge | Erosion | | | | | | | | | | |
| Alongshore currents | Accretion, no change or erosion | | | | | | | | | | |
| Rip currents | Erosion | | | | | | | | | | |
| Underflow | Erosion | | | | | | | | | | |
| Overwash | Erosion | | | | | | | | | | |
| Wind | Erosion | | | | | | | | | | |
| Sediment supply | Accretion or erosion | | | | | | | | | | |
| Inlet presence | Net erosion; high instability | | | | | | | | | | |
| Sea level rise | Erosion | | | | | | | | | | |
| Land subsidence (tectonic) | Accretion or erosion | | | | | | | | | | |
| **Human Activities** | | | | | | | | | | | |
| Dredging | Erosion or accretion | | | | | | | | | | |
| Coastal defence | Erosion or accretion | | | | | | | | | | |
| Vegetation cleaning | Erosion or accretion | | | | | | | | | | |
| Harbour development | Erosion or accretion | | | | | | | | | | |
| River damming | Erosion | | | | | | | | | | |
| Land Reclamation | Erosion | | | | | | | | | | |

*Legend:*

Erosion
Accretion
Erosion or Accretion

beach level. Because of the potential for erosion of the beach downdrift of the last groin in the field, a transition section of progressively shorter Groynes may be provided to prevent the formation of a severe erosion area. Even so, it might be necessary to protect some part of the downdrift beach with a seawall or to nourish a portion of the eroded area with beach material from an alternative source.

Groynes are mainly classified as permeability, height and length. Groynes built of common construction materials can be made permeable or impermeable and high or low in profile. Types of Groynes are classified based on function and material.

#### 4.4.4.6 Based on Function

*Attracting Groynes*

Attracting groynes are constructed in such a way that their head is pointing toward the downstream side of the river, as shown in the figure below. They are built with an angle of 45–60° with the bank since it is inclined towards downstream, the water flow will attract towards the bank on which the groyne is located.

The upstream side of attracting groynes undergoes severe attack by water flow. Hence, it should be constructed with proper protection. They provide safety to the opposite bank, but adjacent banks may get affected by this type of groynes. Silting is also not possible in this type of groyne, which is a disadvantage.

*Repelling Groynes*

Repelling groynes are built with their head towards upstream with an inclination of 60–80° with the bank. They repel the water flow towards the bank on which it is

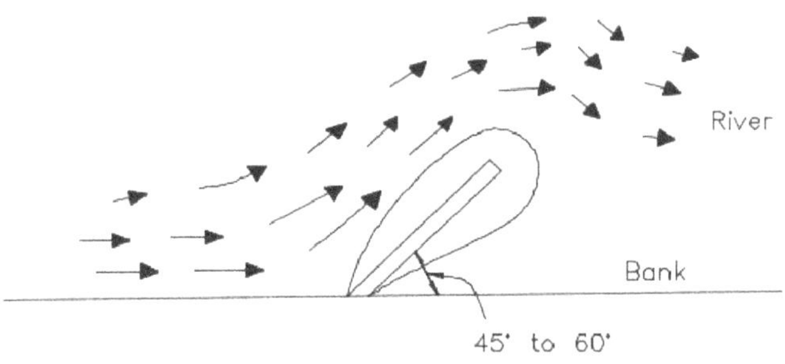


**FIGURE 4.10** Attracting groyne.

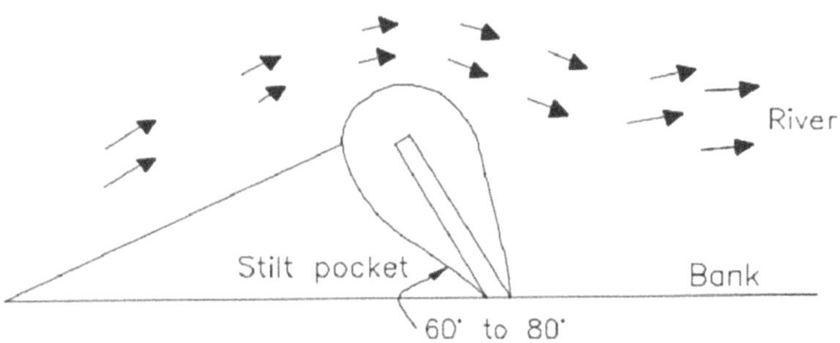


**FIGURE 4.11** Repelling groyne.

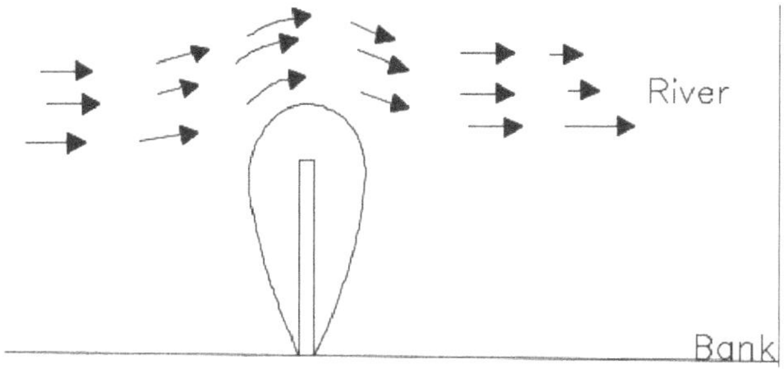


**FIGURE 4.12** Deflecting groynes.

located. The head portion of repelling groynes is under the main attack by the flow. Hence it should be built with strong protection. The sediments carried by the water get deposited in the silt pocket, which is formed at the upstream side of repelling groynes and is used widely for river training and bank protection.

*Deflecting Groynes*

Deflecting groynes are built perpendicular to the bank, and they deflect the water flow without repelling and provide local protection to the banks.

#### 4.4.4.7 Based on Material

*Concrete Groynes*

A concrete groin is an impermeable, prestressed concrete-pile structure with a cast-in-place concrete cap.

*Rubble-Mound Groynes*

Rubble-mound Groynes are constructed with a core of quarry run material, including fine material to make them sand-tight, and covered with a layer of armour stone. The armour stone should weigh enough to be stable against the design wave. If the permeability of a rubble-mound is a problem, the voids between stones in the crest above the core can be filled with concrete or asphalt grout. This seal also increases the stability of the entire structure against wave action.

*Asphalt Groynes*

An example of such a structure is a groin extension placed beyond the low waterline which is composed of a hot asphalt mixture and tends toward early structural failure of the section seaward of the beach berm crest. Failure results from a lack of resistance to normal seasonal variability of the shoreface and consequent undermining of the structure foundation.

The construction of jetties or breakwaters to provide safe navigation conditions at harbour entrances or tidal inlets along sandy coasts usually interrupt the natural land shore transport of sand at the entrance or inlet. The resulting starvation of the down-drift beach can cause severe erosion unless measures are taken to transfer or bypass the sand from the updrift side to the downdrift beaches. The primary method of sand bypassing are as follows: fixed bypassing plants, floating bypassing plants and land-based vehicles or draglines. It provides a low section or weir in the updrift jetty over which sand moves into a sheltered pre-dredged deposition basin. The dredged material is usually pumped across the navigation channel to nourish the downdrift shore.

### 4.4.5 Littoral Transport

When there is a current close to the coast, the combination of wave action on the shore, which loosens material, and the current, which transports beach material, can alter the coastline. This mechanism of sediment transport is called littoral drift. Over a period, littoral drift may occur in one direction and then in the reverse direction, but generally, there will be one direction which predominates. The littoral zone extends from the shoreline to just beyond the seawardmost breakers. It is classified as follows

- Onshore-offshore transport
- Longshore transport

#### 4.4.5.1 Onshore-Offshore Transport

- It has an average net direction perpendicular to the shoreline.
- It is usually the most significant type of transport in the offshore zone, except in regions of strong tidal currents.

#### 4.4.5.2 Longshore Transport

- It has an average net direction parallel to the shoreline.
- Both transport area significantly in the surf zone.

### 4.4.6 Shoreline Evolution

The shoreline or coastline, the boundary between land and sea, keeps changing its shape and position continuously due to dynamic environmental conditions. A wide

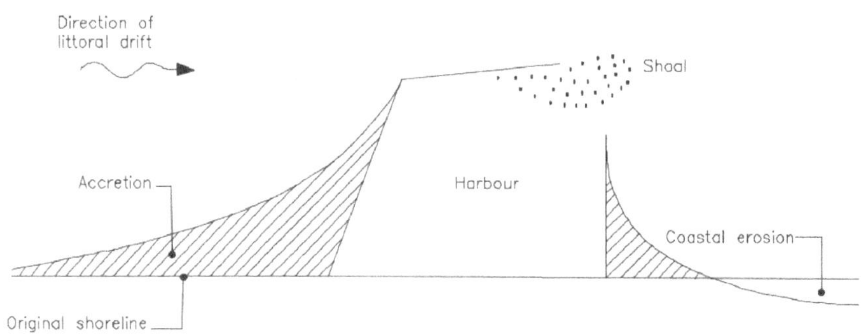


**FIGURE 4.13** Effect of littoral drift on coastal harbour.

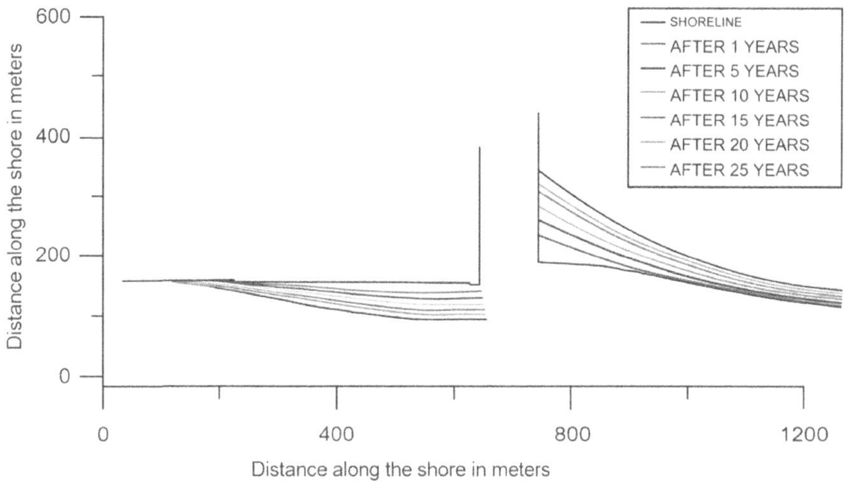


**FIGURE 4.14** Case study picture of numerical model study of shoreline evolution.

variety of numerical models and methods have been proposed to detect the shoreline change and to model the change mathematically. In many cases, the parameters such as surf zone turbulence, the effect of frequency, variations in the directional spread and other wave transformations are also calculated, and the solutions are formulated. A typical case study picture of a numerical model study of shoreline evolution is shown in Figure 4.14.

## 4.5 DESIGN CONSIDERATION

The chief information needed for the design of breakwaters is the wave height and period of the storm waves likely to occur. The standard practice is to choose a design wave that represents the maximum wave measured during a storm which statistically will occur once in so many years, for example, once in 100 years. The likelihood of waves breaking against the structure is also a factor of importance in design. Data about the probable daily maximum wave must also be collected in areas of continuous swell since this may have a significant effect on the cross-sectional design of the breakwater. The determination of the design wave requires the study of available oceanographic and meteorological data. It is preferable that direct recording of wave heights and periods should be made at the port site, but this is not always possible within the time available.

The foundations of breakwaters warrant careful geotechnical investigation. Boreholes must be made, soil and rock samples are taken over an area sufficiently large to allow variations in the position of the breakwater to be considered, the possibility of scouring the sea bed for some distance in front of the breakwater to be investigated. The strength and settlement properties of soils under a proposed breakwater require study to the depth at least equal to the width of the breakwater base since weak soils at depth may be overstressed by a large structure. Weak soils at the surface may have their bearing capacity enhanced by the placing of a blanket of sand or gravel on the surface prior to the construction of the breakwater.

## 4.5.1 Design of Conventional Breakwater

MHWS – Mean high water spring
SS – Storm surge (sea level rise during the storm)
$R_c$ – Crest freeboard
B – Crest width
T – Layer thickness

### 4.5.1.1 Design Data

*Design water level* = mean high water spring (MHWS) + storm surge (SS) + sea level rise (typically 0.3 or 0.45 m for 100 years depending on the location, carbon emission etc.).

*Design water depth (d)* = contour level at one wave length from sea side toe of the breakwater + design water level.

Hs shall be based on one in hundred year return period in deep water (typically 20–30 m).

1. For water depth greater than 2XHs, Hs is the design wave height.
2. For shallow water:

Maximum wave height ($H_{max}$) for shallow water = 0.78 x design water depth (d) (for design water depth less than 2 times Hs; Hs shall be based on one in hundred return period in deep water).

*Significant wave height* (Design wave height for shallow water) $H_s = H_{max} / 1.8$

### 4.5.1.2 Design of Armour Layer

Assuming armour unit is rough quarry stone, two layers, no overtopping.

According to Hudson,

$$W = W_r H_D^{\ 3} / k_D \left(S_r - 1\right)^3 cot, \tag{4.6}$$

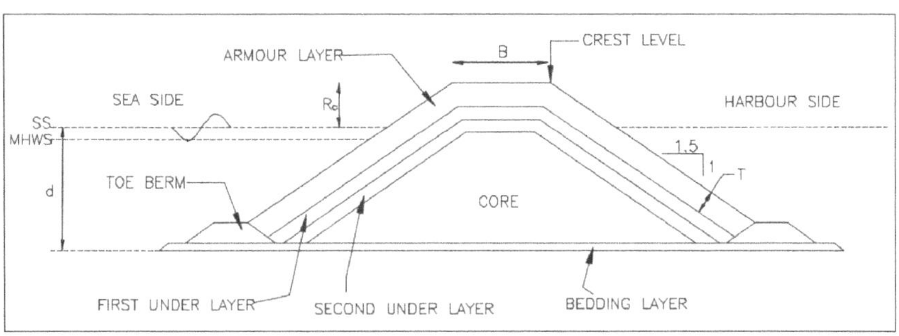


**FIGURE 4.15** Typical cross-section of conventional breakwater.

where
- $W$ = Weight of individual armour unit in Newton
- $H_D$ = Design wave height in 'm'
- $K_d$ = Stability co-efficient (Refer Table 7–8 in Shore Protection Manual Volume II (1984))
- $\theta$ = Angle of structure slope measured with horizontal
- $S_r$ = Specific gravity of armour unit = $W_r/W_w$
- $W_r$ = Unit weight of armour 2.65 T/m$^3$
- $W_w$ = Unit weight of sea water 1.025 T/m$^3$

Nominal diameter ($D_{50}$) of stones corresponding to the armour rubble stone weight

$$D_{50} = \left(\frac{W}{W_r}\right)^{1/3} \tag{4.7}$$

The thickness of the armour layer (T) can be calculated as,

$$(T) = nK_d D_{50} \tag{4.8}$$

$$K_d = \text{Layer Coefficient} = 1 \text{ to } 1.15$$

#### 4.5.1.3 Crest Width

The crest width depends on the maximum of the following:

- Minimum number of stones required
- Width depends on the degree of overtopping
- The functional aspect during the construction and after use which may be equal to the width of the wheel

$$B = nK_d D_{50} \tag{4.9}$$

where
- $B$ = Crest width in m
- $N$ = number of quarry stone (minimum is 3)
- $K_d$ = Layer Co-efficient
- $D_{50}$ = Diameter of armour stone in m

In the case of artificial armour (for example, tetrapod)

B = 2.5 times of tetrapod length on seaside + pavement width + crown wall thickness

### 4.5.1.4 Crest Elevation

The crest elevation above the maximum high-water line

Crest elevation = MHWS + storm surge + run-up + free board

Surf similarity parameter or wave breaker parameter $(\xi)$:

$$Z\,(\text{section modulus}) = \frac{I}{y} \tag{4.10}$$

$$s_{om} = \frac{H_s}{L_{om}} = \frac{2\pi H_s}{gT_m^2} \tag{4.11}$$

Run-up $(R_u)$ (as per CEM Chapter 5 – Part 6)

$$R_{ui\%}/H_s \quad \begin{aligned} &= A\,\xi_{om} && \text{for} \quad 1.0 < \xi_{om} \le 1.5 \\ &= B\,(\xi_{om})^C && \text{for} \quad 1.5 < \xi_{om} \le (D/B)^{1/C} \\ &= D && \text{for} \quad (D/B)^{1/C} \le \xi_{om} < 7.5 \end{aligned}$$

Reduction factor = 0.5 to 0.55

(As per Table VI-5–3 given in CEM for two layers of armour

Reduced wave run-up = run-up calculated x reduction factor

### 4.5.1.5 Secondary Layer

The weight of the stone in the secondary layer should be $W_{50}$ / 10 to $W_{50}$ / 5

$$\text{Diameter of stone (D)} = \left(\frac{W}{W_r}\right)^{1/3} \tag{4.12}$$

where

W = Weight of stone in secondary later in T

$W_r$ = Unit weight stone = 2.65 T/m$^3$

The thickness of the secondary layer $(T)$ can be calculated as

$$T = n\,K_d\,D \tag{4.13}$$

$$K_d = \text{Layer coefficient} = 1 \text{ to } 1.15$$

#### Toe berm

Toe berm should be provided to avoid the scour and also ensure the stability of armour.

The weight of the stone in the toe berm should be twice the weight of under armour

$$\text{Diameter of stone(D)} \left(\frac{W}{W_r}\right)^{1/3}$$

where

W = Weight of toe stone in T

$W_r$ = Unit weight stone = 2.65 T/m$^3$

Toe berm width = $3 \times K_d \times D$

Toe berm height = $2 \times K_d \times D$ (or) half of the width

**Core Layer:**

- The core layer supports the protective armour cover and any other underlayer.
- It prevents sediments passing thought the breakwater.
- Depending on the site condition and construction period, the core shall have wide gradation with sizes from 10 kg to 500–1000 kg.

**Bedding Layer:**

- Bedding layer act as the foundation to support the entire structure.
- Generally, stones of 1–50 kg can be provided as a bedding layer to a thickness of 150 mm.
- If soft soil is present in the site location, it is possible that the bedding layer will submerge into the sea bed. Hence, add 10% additional quantity of stones to the filter and core layer.

**TABLE 4.4**
**Stone Gradation in Different Layers**

| Rock Size | Layer | Rock Size Gradation |
|---|---|---|
| W | Primary cover layer | 75% more than W and 25% more than o.75W for rock armour and uniform size (W) for artificial armour |
| W/5 to W/2.5 | Toe Berm | 50%> W/2.5 |
| W/10 to W/5 | Second underlayer | 50%>W/5 |

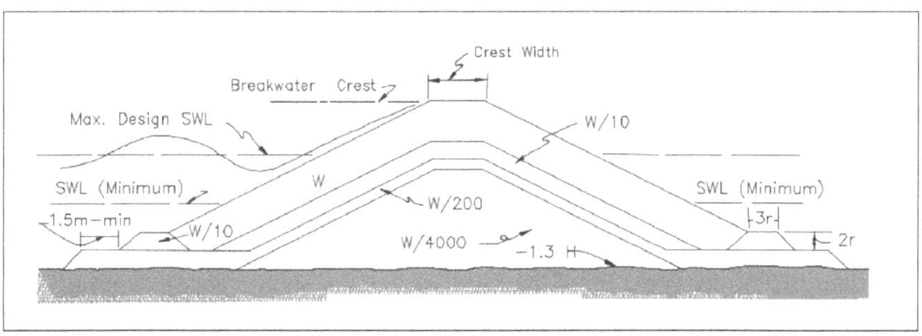


**FIGURE 4.16** Idealized multilayer section.

### 4.5.2 Design of Berm Breakwater

The design of the berm breakwater is based on the Van der Meer formula

$$\text{Size of Armour} = D_{50} = \frac{H_s}{\Delta N_s} \tag{4.14}$$

$D_{50}$ is the size of the primary armour stone,
$H_s$ is the design wave height,
$\Delta$ is relative density,
$N_s$ is stability number.

Size of armour is influenced by significant wave height and stability number
Berm width (B) = 3.5 times of $H_s$
Berm level = design water level + 0.65 times of $H_s$
Berm width and berm level are influenced by significant wave height

$$\text{Toe Design } \frac{H_s}{\Delta N_s} = \left[2 + 6.2\frac{h_t^{2.7}}{h}\right] N_{od}^{0.15} \tag{4.15}$$

where
$h_t$ is the height of the toe
$h$ is water depth
$N_{od}$ is the damage number

## 4.6 PHYSICAL MODEL STUDY

### 4.6.1 Submerged Reef

The experimental investigation of wave transformation and breaking over the submerged reef with triangular steel wedge connected to shore is carried out. The submerged reef was tested for regular and random waves at different water levels in a shallow wave basin at IIT Madras, India. The reef is designed for conditions of the south-east coast of India. The reef is envisaged to trap the sediments from the long shore sediment transport and dissipate the wave energy during cyclones to protect the shoreline.

The submerged reef has a triangular wedge-shaped steel structure with armour stones and concrete cubes, the first of its kind in India. The main part of the submerged reef is triangular shaped steel wedge weighing 900 T (in prototype) with dimensions 60 m × 50 m × 2.5 m with slope on either side resting on rubble stone bed. The model studies were carried out in 1:10 model scale for head sea condition with wave heights 5–15 cm with wave period 1.9–3 s.

The studies indicate:

1. Intensity of the wave breaking increases as the submergence depth of the reef crest is reduced.

2. When crest of reef is submerged long period waves get attenuated and generate dispersive tail wave.
3. The wave with higher incident height shows more shoaling on the upward slope of the submerged reef as compared to the lower incident wave height. Hence wave transformation depends on the incident wave height.

The submerged reef was installed and beach formation was observed.

**FIGURE 4.17** Reef positioning.

**FIGURE 4.18** After implementation of southern reef.

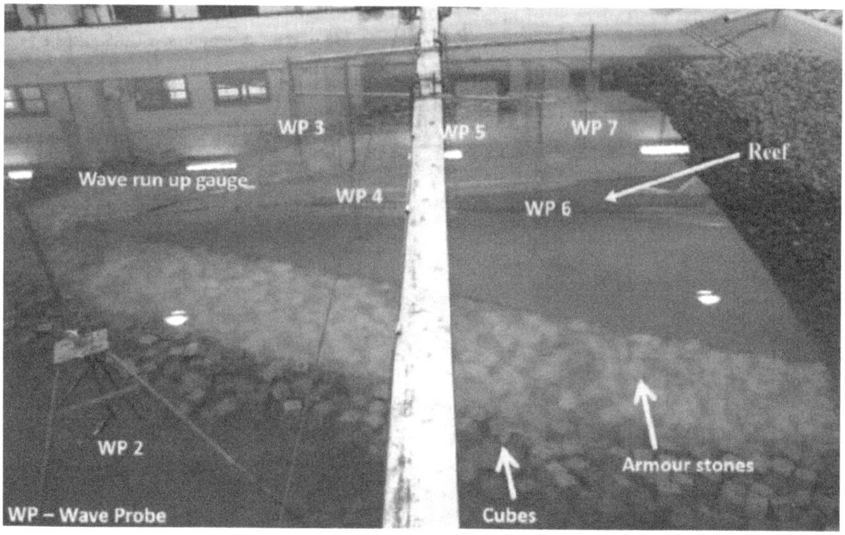


**FIGURE 4.19** Experimental set-up (1:10 scale) at shallow wave basin at IIT Madras.

### 4.6.2 Offshore Breakwater

The 3D model is to verify the roundhead section and 300 m of the trunk section for waves approaching from different direction through the following sub objectives:

- Assessing the wave/water levels at the toe of the structure
- Assessing the wave overtopping at the crest and transmission to the lee side
- Assessing the stability of structure including main armouring, under-layers, toe and toe berm details of the breakwater on seaside and rear side slopes and the crest

A selected roundhead with an adjacent length of the breakwater trunk representing 300 m in the prototype of symmetrical deepwater breakwater has been tested under the effect of oblique waves. In 3D modelling, a selected layout (Figure 4.20) has been modelled, further, the 3D model testing has been done for two dominant wave directions of 150° and 180°. Additionally, modifications were done for some locations and tested until the satisfactory results were obtained.

#### 4.6.2.1 Acceptable Damage Criteria for Accropode

Damage criterion for design wave conditions (Hs):

- No model unit removed
  - Limited settlement
  - Less than 1% of Accropode II units rocking permanently

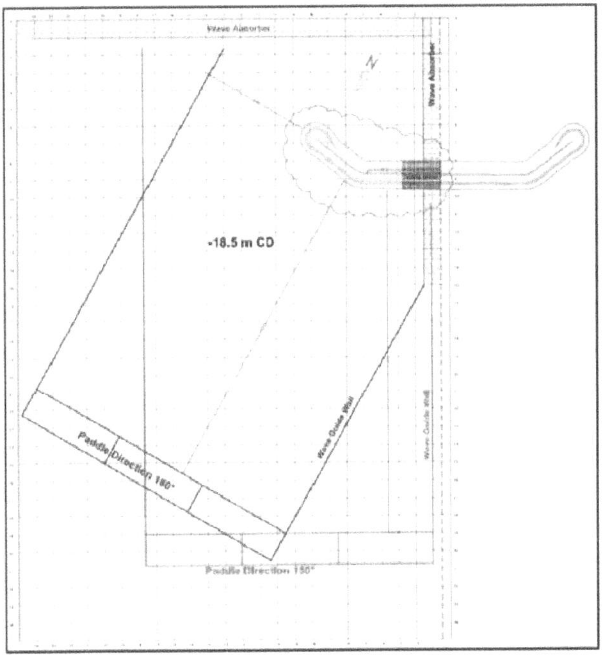


**FIGURE 4.20** Selected layout of 3D model testing.

**TABLE 4.5**
**Wave Condition at −18.5 m CD**

| Classification | Test No. | Water Level m CD | Wave Condition at −18.5 m CD: $H_s$ (m) | $T_p$ (Sec) | Focus |
|---|---|---|---|---|---|
| Initial | SE1 | +1.00 MHWL | 2.00 | 16.0 | Transmission, Stability |
| 100 YR | SE2 | | 3.60 | 9.3 | Overtopping |
| Overload | SE3 | | 4.32 | 10.5 | |
| | SE4 | +1.60 DHWL | 4.68 | 11.2 | |
| Extreme Overload | SE5 | | 5.40 | 12.2 | |
| | | The damage section shall be repaired | | | |
| 100 YR | SE6 | | 3.60 | 9.3 | Toe Stability |
| Overload | SE7 | +0.11 DLWL | 4.68 | 11.2 | |

Damage criterion for an extra load corresponding up to 120% of the design wave conditions (120% Hs from non-depth-limited projects): No model units removed from the layer. After each test, damage levels were identified separately for armour layer and other details.

*Toe Stability*

The toe stability was assessed by the limit for damage number (Noa) which is 0.5 for low water design wave condition and 2.0 for overload condition. $N_{od}$ is the number of displaced units within a strip of width $D_n$ (nominal diameter of rock armour).

*Stability of Rear Armours*

As main rear armour consists with three rock layers, the damage percentage has been estimated considering the inner two layers while assessing the damage of top layer separately. In other words, the topmost layer of the three-layer system can be damaged without compromising the structure and the allowable damage of 0–5% is considered for the inner two layers only as per the design.

1. No dislodgments have been observed for the Accropode II units and hence, can be considered as stable for all the tests in both directions. However, the dislodgement of the units observed during S5 test was due to the continuation of the lee side damage up to the crest and could not be considered as a failure of the main armour.
2. Sea side of the Trunk 1 portion was stable for all the tests done for both directions with and without the crest level modification at the curved portion. Furthermore, the observed damage at the toe berm for SE1-SE5 tests is more or less similar to the damage occurred during the S1-S5 tests done with the unmodified curved portion. Severe damage was observed at the top most layer of 1–3T due to the high overtopping occurred during S5 test and estimated damage for middle/bottom layers was 5% with unmodified section. However, no damage has been observed for repeated S5-ii with the modified section. The damage to Trunk 1 was reduced by the modification done to the curved portion. *Therefore, Trunk 1 is considered as a hydraulically stable section for all the tests in both directions.*
3. Initial tests done with +7.6 m CD crest level indicated that the overtopping created during S4 test was initiated the damage of the main armour at the lee side. Furthermore, the damage has been initiated adjacent to the water level and has been propagated towards the crest during the S5 test. 1–3T armours at the crest was begun to erode and the top part of the curved portion washed away up to −0.2 m CD due to the high overtopping. Since the damage was limited to the curved portion only, the 120 m length at the bend (60 m on either side of the curve) has been modified with crest elevation of +8.4 m CD. Therefore, only S4 and S5 tests were repeated for modified layout with +8.4 m CD crest level at curved portion. *A significant reduction of damage was observed in both repeated tests so that the estimated damage for middle/bottom layer of 1–3T was 2.8%.* However, toe and toe berm in the curved portion were stable during all the tests before and after the modification.
4. Increasing of the crest level (+8.4 m CD) at the curved portion has created a higher overtopping in the adjacent Trunk 2 which has a low crest level

(+7.6 m CD). Hence, damage to both sea side and the lee side was increased at Trunk 2 after the modification. However, *Trunk 2 is a stable section as the damage for the Trunk 2 was within the allowable limits for all the tests.*

5. As far as sea side is concern, the Trunk 3 area gave the highest Nou value for all the tests in 180°N direction. Specially, exceedance of the allowable Nod values were also observed during 180°N direction's design waves of S2 and S6 at DHWL and DLWLs respectively. In both cases, 3.6 m of design waves were directly applied after a swell wave with smaller wave height or a no wave. This might be a reason for exceedance of allowable limit for design waves and remains within the acceptable limit for the overload tests.

## Examples

1. Design a breakwater of mean water depth 4.0 m.

## Data

Mean Water Depth at MHWS = 4.0 m
Tidal Correction = 0.4 m
Sea level rise = 0.4 m
Stability Co-efficient = 2
Unit weight of armour = 2.65 T/m$^3$
Unit weight of sea water = 1.025 T/m$^3$

## Solution

Design Water Depth
= Mean water + Tidal Correction + sea level rise
= 4.0 + 0.4 + 0.4 = 4.8 m
Maximum Wave Height
$H_{max} = 0.78 \times 4.8 = 3.74$ m
Significant wave height (design wave height) for the calculation of Armour weight
$H_s = H_{max}/1.8$
$= 3.74/1.8 = 2.08$ m

## Armour Layer

Assuming armour unit is rough quarry stone, two layers, no overtopping.
According to Hudson,

$$W = W_r H_s^3 / k_D (S_r - 1)^3 cot,$$

Where,
W = Weight of individual armour unit in Newton
H = Design wave height in 'm'
$k_D$ = Stability co-efficient = 2 for breaking wave, from Table 7.8 of Shore protection Manual

$\theta$ = Angle of structure slope measured with horizontal (1:1.5)
$S_r$ = Specific gravity of armour unit = $W_r/W_w$
$W_r$ = Unit weight of armour 2.65 T/m$^3$
$W_w$ = Unit weight of sea water 1.025 T/m$^3$
Cot , = 1.5
$s_r = W_r/W_w = 2.585$
$W = 2.65 \times 2.08^3/2 \times (2.585-1)^3 \times 1.5 = 2.79$ T
Provide Boulder Stone of 3.0 T
Nominal Diameter (D) of stones corresponding to the Armour rubble stone weight

$$D = \left(\frac{W}{W_r}\right)^{1/3} = \left(\frac{3000}{2650}\right)^{1/3} = 1.04$$

## Armour Thickness

The thickness of the armour layer (T) can be calculated as,

$T = nK_dD$

$K_d$= Layer Coefficient = 1.15

T= 2 x 1.15 x 1.04

= 2.392 m

Adopt 2.5 m thick armour layer.

## Crest Width

$$B = nK_dD$$

where
B = Crest width in m
N = number of quarry stone (n = 2)
$K_d$ = Layer Co-efficient
B = $2 \times 1.15 \times 1.04 = 2.392$ m
Provide a crest width of 4.0 m core layer for construction purpose
Provide Stones of weight 10 kg to 500 kg for core

## Bedding Layer

The stones of 1 to 50 kg are provided of thickness 150 mm

## Toe Berm Width and Height

Toe berm width should be Maximum of
$3 \times K_{\Delta} \times D$,
Toe berm height should be Maximum of
$2 \times K \times D$ and
Or 1/2 of berm width

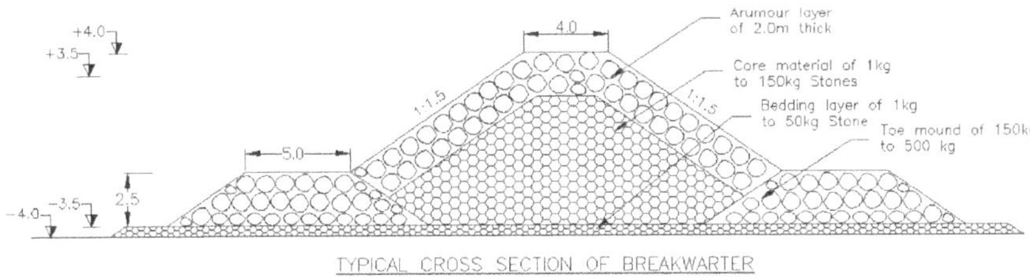


**FIGURE 4.21** Cross-section of breakwater.

# Annexure 1

## A. THE DEEPWATER BREAKWATER ON THE WEST COAST OF INDIA

The upcoming Vizhinjam International Deepwater Multipurpose Seaport it is located on an open coast, a breakwater is being built to provide sufficient tranquility for the safe loading and unloading of cargo from vessels. The length of the rubble mound breakwater is 3100 m, which includes a wave wall to protect the harbour basin from waves and swells. The road at the top of the breakwater has a width of 10 m. The outer part of the breakwater is designed to withstand a significant wave height of 4.7 m.

The breakwater is constructed from the land side by end-on dumping of quarry-run materials for the core portion. The core is then trimmed to the required profile. Similarly, the underlayers/rock armour is also placed and trimmed. Subsequently, the Accropodes are placed, followed by the construction of a wave wall and road made on top.

The geotechnical aspects of the breakwater include slope stability, settlement and liquefaction susceptibility. One of the major inputs for the aforesaid analysis is the density of the quarry run, which forms the major part of the breakwater, i.e. the core. The density of the quarry run depends upon the porosity between rock fragments.

According to the CIRIA manual, the quarry run is defined as a 'volume filling material', whereas it is a cheap material, typically coming from what is left over at a quarry after producing armour rock, which is used to make a core of a structure, this being usually the largest volume of rock required. The Rosin-Rammler method is used to obtain the core gradation for the breakwater construction.

**FIGURE 4.22** Breakwater construction at Vizhinjam Port.

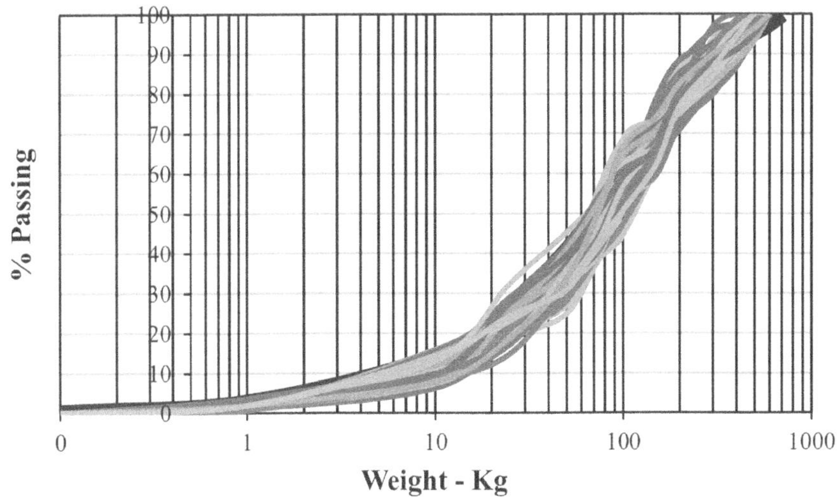


**FIGURE 4.23** Grading curves for various samples, for core 0.1–500 kg.

Well-graded curve contains particles of a wide range of sizes, and it has a good representation of all the sizes of rock. The core materials may consist of either primary rock fill such as quarry run, alternative granular materials. In the core of a structure, notably for underwater placement, the placed bulk density is the most important indicative material parameter, as it relates closely to shear strength and possible settlement as well as to permeability. It is governed by grading width and particle shape.

All the gradation curves satisfy the ratios of $D_{10}/D_5$, $D_{20}/D_{10}$, $D_{30}/D_{15}$, $D_{40}/D_{20}$, $D_{60}/D_{10}$. It is also observed from the test curves that there are no significant gaps in the material gradation. Majority of the curves, the value of grading width ($D_{85}/D_{15}$) is greater than 2.5, which shows that they are of 'very wide gradation'. All the test curves of the core material also satisfy the requirements against other parameters (filter criteria, internal stability, porosity and loss of material) specified. for the core of the breakwater to provide desired strength and stability to the structure.

## B. ACCROPODE PLACEMENT FOR OFFSHORE BREAKWATER

The offshore breakwater in 18 m water depth is 915 m long, with the main trunk portion aligned parallel to the berths and the eastern and western arms at 143° to the main trunk section of the breakwater. The real-time interlocking of armour units is always difficult to achieve in poor underwater visibility conditions. This case study describes the usefulness of the Posibloc software for placing artificial concrete armour units in any water visibility condition using a floating barge in open sea condition. The placement of 4 $m^3$ Accropode from the crane mounted on the floating barge was found to be less than 20 units in a 12-hour duration. To speed up the progress, the 150 T capacity crane is lifted from the floating barge and kept

**FIGURE 4.24** Placing of Accropode from the top of the breakwater.

on top of the emerging breakwater at +3.5 m CD. The width available is 10 m at this level. The Accropodes are transported from the shore to the breakwater and are lifted by the crane, and stored on top of the breakwater. The storage capacity is increased from 30 to 60. Subsequently, the crane is used from the breakwater and with this arrangement, the rate is increased to 40 units from toe level to −2 m CD and 60 units from −2 m CD. The diver assistance is found to be minimal for shore-based crane operation. The quality of placement is perfectly controlled. The placing density achieved is from 95% to 105%.

## C. DESIGN OF CROWN WALL FOR THE BREAKWATER

The crown wall for the breakwaters are designed using Jensen's (1984) & Bradbury et al.'s (1988) method given in CIRIA C683, the Rock Manual, 2007.

The design parameters for the breakwater section at −4.0 m bed level are as below:

Modified design wave height = 4 m
Significant wave height = 3.15 m
Peak wave period = 11 sec
Peak wave length in deep water $L_{op} = gT^2/(2\pi) = 189$ m
Max horizontal force $F_H = (\rho_w g d_c L_{op}) \times (aH_s/R_{ca} - b)$
Density of water = 1030 kg/m$^3$
Height of crest wall face $d_c = 1.25$ m
Crest of armour above the SWL = 4.3 m
$H_s/R_{ca} = 0.732$
$H_s/L_{op} = 0.0167$

Width of the base of the crown wall = 6 m

Empirical coefficients 'a' and 'b' for calculating wave forces on crown wall for cross-section A as per CIRIA C683 Rock Manual, Table 5.49, Page No 637 are as below:

Empirical coefficient, a = 0.051

Empirical coefficient, b = 0.026

Maximum Horizontal Force $F_H = 27{,}110$ N = 27.11 kN

Horizontal Pressure $P_H = F_{H/}d_c = 21.688$ kN/m

Uplift Pressure $P_U = P_H = 21.688$ kN/m

Uplift Force, $F_u = P_U * (B_c/2) = 65.064$ kN

$$F_U = (\rho_w g B_c L_{op}/2) \times (aH_s/R_{ca} - b) \tag{4.16}$$

Weight of concrete per m, $F_G = 123$ kN

| | in kN | Lever Arm (m) | Moment in kNm |
|---|---|---|---|
| W1 | 108 | 2.75 | 297 |
| W2 | 6 | 0.25 | 1.5 |
| W3 | 3 | 0.67 | 2 |
| W4 (Shear Key) | 6 | 0.25 | 1.5 |
| Total | 123 | | 302 |

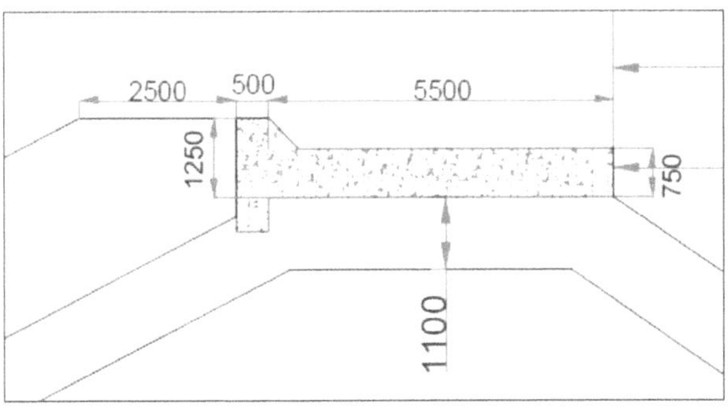


**FIGURE 4.25** Crown wall.

1. Stability against uplift

   Stability against uplift = FG/Fu
   = 1.8905

   > 1.5

   Hence safe

2. Stability against sliding
   As per CIRIA Eqn 5.199,
   $f(F_G - F_U) \geq F_H$
   Total downward weight = 123 kN
   Uplift force Fu = 65.064 kN
   Frictional coefficient = 0.7
   $f(F_G - F_U) = 40.555\ kN > F_H$
   Hence safe
3. Stability against overturning
   As per CIRIA Eqn 5.200,
   $M_G - M_U \geq M_H$
   Resisting Force = 123 kN
   Resisting Moment = 302 kNm
   Horizontal Force = 27.11 kN
   Lever arm = 0.5 m
   Moment due to horizontal force = 13.55 kNm
   Overturning uplift force $P_u * B_c = 119.28$ kN
   Lever arm = 2.00 m
   Overturning moment = 238.57 kNm
   $M_G - M_U = 63.43$ kNm
   $> M_H$
   Hence safe

## D. STABILITY ANALYSIS OF BREAKWATER THROUGH NUMERICAL MODELLING

This study investigates the stability and structural safety of proposed rubble mound breakwaters for fishery harbours using numerical modelling through PLAXIS 2D. Breakwaters are essential coastal structures that protect harbours from oceanic wave forces, often combining wave energy dissipation with berthing facilities. The effectiveness of a breakwater depends on its alignment, material composition, and its resilience under environmental loads such as waves, currents, and soil interaction.

The harbour comprises two breakwaters, are oriented perpendicular to the shoreline, terminating at the −4.0 m depth contour. The length of southern breakwater is 1240 m and northern breakwater is 1150 m. The breakwater layout is shown in Figure 4.26. Subsoil investigations revealed soft clay and silty sand layers extending up to depths of 13 m and 8.8 m respectively, requiring foundation improvements for stability.

The rubble mound breakwater design involves a multilayered system comprising:

- Primary armour units (0.5–1.0T stones)
- Secondary armour layers (0.1–0.2T stones)
- Core layers (20–100 kg stones)
- Filter layers (1–10 kg well-graded stones)
- Toe berms with 3 m width on the sea side and 1.5 m on the port side using large quarry stones

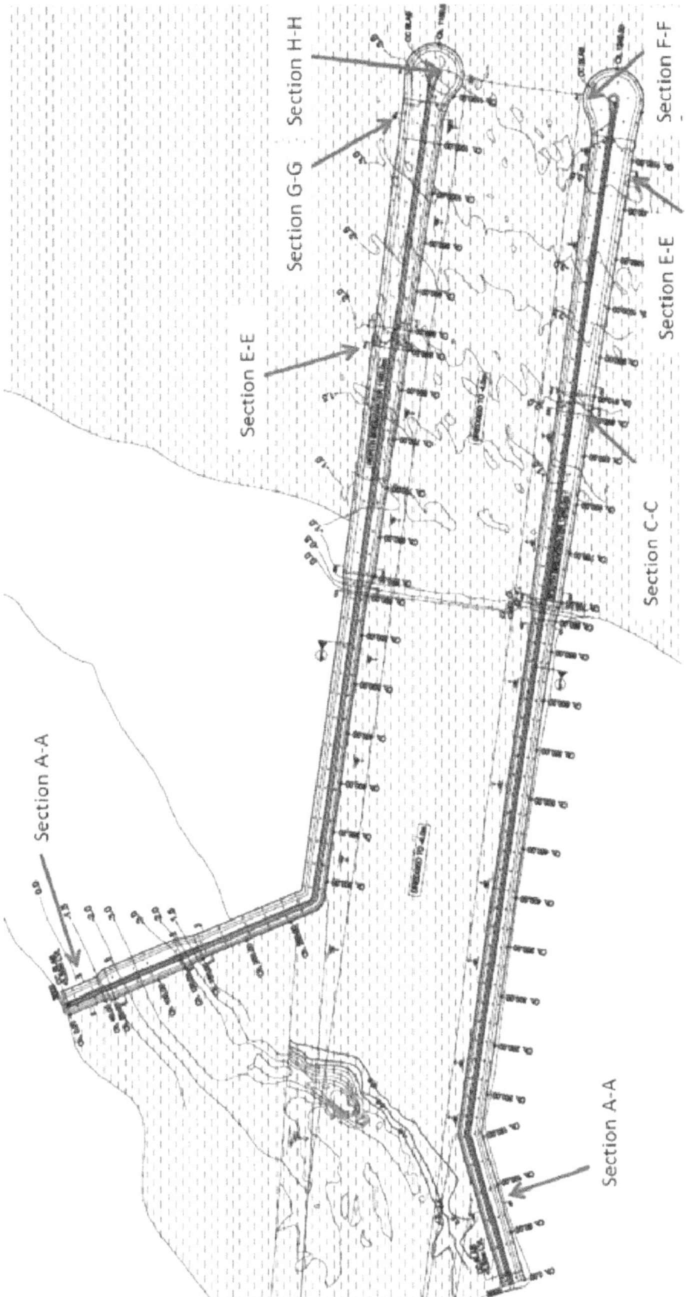


**FIGURE 4.26** Layout of north and south breakwaters at Machilipatnam.

The design methodology includes assessing tidal levels, wave characteristics, crest elevation, toe protection, and layered construction. The slope configurations adopted were 1:2 on the sea side and 1:1.5 on the port side.

## NUMERICAL MODELLING USING PLAXIS 2D

Finite element analysis was conducted using PLAXIS 2D, a specialized geotechnical tool employing the Mohr-Coulomb soil model. This model requires five soil parameters derived from field and laboratory tests. The modelling included various structural phases:

1. Base structure activation
2. Water table simulation
3. Application of live load (50 kN/m²)
4. Final safety analysis

Sections analysed included the trunk and round head portions of both breakwaters. The sections of –3.5 m CD and –4 m CD is resting over a soft clay bed. Hence a uniform settlement of 1 m to 2 m is predicted during the initial construction phase. The settlement and structural deformation of the same will not affect the structural stability of the breakwater. And therefore additional quantity backup needs to be considered during the construction phase to have quality backup in order to compensate the settlement quantity and the same quantity of quarry run material can be add on to the total estimated quantity. Rock should be filled at –3.5 m bed level for the depth of 1 m below the filter unit. Breakwater at –3.5 m bed level is shown in Figure 4.27. At –4 m bed level, rock should be filled for 2 m depth. PLAXIS analysis is carried out with the improved soil condition. It gives the satisfactory results.

## RESULTS AND CONCLUSION

At Machilipatnam, sections at 0 m, –2 m, –3.5 m, and –4 m bed levels were analysed. Even with soft foundation conditions, the rock replacement strategy ensured

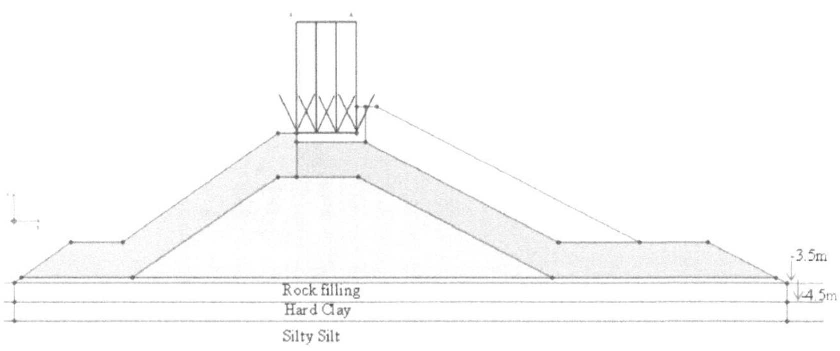


**FIGURE 4.27** PLAXIS model at –3.5 m CD with rock filling in north breakwater.

**TABLE 4.6**
**Summary of Breakwater Analysis in PLAXIS 2D**

| Cross-Section | Factor of Safety | Total Displacement |
|---|---|---|
| North Breakwater – Section AA South Breakwater – Section AA (0 m bed level) | 2.02 | 109.42 mm |
| North Breakwater – Section HH South Breakwater – Section CC (−2 m bed level) | 1.7 | 151.09 mm |
| North Breakwater Section GG South Breakwater – Section EE (−3.5 m bed level) | 1.66 | 158.68 mm |
| North Breakwater Section HH South Breakwater – Section FF (−4 m bed level) | 1.85 | 183.33 mm |

satisfactory safety margins, with factors of safety ranging from 1.66 to 2.02 and displacements between 109 mm and 183 mm.

The study concludes that the breakwater designs at both locations are geotechnically and structurally stable under the given environmental and operational conditions. The use of PLAXIS 2D modelling provided reliable insights into behaviour under real-world load conditions, guiding design optimization and material specification.

## REVIEW QUESTIONS

1. What is a breakwater? Explain it with a cross-section.
2. Write the purpose of a breakwater in brief.
3. What is tranquility?
4. Name the different types of breakwater.
5. Explain berm breakwater with a cross-section.
6. Define erosion and accretion with a sketch.
7. What are the causes of coastal erosion?
8. Describe littoral transport with a sketch.
9. Name some coastal protection methods.
10. Draw the typical cross-section of a seawall and explain it.
11. What are Groynes? Explain different types of Groynes with a diagram.
12. Design a breakwater for the following given data.
    Mean sea level = 0.34 m
    MHWS = 0.76 m
    Contour level = 0
    Storm surge = 1
    Design high water level = 1.76 m
13. Design a berm breakwater of significant wave height 2 m and stability number 2.
14. How is the physical model study of breakwaters done?

## REFERENCES

**Coastal Engineering Manual (Part VI) (2011)** *Department of Army U.S. Army Corps of Engineers*

**Durusoju Hari Prasad and Nandyala Darga Kumar (2014)** Coastal erosion studies – A review, *International Journal of Geosciences*, Volume 5, pp. 341–345, http://dx.doi.org/10.4236/ijg.2014.53033

**Gegar Prasetya Thematic Paper**, *The Role of Coastal Forests and Trees in Protecting against Coastal Erosion*. Available at www.fao.org

**IRC 75 (1979)** *Guidelines for the Design of High Embankments*, Indian Road Congress

**S. Karthik, S.A. Sanniasiraj, and V. Sundar (2020)** *Hydrodynamic Characteristics of Curved and Vertical Front Face Pile-supported Breakwaters in Regular Waves*, https://doi.org/10.1016/j.oceaneng.2020.108105

**Shore Protection Manual: Volume II (1984)** *Coastal Engineering Research Centre Department of the Army Waterways Experiment Station*, Corps of Engineers

**UNCTAD (1985)** *Port Development: A Handbook for Planners in Developing Countries*, United Nations Conference on Trade and Development, United Nations, New York

# 5 Berthing Structures

## 5.1 BERTHING STRUCTURES

Berthing structures are constructed for the berthing and mooring of vessels to enable the loading and unloading of cargo and for embarking and disembarking passengers and vehicles. The design of berthing structures depends on various factors. However, the vessel characteristics govern the design of berthing structures.

Several types of berthing structures, load combination for analysis, fixity and soil springs calculation for foundation analysis, limit state of collapse and serviceability for design are described in this chapter. Design of concrete and steel piles and design of deck with precast and cast in-situ construction of berthing structures are also explained in this chapter. In addition to these, the pile load test is illustrated in this chapter.

## 5.2 GENERAL

### 5.2.1 Types of Berthing Structures

A berthing structure is a facility where a vessel may be safely moored. Berthing arrangements are classified as alongside type, open dolphin type and Ferry type.

Berthing structures are classified as follows

- Based on configuration – quay, wharf, pier or jetty
- Based on the type of cargo handled – passenger, general cargo, bulk cargo, container and liquid cargo berth
- Based on the face of the structure – vertical type and open type
  Vertical type includes caisson, sheet pile wall, block wall and cellular bulkhead.
  Open type structures are represented by open piled structures.
- Based on load transfer – gravity and flexible
  Gravity structures – masonry wall, concrete block walls and concrete caissons
  Flexible structures
  Steel sheet piles – tie back, cantilever
  Diaphragm walls – cantilever, relieving platform, tie back
  Jetties – consisting of berthing and mooring dolphins, jetty head and approach jetty

The minimum length of a berthing structure should be sufficient for mooring the longest ship expected to arrive. The minimum depth includes a bottom clearance equivalent to 10% of the draught of the largest vessel using the terminal. The top

DOI: 10.1201/9781003361275-5

surface of the berthing structure should be built above the highest high-water level. The dimension of berth, dredge level and deck top-level shall be derived as per IS 4651 (Part V) (1980) (discussed in Chapter 2). The dimension of the berth is also given in Appendix 6.1 of IS 4651 (Part V) – 1980 for various sizes of passenger ships, freighters, tankers, ore carriers and large fishing vessels.

#### 5.2.1.1 Quay

A quay is a stretch of paved bank or a solid artificial landing place parallel to the navigation waterway for loading and unloading vessels. Quays are constructed as sheet pile wall, diaphragm wall, open piled structure or gravity structures. A quay is a continuous berth where fenders and bollards are provided at different points along the berth to facilitate the berthing of ships.

#### 5.2.1.2 Wharf

A wharf is a structure built on the shore of a harbour, river or canal so that vessels may lie alongside to receive and discharge cargo and passengers. It is constructed as an open structure supported on piles.

#### 5.2.1.3 Pier

A pier is a structure usually of open construction, extending out into the water from the shore to serve as a landing place, recreational facility, etc., rather than to afford coastal protection.

#### 5.2.1.4 Jetty

A jetty is a structure that projects into the water on open sea-coasts. A finger jetty will have berths on two sides. A jetty consists of a number of structures such as berthing dolphin, mooring dolphin, loading platform and approach trestle (loading platform to shoreline), each of which has a special type of function.

Berthing dolphins support fenders which absorb berthing impacts. Berthing dolphins should be placed as wide apart as possible. The distance should neither exceed the length of the straight side of the smallest vessel nor be less than approximately one-third of the maximum length of the largest vessel.

Mooring dolphins pick up the pull from the hawsers. Mooring dolphins for breast lines are located at the bow and stern at a distance (about the beam of the ship) from the berth line, which will not make the moorings too steep.

Loading platforms support special loading or unloading equipment but normally, no horizontal forces apart from wind loads will act on the loading platforms.

### 5.2.2 Factors Governing the Selection of Berthing Structure

- Purpose of berth
- Average soil type below the sea bed
- Dead weight tonnage and type of vessel
- Tide levels
- Level of rock/dense sand/very stiff clay

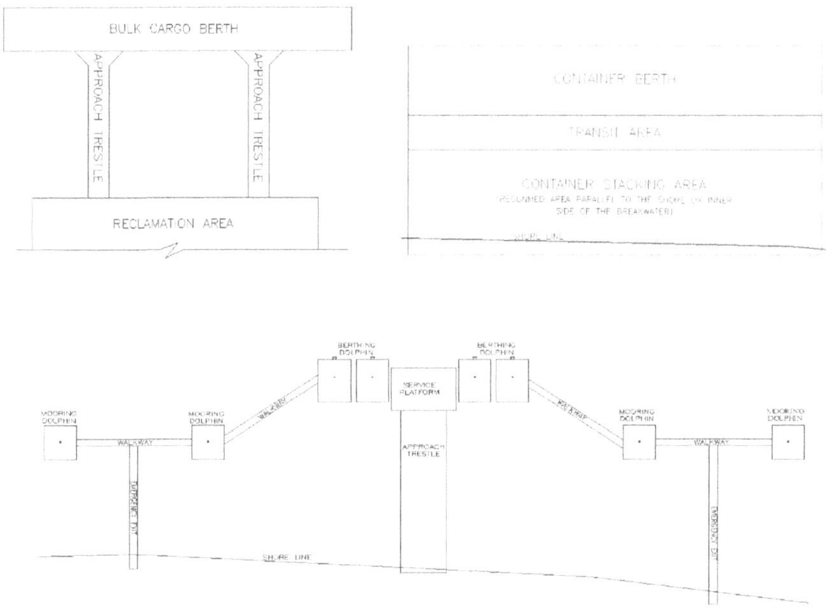


**FIGURE 5.1** Layout of bulk cargo berth, container berth and an oil jetty.

### 5.2.3 Selection of Type of Berthing Structure

- Selection depends on the magnitude and nature of loading hydraulic conditions such as wave action and currents.
- For bulk cargo berth, open-type constructions with approach trestle are preferable. A layout of an example of a bulk cargo berth with an approach trestle is shown in Figure 5.1.
- For an oil jetty, an open type with berthing and mooring dolphins is preferable. A layout of an oil jetty with berthing and mooring dolphins is shown in Figure 5.1.
- For a container berth, an open or vertical face type adjacent to the shore or breakwater is preferable.
- If the sea bed is rock/dense sand/very stiff clay, gravity structures like caissons are preferable. The cross-section of the typical caisson structure is shown in Figure 5.2.
- Soil conditions will have an essential bearing on the type of deck selected.

### 5.2.4 Open-Face and Vertical-Face Berthing Structures

The berthing structure can also be classified depending upon its infrastructures such as open-face or vertical-face structures.

1. Open-face structures
   - Jetties – these consist of berthing and mooring dolphins, jetty head and approach jetty.

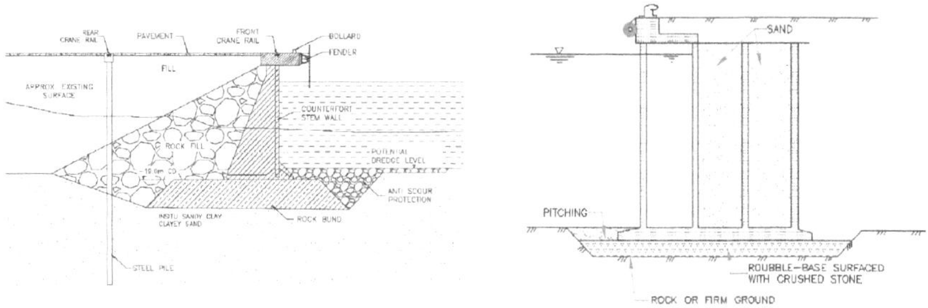


**FIGURE 5.2** Block wall and caisson berth.

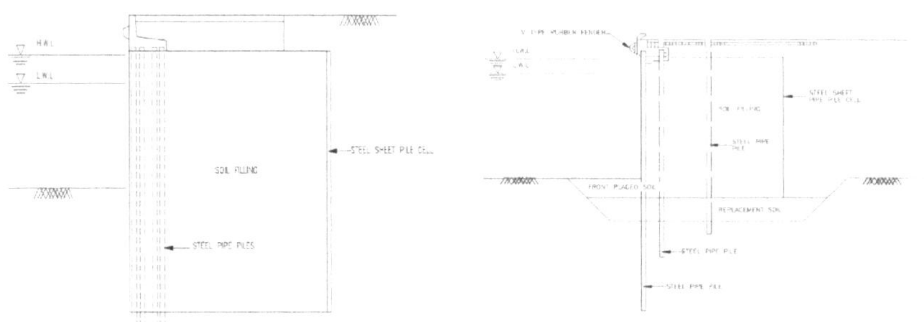

**FIGURE 5.3** Embedded type steel cellular bulkhead quay wall.

2. Vertical-face structures
   - Block wall (Figure 5.2)
   - Caissons (Figure 5.2)
   - Cellular bulkhead (Figure 5.3)
   - Touch pile sheet pile (tie back or cantilever)
   - Diaphragm wall (tie back or cantilever) (Figure 5.4)

The touch pile sheet pile or diaphragm wall structure will have a vertical face at the berthing side or rear side, depending upon the requirement of the structure.

### 5.2.4.1 Sheet Pile Wall

Sheet piles are usually earth retention techniques to retain soil. They can also be used to avoid water percolation for marine construction work. They are installed in sequence to the designed depth along a planned alignment. The sheet piles act as a permanent or temporary wall to resist lateral earth pressure.

The sheet pile wall is unable to resist the horizontal loads from backfill, moving forces, etc. A part of the horizontal force is absorbed by passive earth pressure in front of the sheet piles through their fixations in the sea bed, and tied anchor rods to an anchor structure behind the sheet pile wall will transmit part of the force.

Sheet pile materials can be wood, reinforced concrete or steel. Today, steel sheet piles are the most widely used sheet wall elements in berth structures.

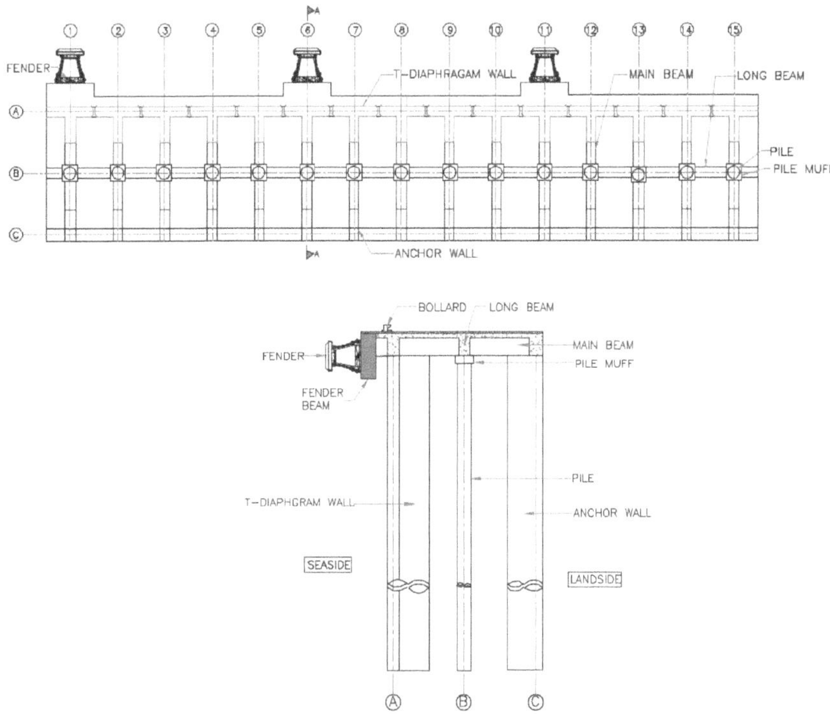


**FIGURE 5.4** Plan and cross-section of diaphragm wall berthing structure.

Sheet pile walls are divided into three groups depending on how the load-bearing elements are designed:

1. Simple sheet pile wall berths
2. Solid platform berths
3. Semi-solid platform berths

### *Simple Sheet Pile Wall Berths*

- These are used when water depth at the berth front is moderate and the soil conditions are favorable for a simple sheet pile wall.
- The sheet piles are driven down to sufficient depth, one after the other.
- The horizontal anchor force is transmitted from the wall (anchoring beam) by way of a tie rod to a retaining plate or another type of anchorage.
- The most economic free wall height of a simple sheet pile wall berth with a simple anchoring system varies from about 7 m to 10 m, depending on soil conditions, price of steel, driving equipment and the useful load on the quay deck.

### *Solid Platform Berths*

- When the height of the berth front exceeds about 8–10 m, the simple sheet pile wall structure will typically not be the most economical solution.

- Better solutions can be obtained by using one or more of the following adaptations:
  1. Tie rods are placed at two different levels. This can also be done where the tidal variations are significant.
  2. Use of special sheet pile profiles designed to resist great bending moments.
  3. Use of lightweight fill material behind the sheet pile wall.
  4. A relieving plate, placed on the wall and on piles behind the wall, transmits useful loads and the weight of fill on top as axial loads to the wall and the piles and reduces the horizontal load acting on the wall. This type of structure is called the solid platform type of berth.

Horizontal forces are absorbed either by batter piles under the relief plate or by anchorage further back. This type is economical in cases where live load and/or the quay front height are relatively high.

*Semi-solid Platform Berth*

- If a greater quay front height than about 14–18 m is needed, this type of berth structure is used.
- In principle, it is very much like the solid platform type, but the soil pressure against the sheet pile wall is further reduced by omitting some of the fill under the platform. This implies that a formwork has to be made for the platform, instead of casting it directly on the soil, as for the solid quays.
- The main advantage of semi-solid type compared with the solid type is that the berth front height can be made substantially higher.
- The semi-solid platform berth may be characterized as a transition from the solid berth structure to open column and lamella berth structures.

#### 5.2.4.5 Diaphragm Wall

A diaphragm wall is a structural concrete wall constructed in a deep trench excavation, either cast in-situ or using precast concrete components. Diaphragm walls are often used on congested sites, close to existing structures, where there is restricted headroom, or where the excavation is of a depth that would otherwise require the removal of much greater volumes of soil to provide stable battered slopes.

Diaphragm walls are suitable for most subsoils, and their installation generates only a small amount of vibration and noise, which increases their suitability for works carried out close to existing structures. In addition, floor slab connections and recessed formwork can be incorporated into the walls.

The walls generally range in thickness from 0.5 m to 1.5 m and can be excavated to depths of over 50 m. Excavation is typically carried out using rope-suspended mechanical or hydraulically operated grabs.

Specific ground conditions or greater depths may require hydro mills (hydraulically operated reverse circulation trench cutters) to penetrate hard rock by 'cutting' rather than 'digging'. Hydro mills can achieve depths of up to 80 m.

Excavation stability is maintained using a drilling fluid, usually a bentonite slurry. This is a controlled mixture that has thixotropic properties, meaning that it exerts pressure in excess of the earth and hydrostatic pressures on the sides of the excavation. The walls are constructed using reinforced or unreinforced concrete, in discrete panel lengths generally ranging between 2.5 m and 7 m. Purpose-made stop ends can be used to form the joints between adjacent panels, with a water bar incorporated across the joints.

More complicated arrangements such as L- or T-shaped panels can be constructed where additional bending moment capacity or wall stiffness is required.

### 5.2.5 Offshore Berthing Structures

Offshore berthing structures are used for liquid cargo (oil or gas) or for dry cargo, for iron ore, coal, sugar, phosphates or grains. The design for offshore berthing structures should consider the following:

- Single type of cargo
- Rapid loading and unloading (10,000 T of iron ore per hour or 60,000 bbl (barrels) of oil per hour)
- Sufficient storage onshore
- Open sea and exposed to winds, waves and currents
- Construction practicability
- DWT of the vessel in the range of 0.1–0.3 million T

The type of ship loader generally governs the design of offshore berth. The three main types of ship loader are (1) fixed type, (2) travelling gantry type and (3) slewing telescopic boom loaders. The fixed loader is used in small ships. The travelling gantry loader is expensive since the loader is to be supported by a berth which is continuous. These three types of loaders are to be critically evaluated for dry bulk cargo terminals, whereas for liquid cargo, the loading system does not influence the offshore berthing structure. The approach jetty to the offshore berthing structure is the critical component and governs the total cost of the facility. The mooring and berthing force in the offshore berth is to be critically evaluated for the safe design of the offshore berthing structure.

## 5.3 LOAD COMBINATION

The combination of loadings for design is dead load, vertical live loads, plus either berthing load, or line pull or earthquake or wave pressure, for open type berthing structure. The governing combination should be taken for the design. In addition to the above load, earth pressure and differential water pressure shall be considered for vertical force typing structures.

For a given load combination, partial safety factors are considered for reliability-based design of structural components.

The partial safety factors for different types of loads in limit state design are given in IS 4651 (Part 4) (2023).

The points to be considered while using the partial safety factor for load combination as per IS 4651 (Part 4) (2023) are as follows:

- All members shall be designed to sustain safely the effect of the combination of various loads, forces and stresses that can possibly coexist.

- All calculations shall distinctly tabulate the various combinations of above loads and stresses covered by design.
- Crane/machine operating loads and along with dead load combined with wind load specified by the manufacturer or the normal wheel load with maximum wind load as specified in IS 875 (Part 3) (2015), whichever is more severe, should be taken for design purposes.
- The berthing force and crane/machine operating load may not be considered to occur simultaneously.
- Wind load and seismic forces need not be deemed to act simultaneously.
- Limit state of serviceability load combinations shall be given as below:
  1. Short-term effect for crack width check only
     a. $1.0(DL) + 1.1(DyL) + 1.0(LL + EP + HyF + WL + CL + BF + WWiF + SS)$
     b. $1.0(DL) + 1.1(DyL) + 1.0(LL + EP + HyF + WL + CL + MF + WWiF + SS)$
  2. Short-term effect for deflection check near expansion joint only
     a. $1.0(DL) + 1.1(DyL) + 1.0(LL + EP + HyF + WL + CL + EL + SS)$
- Limit state of serviceability load combinations (long-term).
  1. $1.0(DL) + 0.5(DyL + LL) + 1.0(EP + Shrinkage + Creep + TempL)$

     For shrinkage, creep and temperature loads, IS 456 (2000) and IS 875 (Part 5) (1987) may be referred.
- For normal loading berthing energy shall be considered for the BF calculation, whereas for extreme loading, ultimate berthing energy shall be considered for the BF calculation.
- Limit state of collapse load combinations.
  1. Normal load condition
     a. $1.5(DL + DyL + LL) + 1.2(EP) + 1.0(HyF) + 1.2(WL + CL) + 1.5(BF) + 1.0(WWiF)$
     b. $1.5(DL + DyL + LL) + 1.2(EP) + 1.0(HyF) + 1.2(WL + CL) + 1.5(MF) + 1.0(WWiF)$
  2. The seismic load combination in limit state design of RCC and prestressed concrete structures are
     a. $1.5(DL + IL)$
     b. $1.2(DL + IL \pm EL)$
     c. $1.5(DL \pm EL)$

     where IL is imposed load such as weight of stairways, handrails, equipment reactions and super structures, etc.
  3. Extreme/survival load condition
     a. $1.2(DL + DyL + LL) + 1.0(EP) + 1.0(HyF + WL + CL) + 1.0(BF)$
     b. $1.2(DL + DyL + LL) + 1.0(EP) + 1.0(HyF + WL + CL) + 1.2(EL)$

c. $1.2(DL + DyL + LL) + 1.0(EP) + 1.0(HyF + WL + CL) + 1.2(TL)$

d. $1.2(DL + DyL + LL) + 1.0(EP) + 1.0(HyF + WL + CL) + 1.2(EWiF)$

4. Temporary load condition
   a. $1.2(DL + DyL + LL) + 1.0(EP) + 1.0(HyF + WL + CL)$
5. Reversal load condition
   a. $0.9(DL + DyL + LL) + 1.0(EP) + 1.0(HyF + WL + CL) + 1.5(BF \text{ or } MF)$

   b. $0.9(DL + DyL + LL) + 1.0(EP) + 1.0(HyF + WL + CL) + 1.5(EWiF)$

   c. $0.9(DL + DyL + LL) + 1.0(EP) + 1.0(HyF + WL + CL) \pm 1.5(EL)$

Increases in permissible stresses for different combinations of loads are given in Table 5.1.

**TABLE 5.1**
**Increases in Permissible Stresses**

| Sl. No. | Combination of Loads | Increase in Permissible Stress: Reinforced Concrete | Increase in Permissible Stress: Other Materials Such as Steel and Timber | Increase in Allowable Bearing Pressure |
|---|---|---|---|---|
| 1 | DL + LL + impact of breaking or traction or vehicles + centrifugal forces of vehicles | Nil | Nil | Nil |
| 2 | DL + LL with impact, breaking or tractive and centrifugal forces + earth pressure, percent | 15 | 15 | 15 |
| 3 | DL with/without LL including impact, breaking or tractive and centrifugal forces + earth pressure + hydrodynamic and hydrostatic forces + berthing and mooring forces, percent | 25 | 33 1/3 | 25 |
| 4 | Wind forces on structures + load combination of (1) + (2) or (3) | | | |
| 5 | Seismic forces + load combination of (1), (2) or (3), percent | | | |
| 6 | Secondary stress + load combination of (1), percent | 15 | 15 | 15 |
| 7 | Erection stage stresses with DL and appropriate LL + earth pressure + hydrodynamic forces + wind forces, percent | 15 | 33 1/3 | 25 |

**TABLE 5.2**
**Factor of Safety for Load Combination as per BS 6349 (Part 2) (2019)**

| | | Action | Symbol (from BS EN 1990) | EQU (Set-A) | | STR/GEO (Set B) | | STR/GEO (Set C) | |
|---|---|---|---|---|---|---|---|---|---|
| | | | | Unfavorable | Favorable | Unfavorable | Favorable | Unfavorable | Favorable |
| Permanent actions | | Permanent actions including geo-technical actions (unfavourable) | $\gamma_{Gsup}$ | 1.05 | – | 1.35 | – | 1.0 | – |
| | | Permanent actions including geo-technical actions (unfavourable) | $\gamma_{Ginf}$ | – | 0.95 | – | 1.0 | – | 1.0 |
| | | Deck surfacing | $\gamma_{Gsup}$ | 1.75 | 0 | 1.75 | 0 | 1.0 | 1.0 |
| Variable actions | Persistent | Cargo loads | $\gamma_Q$ | 1.5 | 0 | 1.5 | 0 | 1.3 | 0 |
| | | Ship ramp loads | $\gamma_Q$ | 1.2 | 0 | 1.2 | 0 | 1.2 | 0 |
| | | Road and traffic actions (including pedestrians) | $\gamma_Q$ | 1.35 | 0 | 1.35 | 0 | 1.15 | 0 |
| | | Crane loads "A" | $\gamma_Q$ | 1.35 | 0 | 1.35 | 0 | 1.15 | 0 |
| | | Wind loads | $\gamma_Q$ | 1.4 | 0 | 1.4 | 0 | 1.3 | 0 |
| | | Temperature | $\gamma_Q$ | 1.3 | 0 | 1.3 | 0 | 1.3 | 0 |
| | | Differential settlement | $\gamma_Q$ | 1.2 | 0 | 1.2 | 0 | 1.0 | 0 |
| | | Snow load | $\gamma_Q$ | 1.4 | 0 | 1.4 | 0 | 1.3 | 0 |
| | | Water current loads | $\gamma_Q$ | 1.4 | 0 | 1.4 | 0 | 1.3 | 0 |
| | | Wave loading | $\gamma_Q$ | 1.4 | 0 | 1.4 | 0 | 1.3 | 0 |
| | | Ship berthing loads | $\gamma_Q$ | 1.4 | 0 | 1.4 | 0 | 1.3 | 0 |
| | | Mooring loads | $\gamma_Q$ | 1.4 | 0 | 1.4 | 0 | 1.3 | 0 |
| | | Forces from ship propulsion | $\gamma_Q$ | 1.5 | 0 | 1.5 | 0 | 1.3 | 0 |

| | | | | | | | | | |
|---|---|---|---|---|---|---|---|---|---|
| | | Earth and geotechnical water pressures | $\gamma_Q$ | 1.5 | 0 | 1.5 | 0 | 1.3 | 0 |
| | Transient | Abnormal berthing loads | $\gamma_Q$ | 1.2 | 0 | 1.2 | 0 | 1.2 | 0 |
| | | Erection case | $\gamma_Q$ | 1.2 | 0 | 1.2 | 0 | 1.2 | – |
| | | Earth and geotechnical water pressures | $\gamma_Q$ | 1.5 | 0 | 1.5 | 0 | 1.3 | 0 |
| Pre-stress | | | $\gamma_p$ | See BS EN (1992) or relevant Eurocode | – | See BS EN (1992) or relevant Eurocode | – | See BS EN (1992) or relevant Eurocode | – |

The cases for UPL and HYD limit states are dealt with in BS EN (1997). The cases for FAT limit states are dealt with in BS EN (1993-1-9) and BS EN (1993-2).

**TABLE 5.3**
**$\psi$ Factors for Combinations as per BS 6349 (Part 2) (2019)**

| Action | | $\psi_0$ Factor for the Combination Value of a Variable Action | $\psi_1$ Factor for the Frequent Value of a Variable Action | $\psi_2$ Factor for the Quasi Permanent Value of a Variable Action |
|---|---|---|---|---|
| General port cargo loadings etc. | Cargo loads | 0.7 | 0.50 | 0.3 |
| | Ship ramp loads | 0.7 | 0.5 | 0 |
| Traffic actions (including pedestrians) | Port vehicle loadings | 0.75 | 0.75 | 0 |
| | Tandem system A) | 0.75 | 0.75 | 0 |
| | Uniformly distributed load A) | 0.75 | 0.75 | 0 |
| | Pedestrian, cycle tracks | 0.4 | 0.4 | 0 |
| | Crane loads | 0.75 | 0.75 | 0 |
| | Unusual vehicles B) | 0 | 0 | 0 |
| Environmental loads | Wind loads | 0.6 | 0.2 | 0 |
| | Temperature | 0.6 | 0.6 | 0.5 |
| | Snow load | 0.5 | 0.2 | 0 |
| | Hydrostatic loads (excluding ground water effects) C) | 0.7 | 0.5 | 0.3 |
| | Water current loads | 0.6 | 0.2 | 0 |
| | Wave loading | 0.6 | 0.2 | 0 |
| Ship and mooring loads | Ship berthing loads | 0.75 | 0.75 | 0 |
| | Mooring loads | 0.6 | 0.2 | 0 |
| | Forces from ship propulsion | 0.75 | 0.75 | 0 |
| Others | Earth and ground water pressures | 1.0 | 1.0 | 1.0 |
| | Erection case | 1.0 | – | 1.0 |
| | Pre-stress | See BS EN (1992) or relevant Eurocode | See BS EN (1992) or relevant Eurocode | See BS EN (1992) or relevant Eurocode |

A) Grla is a load combination that comprises the TS and UDL systems in accordance with BS EN (1991-2).
B) This assumes that special vehicles will be loading the structures in controlled conditions.
C) Hydrostatic loads for floating structures are regarded as permanent loads and therefore do not feature here.

**TABLE 5.4**
**Formulae for Combinations of Actions as per BS 6349 (Part 2) (2019)**

| Design Situation | Limit State | Permanent Actions | Pre-stress | Leading Accidental or Seismic Action | Leading Variable Action A) | Accompanying Variable Actions A) | Eq. no. B) |
|---|---|---|---|---|---|---|---|
| **Ultimate Limit States** | | | | | | | |
| Persistent and transient | Unfavorable | $\Sigma\gamma_{Gj.sup}G_{kj.sup}$ | $+\gamma_p$ | – | $+\gamma_{Q,1}Q_{k,1}$ | $+\Sigma\gamma_{Q,i}\Psi_{0,i}Q_{k,i}$ .....[C)] | 6.10 |
| | Favorable | $\Sigma\gamma_{Gj.inf}G_{kj.inf}$ | $+\gamma_p$ | – | $+\gamma_{Q,1}Q_{k,1}$ | $+\Sigma\gamma_{Q,i}\Psi_{0,i}Q_{k,i}$ .....[C)] | 6.10 |
| Accidental | Unfavorable | Either: $\Sigma G_{kj.sup}$ | $+P$ | $+A_d$ | $+\Psi_{1,1}Q_{k,1}$ | $+\Sigma\Psi_{2,i}Q_{k,i}$ | 6.11b |
| | | or $\Sigma G_{kj.sup}$ | $+P$ | $+A_d$ | $+\Psi_{2,1}Q_{k,1}$ | $+\Sigma\Psi_{2,i}Q_{k,i}$ | 6.11b |
| | Favorable | Either: $\Sigma G_{kj.inf}$ | $+P$ | $+A_d$ | $+\Psi_{1,1}Q_{k,1}$ [D)] | $+\Sigma\Psi_{2,i}Q_{k,i}$ | 6.11b |
| | | or $\Sigma G_{kj.inf}$ | $+P$ | $+A_d$ | $+\Psi_{2,1}Q_{k,1}$ [D)] | $+\Sigma\Psi_{2,i}Q_{k,i}$ | 6.11b |
| Seismic | Unfavorable | $\Sigma G_{kj.sup}$ | $+P$ | $+A_{Ed}$ [E)] | – | $+\Sigma\Psi_{2,i}Q_{k,i}$ | 6.12b |
| | Favorable | $\Sigma G_{kj.inf}$ | $+P$ | $+A_{Ed}$ [E)] | – | $+\Sigma\Psi_{2,i}Q_{k,i}$ | 6.12b |
| Accidental construction (execution) | Unfavorable | $\Sigma G_{kj.sup}$ | $+P$ | $+A_d$ | – | $+\Sigma\Psi_{2,}Q_{c,k}$ [F)] | A2.2 |
| | Favorable | $\Sigma G_{kj.inf}$ | $+P$ | $+A_d$ | – | $+\Sigma\Psi_{2,}Q_{c,k}$ [F)] | A2.2 |

*(Continued)*

**TABLE 5.4** *(Continued)*

**Formulae for Combinations of Actions as per BS 6349 (Part 2) (2019)**

| | | | | | | | |
|---|---|---|---|---|---|---|---|
| **Serviceability Limit States** | | | | | | | |
| Characteristic | $c_1$ Unfavorable $c_1$ | $\Sigma G_{kj.sup}$ | $+P$ | – | $+Q_{k,1}$ | $+\Sigma\Psi_{0,i}Q_{k,i}$ | 6.14b |
| | $c_1$ Favorable $c_1$ | $\Sigma G_{kj.inf}$ | $+P$ | – | $+Q_{k,1}$ | $+\Sigma\Psi_{0,i}Q_{k,i}$ | 6.14b |
| Frequent | $c_1$ Unfavorable $c_1$ | $\Sigma G_{kj.sup}$ | $+P$ | – | $+\Psi_{1,1}Q_{k,1}$ | $+\Sigma\Psi_{2,i}Q_{k,i}$ | 6.15b |
| | $c_1$ Favorable $c_1$ | $\Sigma G_{kj.inf}$ | $+P$ | – | $+\Psi_{1,1}Q_{k,1}$ | $+\Sigma\Psi_{2,i}Q_{k,i}$ | 6.15b |
| Quasi-permanent | $c_1$ Unfavorable $c_1$ | $\Sigma G_{kj.sup}$ | $+P$ | – | – | $+\Sigma\Psi_{2,i}Q_{k,i}$ | 6.16b |
| | $M_r = 0.87\ f_y A_{st} d\left[\rightarrow -\frac{A_{st} f_y}{bdf_{ck}}\right]$ Favorable $c_1$ | $\Sigma G_{kj.inf}$ | $+P$ | – | – | +£¤ $_{2,i}Q_{k,i}$ | 6.16b |

**Key:**

+ To be combined with

£ The combined effect of

A) Variable actions are listed in the table. Any one of the variable actions in a design situation could be classified as 'Leading', the others then being regarded as 'Accompanying'.

B) Refers to equation numbers in BS EN (1990:2002+A1:2005).

C) For the case where wind and waves are acting simultaneously on the structure for the Set A, Set B and Set C ULS cases, $\psi_0$ should be changed to $\psi_1$ to allow for the low probability that the actual maximum wave will coincide with the actual design gust levels on which wind loads in BS EN (1991-1-4) are based. An additional check using the characteristic values of both loads should be performed but treated as an accidental load case.

D) The choice between $\psi_{1,1}$ or $\psi_{2,1}$ should be related to the relevant accidental design situation.

E) $A_{Ed} = \gamma_I A_{Ek}$, where the seismic effects are as defined in BS EN (1998).

F) $Q_{c,k}$ is the characteristic value of the construction loads as defined in BS EN (1991-1-6).

The factors of safety for loads, factors for combination of loads and formulae for combination of loads are given in Tables 5.2, 5.3 and 5.4, respectively, as per BS 6349 (Part 2) (2019).

## 5.4 FIXITY DEPTH, SOIL SPRINGS AND PILE CAPACITY, AND PILE TERMINATION CRITERIA

### 5.4.1 Fixity Depth Calculation

The point of fixity needs to be calculated to ensure suitable embedment when lateral loads apply. For reinforced concrete piles, this point is required to determine the extent of additional reinforcement at the top of the pile.

The point of fixity is based on the load, pile type, size and soil condition.

As per IS 2911 (Part I/Sec 2) (2010), the first step is to calculate the stiffness factor R or T for a particular combination of pile and soil to determine if the pile behaves as a short, rigid unit or as an infinitely long flexible member.

Having calculated the stiffness factor, the criteria for behaviour as a short rigid pile or as a long elastic pile is related to the embedded length L of the pile. The depth from the ground surface to the point of virtual fixity is then calculated (Figure 5.5) and used in the conventional elastic analysis for estimating the lateral deflection and bending moment.

Stiffness factor for piles in sand and normally loaded clays:

$$T=5\sqrt{\frac{EI}{n_h}}\text{ for sand} \tag{5.1}$$

$$T=5\sqrt{\frac{EI}{K_1}}\text{ for normally loaded clay} \tag{5.2}$$

where

$E$ = Young's modulus of the pile material in kg/cm$^2$

$I$ = Moment of inertia of the pile cross-section in cm$^4$

$\eta_h$ = Modulus of subgrade reaction for sand (Table 5.5)

$k_1$ = Modulus of subgrade reaction for normally loaded clay (Table 5.6)

This figure is valid for long, flexible piles where the embedded length $L_e$ is ≥ 4R or 4T.

### 5.4.2 Spring Constant Calculations

Soil-structure interaction analysis is performed by springs as a representation of soil.

#### The spring stiffness is calculated based on the following procedure.

- The analysis of the structure has been carried out using STAAD.Pro, by idealizing the soil support using springs for both the diaphragm wall and piles.
- The springs are placed at 1 m intervals.

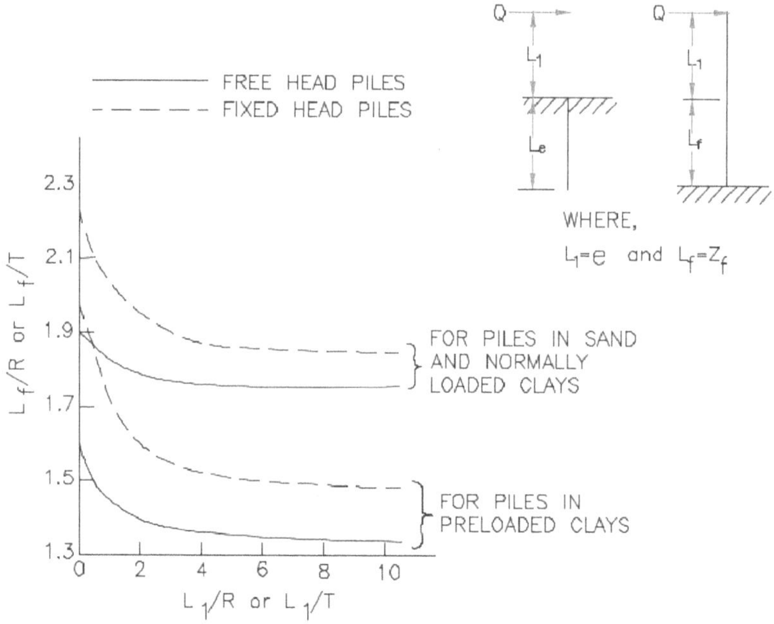


**FIGURE 5.5** Determination of fixity depth.

**TABLE 5.5**
**Modulus of Subgrade Reaction for Sand, $\eta_h$, in kN/m³**

| Sl. No. | Soil Type | *N* (Blows/30 cm) | Range of $K_1$ ($kN/m^3 \times 10^3$) | |
|---|---|---|---|---|
| | | | Dry | Submerged |
| 1 | Very loose sand | 0–4 | <0.4 | <0.2 |
| 2 | Loose sand | 4–10 | 0.4–2.5 | 0.2–1.4 |
| 3 | Medium sand | 10–35 | 2.5–7.5 | 1.4–5.0 |
| 4 | Dense sand | >35 | 7.5–20.0 | 5.0–12.0 |

The $K_1$ values may be interpolated for intermediate standard penetration values, *N*.

- The spring spacing shall be nearly equal to the thickness of the diaphragm wall or the pile diameter for effective modelling of soil support in finite element analysis.
- The spring constants at each node are calculated as the reaction offered by the soil in the region 0.5 m above the node and 0.5 m below the node.

## TABLE 5.6
## Modulus of Subgrade Reaction for Cohesive Soils, $k_1$, in kN/m$^3$

| Sl. No. | Soil Type | Unconfined Compression Strength, $q_u$ (kN/m$^2$) | Range of $k_1$ (kN/m$^3$ × 10$^3$) |
|---|---|---|---|
| 1 | Soft | 25–50 | 4.50–9.00 |
| 2 | Medium stiff | 50–100 | 9.00–18.00 |
| 3 | Stiff | 100–200 | 18.00–36.00 |
| 4 | Very stiff | 200–400 | 36.00–72.00 |
| 5 | Hard | >400 | >72.00 |

For $q_u < 25$, $k_1$ may be taken as zero, which implies that there is no lateral resistance.

## TABLE 5.7
## Criteria for Behaviour of Pile Based on Its Embedded Length

| | | Relation of Embedded Length With Stiffness Factor | |
|---|---|---|---|
| Sl. No. | Soil Type | Linearly Increasing | Constant |
| i) | Short (rigid) pile | $L \leq 2T$ | $L \leq 2R$ |
| ii) | Long (elastic) pile | $L \geq 4T$ | $L \geq 3.5R$ |

The intermediate $L$ indicates a case between rigid pile behaviour and elastic pile behaviour.

## TABLE 5.8
## Fixity Depth for Piles in Clayey Soil – Maximum Free Length

| Clay Description | Shear Strength* (kN/m$^2$) | Compressive Strength** (kN/m$^2$) | Fixity Depth in Terms in Diameter |
|---|---|---|---|
| Soft | 12.5 | 25 | 5.83 D |
| Soft to medium stiff | 25 | 50 | 5.05 D |
| Medium stiff to stiff | 50 | 100 | 5.25 D |
| Stiff to very stiff | 100 | 200 | 3.79 D |
| Very stiff to hard | 200 | 400 | 3.26 D |

* Shear strength for clay = 6 × SPT value.

** Compressive strength for clay = 12 × SPT value.

**TABLE 5.9**
**Fixity Depth for Piles in Clayey Soil – Zero Free Length**

| Clay Description | Shear Strength* (kN/m$^2$) | Compressive Strength** (kN/m$^2$) | Fixity Depth in Terms in Diameter |
|---|---|---|---|
| Soft | 12.5 | 25 | 6.82 D |
| Soft to medium stiff | 25 | 50 | 5.94 D |
| Medium stiff to stiff | 50 | 100 | 5.17 D |
| Stiff to very stiff | 100 | 200 | 4.51 D |
| Very stiff to hard | 200 | 400 | 3.92 D |

* Shear strength for clay = 6 × SPT value.

** Compressive strength for clay = 12 × SPT value.

**TABLE 5.10**
**Fixity Depth for Piles in Sand – Maximum Free Length**

| Sand Description | *N* Value | Modulus of Subgrade Reaction for Granular Soils (Submerged) (kN/m$^3$) | Fixity Depth in Terms in Diameter |
|---|---|---|---|
| Very loose to loose sand | 4 | 200 | 10.99 D |
| Loose to medium sand | 10 | 1400 | 7.36 D |
| Medium to dense sand | 35 | 5000 | 5.67 D |
| Dense sand | >35 | 12000 | 4.74 D |

**TABLE 5.11**
**Fixity Depth for Piles in Sand – Zero Free Length**

| Sand Description | *N* Value | Modulus of Subgrade Reaction for Granular Soils (Submerged)(kN/m$^3$) | Fixity Depth in Terms in Diameter |
|---|---|---|---|
| Very loose to loose sand | 4 | 200 | 12.72 D |
| Loose to medium sand | 10 | 1400 | 8.62 D |
| Medium to dense sand | 35 | 5000 | 6.70 D |
| Dense sand | >35 | 12000 | 5.61 D |

**TABLE 5.12**
**Fixity Depth for Piles in Rock – Maximum Free Length**

| Rock Description | Shear Strength (MPa) | Compressive Strength (MPa)* | Fixity Depth in Terms in Diameter |
|---|---|---|---|
| Grade A | 40 | 80 | 1.24 D |
| Grade B | 10 | 20 | 1.76 D |
| Grade C | 4 | 8 | 2.21 D |
| Grade D | 2 | 4 | 2.63 D |
| Grade E | 1 | 2 | 3.12 D |
| Grade F | 0.6 | 1.2 | 3.55 D |

* Compressive strength for rock = 2 × shear strength.

**TABLE 5.13**
**Fixity Depth for Piles in Rock – Zero Free Length**

| Rock Description | Shear Strength (MPa) | Compressive Strength (MPa)* | Fixity Depth in Terms in Diameter |
|---|---|---|---|
| Grade A | 40 | 80 | 1.71 D |
| Grade B | 10 | 20 | 2.42 D |
| Grade C | 4 | 8 | 3.05 D |
| Grade D | 2 | 4 | 3.62 D |
| Grade E | 1 | 2 | 4.31 D |
| Grade F | 0.6 | 1.2 | 4.90 D |

* Compressive strength for rock = 2 × shear strength.

## Modulus of Subgrade Reaction

- The modulus of subgrade reaction is a conceptual relationship between soil pressure and deflection that is widely used in the structural analysis of foundation members.
- It is used for continuous footings, mats and various types of piling.

## Calculation of Subgrade Modulus

The subgrade modulus is calculated as per Equation (5.3):

$$k_s = \frac{1.3}{D}\sqrt[12]{\frac{E_s B^4}{E_h I}}\,\frac{E_s}{1-\mu^2} \tag{5.3}$$

where

Es = Elastic modulus of the soil
Eb = Elastic modulus of the pile
I = Moment of inertia
D = Diameter of pile/thickness of diaphragm wall
μ = Poisson's ratio

## Calculation of Soil Spring Values for Pile

Equations (5.4) to (5.6) are used to calculate individual spring constants:

Top spring value

$$K_t = \frac{BL}{24}\left(7Ks_1 + 6Ks_2 - Ks_3\right) \tag{5.4}$$

Bottom spring value

$$K_b = \frac{BL}{24}\left(7Ks_n + 6Ks_{n-1} - Ks_{n-2}\right) \tag{5.5}$$

Intermediate spring value

$$K_i = \frac{BL}{12}\left(10Ks_{i-1} + Ks_i - Ks_{i+1}\right) \tag{5.6}$$

where

B = Lateral dimension of pile
L = Segment length of pile
$K_s$ = Modulus of subgrade

Equations (5.7) to (5.9) are used to calculate soil spring values for a diaphragm wall:

Top spring value

$$K_1 = \frac{DL}{24}\left(7Ks_{[1]} + 6Ks_{[2]} - Ks_{[3]}\right) \tag{5.7}$$

Intermediate spring value

$$K_i = \frac{DL}{12}\left(Ks_{[i-1]} + 10Ks_{[i]} + Ks_{[i+1]}\right) \tag{5.8}$$

Bottom spring value

$$K_n = \frac{DL}{24}\left(7Ks_{[n]} + 6Ks_{[n-1]} - Ks_{[n-2]}\right) \tag{5.9}$$

## TABLE 5.14
## Range of Modulus Grade Reaction $K_s$

| Soil | | $K_s$(kN/m$^3$) |
|---|---|---|
| Loose sand | | 4800–16,000 |
| Medium dense sand | | 9600–80,000 |
| Dense sand | | 64,000–128,000 |
| Clayey medium dense sand | | 32,000–80,000 |
| Silty medium dense sand | | 24,000–48,000 |
| Clayey soil | | |
| 200 < | $q_a \le 200$ kPa | 12,000–24,000 |
| | $q_a \le 800$ kPa | 24,000–48,000 |
| | $q_a > 800$ kPa | >48,000 |

Use values as a guide and for comparison when using approximate equations.

## TABLE 5.15
## Poisson's Ratio for Selected Materials

| Material | Poisson's Ratio |
|---|---|
| Sand | |
| Dense | 0.3–0.4 |
| Loose | 0.2–0.35 |
| Fine | 0.25 |
| Coarse | 0.15 |
| Rock (basalt, granite, limestone, sandstone, schist, shale) | 0.1–0.4<br>Depending on rock type, density and quality, commonly 0.15–0.25 |
| Clay | |
| Wet | 0.1–0.3 |
| Sandy | 0.2–0.35 |
| Silt | 0.3–0.35 |
| Saturated clay on silt | 0.45–0.5 |
| Glacial silt (wet) | 0.2–0.4 |
| Loess | 0.1–0.3 |
| Ice | 0.36 |
| Concrete | 0.15–0.25 |
| Steel | 0.28–0.31 |

*Source*: Foundation analysis and design by Joseph E. Bowles.

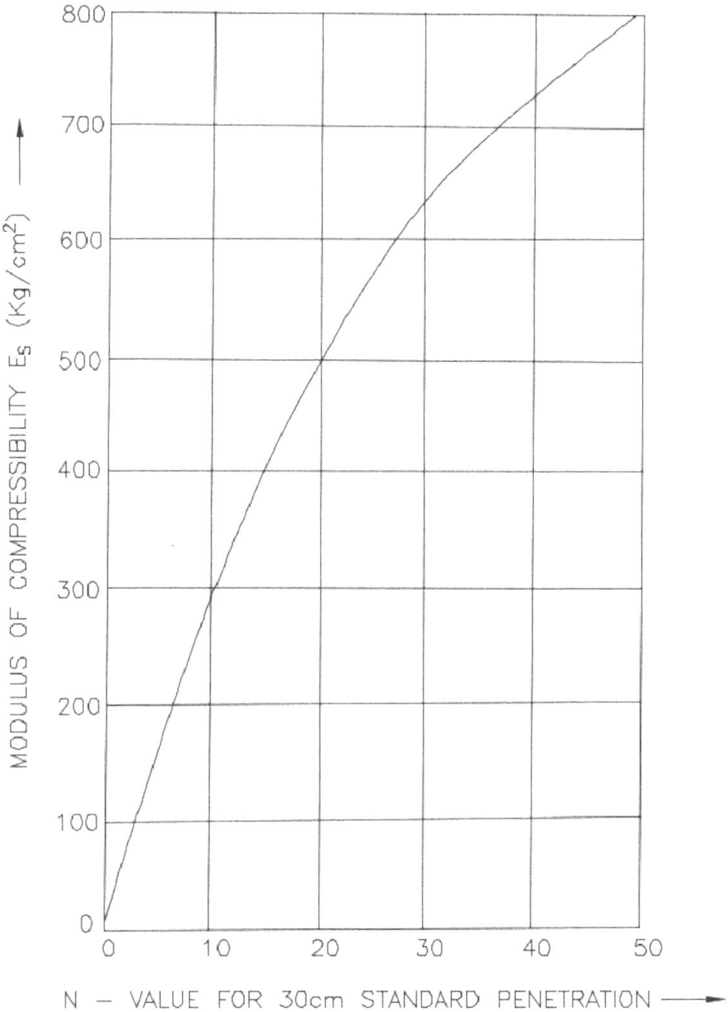


**FIGURE 5.6** Graph representing range of stress-strain relationship based on *N* value.

**Source: Foundation analysis and design by Joseph E. Bowles.**

where

$\gamma_{sub} = \gamma - \gamma_{water}$

$h_1$ and $h_2$ = Height of soil layers

$\delta$ = Angle of internal friction between pile and so

$$A_s - \text{Surface area of pile m}^2 = \pi D h_i \tag{5.12}$$

**TABLE 5.17**
**Range of Properties for Selected Rock Groups**

| Type of Rock | Typical Unit Weight (kN/m³) | Modulus of Elasticity, E (MPa × 10³) | Poisson's Ratio | Compressive Strength (MPa) |
|---|---|---|---|---|
| Basalt | 28 | 17–103 | 0.27–0.32 | 170–415 |
| Granite | 26.4 | 14–83 | 0.26–0.30 | 70–276 |
| Schist | 26 | 7–83 | 0.18–0.22 | 35–105 |
| Limestone | 26 | 21–103 | 0.24–0.45 | 35–170 |
| Porous limestone | | 3–83 | 0.35–0.45 | 7–35 |
| Sandstone | 22.8–23.6 | 3–42 | 0.20–0.45 | 28–138 |
| Shale | 15.7–22 | 3–21 | 0.25–0.45 | 7–40 |
| Concrete | 15.7–23.6 | Variable | 0.15 | 15–40 |

*Source:* Foundation analysis and design by Joseph E. Bowles.

where

$A_p$ – Cross section of the pile $m^2$
D – Diameter of pile m
$\gamma$ – Unit weight of soil $kN/m^3$
$N_c, N_\gamma$ & $N_q$ – Bearing capacity factors
$\alpha$ – Adhesion factor
Weathered rock – 0.9
Cohesive soil – Clay – Refer IS 2911 (Part 1/Sec 2) (2010) Annex B – Fig 2
c – Cohesion $kN/m^2$

**FACTOR OF SAFETY:**

Minimum considered while using static formula: 2.5
For piles in weathered or soft rock: 3

*Pile capacity as per API RP 2A-WSD 2000 (Steel pile)*

The ultimate bearing capacity of piles, including belled piles, $Q_d$ is calculated as follows:

$$Q_d = Q_f + Q_p = fA_s + qA_p \tag{5.13}$$

where

$Q_f$ = skin friction resistance, lb $(kN)$
$Q_p$ = total end bearing, lb $(kN)$
$f$ = *unit skin friction capacity*, $lb/ft^2\,(kPa)$

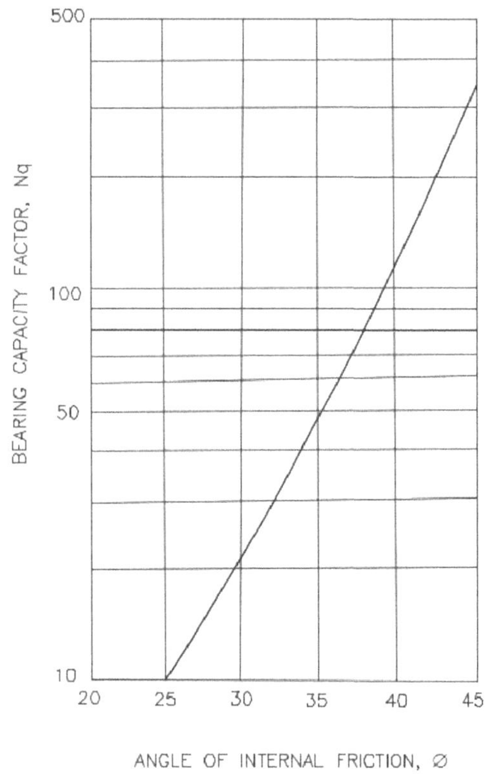


**FIGURE 5.7** Bearing capacity factor $N_q$ for bored piles.

$A_s$ = *side surface area of pile*, $ft^2 \left(m^2\right)$
$q$ = *unit end bearing capacity*, $lb/ft^2 \left(kPa\right)$
$A_p$ = *gross end area of pile*, $ft^2 \left(m^2\right)$

For pipe piles in cohesive soils, the shaft friction, *f*, in lb/ft$^2$ (kPa) at any point along the pile is calculated by the following equation:

$$f = \alpha c$$

where
$\alpha$ = *a dimensionless factor*
$c$ = *undrained shear strength of the soil*

The factor, –, can be computed by the equations:

$$\alpha = 0.5\,\Psi^{-0.5}\Psi \leq 1.0$$

– = 0.5 ¤ $^{-0.25}$¤ > 1.0 with the constraint that $\alpha \leq 1.0$

where

$\Psi = c/p'_o$

$p'_o$ = effective overburden pressure $lb/ft^2$ (kPa)

For piles end bearing in cohesive soils, the unit end bearing $q$, in $lb/ft^2$ (kPa) is computed by the following equation:

$$q = 9c$$

For pipe piles in cohesionless soils, the shaft friction, $f$, in $lb/ft^2$ (kPa) is calculated by the following equation:

$$F = Kp_o \tan \delta$$

where

$K$ = Coefficient of lateral earth pressure (ratio of horizontal to vertical normal effective stress)

$p_o$ = Effective overburden pressure $lb/ft^2$ (kPa)

$\delta$ = Friction angle between the soil and pile wall

For open-ended pipe piles driven unplugged, it is usually appropriate to assume $K$ as 0.8 for both tension and compression loadings. Values of $K$ for full displacement piles (plugged or closed end) may be assumed to be 1.0.

For piles end bearing in cohesionless soils, the unit end bearing $q$ in $lb/ft^2$ (kPa) is computed by the following equation:

$$q = p_o N_q \tag{5.14}$$

where

$p_o$ = Effective overburden pressure $lb/ft^2$ (kPa) at the pile tip

$N_q$ = Dimensionless bearing capacity factor

### 5.4.4 Pile Termination Criteria

Pile founding level shall be calculated based on the allowable capacity of the soil strata for axial load acting on the pile. The allowable capacity of the bored cast in-situ piles shall be calculated as per IS 2911 (Part 1/Sec 2) (2010) and is illustrated in section 5.3.3.

The pile termination level shall be at least 1 times the diameter of the pile into hard rock/stiff clay/very dense sand. For soil classification, see section 5.3.1. Pile fixity depth shall be referred.

In practice, the hard rock shall be identified by pile penetration criteria as given below:

3T × 600 mm drop height × 20 blows/min for 1 hour of chiselling £ pile penetration of 200 mm/hr

**TABLE 5.18**
**Design Parameters for Cohesionless Siliceous Soil as per API**

| Density | Soil Description | Soil-Pile Friction Angle, δ Degrees* | Limiting Skin Friction Values kips/ft²(kPa) | $N_q$ | Limiting Unit End Bearing Values Kips / ft²(MPa) |
|---|---|---|---|---|---|
| Very loose | Sand | 15 | 1.0 (47.8) | 8 | 40 (1.9) |
| Loose | Sand-silt** | | | | |
| Medium | Silt | | | | |
| Loose | Sand | 20 | 1.4(67.0) | 12 | 60 (2.9) |
| Medium | Sand-Silt** | | | | |
| Dense | Silt | | | | |
| Medium | Sand | 25 | 1.7 (81.3) | 20 | 100 (4.8) |
| Dense | Sand-Silt** | | | | |
| Dense | Sand | 30 | 2.0 (95.7) | 40 | 200 (9.6) |
| Very dense | Sand-silt** | | | | |
| Dense | Gravel | 35 | 2.4 (114.8) | 50 | 250 (12.0) |
| Very dense | Sand | | | | |

* Theparameters listed in this table are intended as guidelines only. Where detailed information such as in-situ cone tests, strength tests on high quality samples, model tests, or pile driving performance is available, other values may be justified.

** Sand-silt includes those soils with significant fractions of both sand and silt. Strength values generally increase with increasing sand fractions and decrease with increasing silt fractions.

### 5.4.5 Pile Liner Termination Criteria

Pile liner shall be used to prevent the sliding of soil inside the pile while boring from the pile cut-off level to bed level and up to the required pile depth. It helps to reduce the corrosion in the pile cage rebar and helps to reduce the concrete deterioration by water. Pile liner may be considered as a structural component depending upon the requirement of infrastructure. But in most of the port and harbour structures, a pile liner will act as a non-structural member.

In practice, a pile liner of thickness 6 mm or 8 mm will be proposed.

If sheet pile is connecting to the pile liner, pile liner of thickness up to 12 mm shall be provided.

Steel plates rolled up to form a tubular section shall be used as pile.

The steel of grade varies from 240 N/mm² to 430 N/mm² and shall be used based on the site limitations conforming to the EN 10248 (Part 1)/IS 2062 (2011).

The pile liner shall be terminated up to refusal level or up to founding level, whichever is earlier as given below:

| Soil strata | Pile liner termination criteria |
|---|---|
| Rock (fresh/hard/weathered) | Up to rock touch level |

Clay — Up to refusal level or up to very stiff clayey level in which SPT N should be >50 or up to pile founding level, whichever is earlier.

Sand — Up to refusal level or up to very dense sand level in which SPT N should be >35 or up to pile founding level, whichever is earlier.

### 5.4.6 Pile Termination

In practice, the strong rock (with shear strength = 40 MN/m$^2$) using chisel method of bored cast in-situ pile shall be identified by pile penetration criteria as given below:

> 3T hammer with 600 mm drop height and 15 blows/min for 1 hour of chiselling shall have pile penetration < 200 mm/hr

The pile penetration rate (PPR) for this criteria can be calculated for a 1 m diameter pile as given below:

$$W = \text{Weight of chisel } (\text{in MT}) \ : \ 3.0 \text{ MT}$$

$$h = \text{Fall of chisel } (\text{in m}) \ : \ 0.60 \text{ m}$$

$$n = \text{Number or blows of hammer} \ : \ 15 \text{ blows/min}$$

$$: \ 900 \text{ blows/hr}$$

$$A = \text{Area of pile } (\text{in m}^2) \ : \ (\pi \times D2)/4 = (22/7) \times 1.0 \times 1.0$$

$$: \ 0.786 \text{ m}^2$$

$$p = \text{Penetration } (\text{in cm}) \ : \ 20 \text{ cm}$$

$$\text{Pile penetration ratio } (\text{PPR}) = W \times h \times n/(A \times P) \qquad (5.15)$$

$$= 3.0 \times 0.6 \times 900/(0.786 \times 20) \text{ ton-meter/m}^2\text{/cm}$$

$$= 1620/15.72 \text{ ton-meter/m}^2\text{/cm}$$

$$\text{PPR} = 103.05 \text{ ton-meter/m}^2\text{/cm}$$

The equivalent pile penetration criteria for hydraulic rig can be calculated as follows:

n = Revolutions per minute: 19

$$\text{Penetration (P)} = 2\pi \times n \times T \times t/(A \times \text{PPR})$$

$$= 2 \times (22/7) \times 19 \times 18.49 \times 1/(0.786 \times 103.05)$$

$$= 2207.35/81.0$$

$$\text{Penetration (P)} = 27.25 \text{ cm}$$

## 5.5 LIMIT STATE OF COLLAPSE AND SERVICEABILITY

### 5.5.1 Method of Design

Structures are designed by any of the following methods:

- Limit state method
- Working stress method

RCC and pre-stressed concrete members can be designed by any of the above two methods, whereas for designing structures with other material, the working stress method should be adhered to.

### 5.5.2 Limit State Method

In the limit state method of design, a structure is considered unfit for use when it reaches a particular state called the *limit state*, at which it ceases to fulfil the function or satisfy the conditions for which it is designed.

The structure shall be designed to withstand safely all loads liable to act throughout its life, and it shall also satisfy the serviceability requirements such as deflection and cracking.

#### 5.5.2.1 Partial Safety Factor for Material Strength

When assessing the strength of a structure or structural member for the limit state of collapse, the values of partial safety factors shall be taken as 1.5 for concrete and 1.15 for steel.

### 5.5.3 Working Stress Method

The working stress method or permissible stress method may also be adopted in the designs till such time the complete changeover to the limit state method in made in other relevant Indian Standards on the subject.

However, as the limit state method is more rational and adoptable, the designs may be carried out by the limit state method.

Limit state method affords easy calculation in the case of circular columns, which involve a complicated procedure under the working stress method.

## 5.6 DESIGN OF SUBSTRUCTURE COMPONENTS

### 5.6.1 Design of Concrete Pile as per IS Code

Piles must be designed to withstand stresses caused during their installation and subsequently when they function as supporting members in a foundation structure.

Piles are constructed in both RCC and steel type. Pile capacity, fixity value and spring value according to the soil condition are considered for designing of piles.

The design procedure for the design of concrete piles is as follows:

- Calculate the maximum axial load (P), the corresponding bending moment (M), the maximum bending moment (M) and the corresponding axial force (P) from STAAD analysis or manually.
- Calculate the total moment for both the cases i.e., total moment = factored moment + moment due to slenderness.

$$\text{Factored moment} = \sqrt{M_y^2 + M_z^2} \tag{5.16}$$

Moment due to slenderness $= P_u \times e$

- Calculate the percentage of pile reinforcement.

Calculate $d'/D$

where

$d'$ = Clear cover + (diameter of bar/2)
$D$ = diameter of pile

- According to SP–16 (1980), use the chart comparison with bending-circular section. Select the chart number by using $d'/D$, $f_{ck}$ and fy. The charts are given below:
- Take the value of $p/f_{ck}$
- Calculate the following equations (1) and (2) for both cases

(1) $P_u / f_{ck} \times D^2$

where

$P_u$ = Axial force in kN
$D$ = Diameter of pile in mm

(2)Area of steel reinforcement provided = number of bars provided $\times$ area of one bar

where

$M_u$ = Bending moment
$D$ = Diameter of pile

- The percentage of reinforcement considered for design of governing cases both the condition is $(P_t) = (p/f_{ck}) * f_{ck}$
- Calculate the area of steel required, $A_{st} = P_t \times \pi/4 \times D^2$
- Calculate the area of steel provided:
- Area of one bar = $\pi/4 * d^2$
- No. of bar required = Area of steel/area of one bar
- Provide the number of bars more than required

- Area of steel reinforcement provided = number of bars provided × area of one bar
- Spacing of bars calculated as per IS 456 (2000) Cl.26.5.3.2(d)

#### 5.6.1.1 Helical Reinforcement

1. Pitch-helical reinforcement shall be of regular formation with the turns of the helix spaced evenly, and its ends shall be anchored properly by providing one and a half extra turns of the spiral bar. Where an increased load on the column on the strength of the helical reinforcement is allowed for, the pitch of helical turns shall be not more than 75 mm, nor more than one-sixth of the core diameter of the column, nor less than 2.5 mm, nor less than three times the diameter of the steel bar forming the helix. In other cases, the requirements of IS 456 (2000) Cl.26.5.3.2 shall be complied with.
2. The diameter of the helical reinforcement shall be in accordance with IS 456 (2000) Cl.26.5.3.2(c)(2).
   Clear distance between longitudinal bars = π × (D – (2 × C.C) – (2 × diameter of helical reinforcement) – diameter of bar)/no. of bars provided

- Pitch and diameter of lateral ties, according to IS 456 (2000) Cl.26.5.3.2(c)
  1. *Pitch*: The pitch of transverse reinforcement shall be not more than the least of the following distances:
     a. The least lateral dimension of the compression members
     b. Sixteen times the smallest diameter of the longitudinal reinforcement bar to be tied
     c. 300 mm
  2. *Diameter*: The diameter of the polygonal links or lateral ties shall be not less than one-fourth of the diameter of the largest longitudinal bar and in no case less than 16 mm.

- Determination of crack width
  1. The crack width of the pile has been evaluated by the following Gerely-Lutz equation, given in ACI 318R-95 Commentary of Building code requirements for structural concrete published by the American Concrete Institute since in IS code crack width calculation for rectangular section only provided.

$$cw = \frac{0.076\ X\ \gamma\ X\ \sigma_{st X \sqrt[3]{d_{c\ X\ A}}}}{1000} \tag{5.17}$$

$$\gamma = \frac{\textit{Distance of extreme tension fibre from Neutral Axis } (h_2)}{\textit{Distance of centroid of tension steel from neutral axis } (h_1)} \tag{5.18}$$

where

A = Effective tension area of concrete surrounding the flexural tension reinforcement and having the same centroid, as that reinforcement, divided by the total number of bars

$$A = \frac{\pi X(D - 2Xd - \theta) \times (2Xd + \theta)}{N_b} \tag{5.19}$$

2. As per IS 456 (2000), crack width shall be also be checked using

$$W_{cr} = \frac{3a_{cr} \in_m}{1 + 2(a_{cr} - C_{min})/(D - x)} \tag{5.20}$$

where

$a_{cr}$ = Distance from the point considered to the surface of the nearest longitudinal bar

$C_{min}$ = Minimum cover to the longitudinal bar

$\varepsilon_m$ = Average steel strain at the level considered

D = Overall depth of the member

x = Depth of the neutral axis

$$\varepsilon_m = \varepsilon_{1-} \frac{b(h - x)(a - x)}{3E_s A_s (d - x)} \tag{5.21}$$

where

$A_s$ = Area of the tension reinforcement

b = Width of the section at the centroid of the tension steel

$\varepsilon_1$ = Strain at the level considered, calculated ignoring the stiffening of the concrete in the tension zone

a = Distance from the compression face to the point at which the crack width is being calculated

d = Effective depth

### 5.6.2 Design of Steel Pile as per IS Code Working Stress Method

- Calculate the bending moment by STAAD analysis or manually. (Take the maximum bending moment.)
- Check for thickness as per IS 800 (2007).

$f_{abc}$ or $f_{abt}$ > $0.75 \times f_y$

$f_y$ = 250 N/mm$^2$

Bending stress = M/Z

where

$$Z \text{ (section modulus)} = \frac{I}{y}$$

I = Moment of inertia for circular section

$$y = \frac{D}{2}$$

- If allowable bending stress is greater than provided bending stress, thickness provided is safe.
- Check for deflection as per IS 800 (2007).
- Allowable deflection = L/150
- Actual deflection can be taken from STAAD analysis or calculated manually, and it should be less than the allowable deflection.

### 5.6.3 Design of Steel Sheet Pile as per IS Code Working Stress Method

- Calculate the total load which includes the dead load, active earth pressure and differential water pressure.
- Calculate the factored load, which is total load multiplied by factor of safety.
- Calculate the bending moment assuming the end condition as simply supported:

$$M = \frac{Wl^2}{8}$$

- Check for thickness as per IS 800 (2007).

$$f_{abc} \text{ or } f_{abt} \quad > 0.75 \times f_y$$
$$f_y \quad = 250 \quad \text{N/mm}^2$$

Bending stress = M/Z

where

$$Z\ (\text{section modulus}) = \frac{bd^2}{6} \text{ for rectangular section}$$

- If allowable bending stress is greater than provided bending stress, thickness provided is safe.

### 5.6.4 Design of Diaphragm Wall (T Shaped)

- Calculate the bending moment and shear force (from STAAD or manually).
- Assume the diameter of bar and diameter of the stirrup to calculate the effective depth:

Effective depth d = (D (overall depth) – clear cover – diameter of stirrup – diameter of bar)/2

- Calculate the area of main reinforcement on tension face:

As per IS 456 (2000):

$$R = \frac{M_u}{bd^2} \tag{5.22}$$

$$\frac{P_t}{100} = \frac{f_{ck}}{2 \times f_y}\left[1 - \sqrt{1 - \frac{4.598 \times R}{f_{ck}}}\right] \tag{5.23}$$

where

$M_u$ = Bending moment in kNm
d = Effective depth in mm
b = Width in mm
$P_t$ = Percentage of steel
$f_{ck}$ = Characteristic compressive strength in N/mm²
$f_y$ = Grade of steel in N/mm²
Area of main reinforcement $= P_t \times b \times d/100$
Area of one bar = (π/4) × d²,
d = Diameter of bar

Calculate the area of steel provided, provide number of bars more than number of bars required

- Calculate the area of reinforcement in compression face:
  As per IS 456 (2000)
  $A_{st}/bd = 0.85/f_y$
  Provide number of bars greater than required.
- Calculate the side reinforcement:
  As per IS 456 (2000)
  $A_{st} = (0.05/100) \times B \times D$
  Number of bars provided
  where d = Diameter of bar
- Calculate shear reinforcement:
  Calculate the nominal shear stress
  Calculate the percentage of reinforcement = (Area of main reinforcement × 100)/(effective depth × width of beam)
  Take design shear stress $\tau_c$ from IS 456 (2000) Table 19 by using percentage of reinforcement calculated
  Maximum shear stress $\tau_{cmax}$ from IS 456 (2000) Table 20
  Calculate the area of stirrups $A_{sv}$ = (Number of legs × (π/4) × d²))/4, where d is the diameter of the stirrup
  Spacing of stirrup $S_v = (0.87 \times f_y \times A_{sv})/(b \times 0.4)$ as per IS 456 (2000)
  $\tau_c > \tau_v$, provide minimum shear reinforcement
  Spacing of stirrups $S_v = (0.87 \times f_y \times A_{sv} \times d)/V$ when $\tau_v > \tau_c$ as per IS 456 (2000)
- Check for shear stress in steel:

$$\text{Stress in steel } f_{st} = \frac{M}{\left(d_{c.g} - \frac{x}{3}\right) A_{st}} \tag{5.24}$$

$$\text{Stress in steel in extreme fibre } f_{st} = \left(\frac{f_{st.cg}}{d_{c.g} - x}\right) d_e - x \tag{5.25}$$

where

M = Bending moment

$A_{st}$ = Area of reinforcement provided

$d_{cg}$ = Effective depth = total depth – $Y_{c.g}$

$$Y_{c.g} = \frac{A_{st1}Y_1 + A_{st2}Y_2 + A_{st3}Y_3 + \ldots + A_8Y_8}{A_{st1} + A_{st2} + A_{st3} + \ldots + A_{st8}} \tag{5.26}$$

x- depth of neutral axis $\frac{bx^2}{2} = m.A_{st}(d - x)$

'm' modular ratio = $E_s / E_c$

Es – Young's modulus of steel

$E_c$ – Modulus of concrete = $5000\sqrt{f_{ck}}$

- Check for crack width:

$$\text{Crack width } W_{cr} = \frac{3a_{cr}\varepsilon_m}{1 + \frac{2(a_{cr} - C_{min})}{h - x}} \tag{5.27}$$

$$\varepsilon_1 = \frac{f_s}{E_s}\frac{(h - x)}{(d - x)} \tag{5.28}$$

$$\varepsilon_m = \varepsilon_1 - \frac{b(h - x)(a - x)}{3E_sA_s(d - x)} \tag{5.29}$$

$$a_{cr} = \sqrt{\left(\frac{s}{2}\right)^2 + (d_c)^2} - \frac{d_b}{2} \tag{5.30}$$

where

S = Centre to centre spacing of bar

$d_c$ = Effective cover

$d_b$ = Diameter of longitudinal bar

x = Depth of neutral axis

$\varepsilon_m = \varepsilon_1$ = Average strain

$E_s$ = Young modulus of steel

$A_s$ = Area of reinforcement provided

b = Width

h = Total depth

$C_{min}$ = minimum cover to reinforcement

Design of anchor wall is also similar to above procedure.

## 5.7 DESIGN OF DECK WITH PRECAST AND CAST IN-SITU CONSTRUCTION

Due to the unfavorable condition of the location for the shuttering of RCC members, the precast concrete members are preferred, and at the junction, cast in-situ concrete is laid. The structure is modelled in STAAD.Pro software. The necessary loads as per Chapter 3 are assigned, and the model is analysed as per the load combination given in section 5.3. The typical design procedure of precast deck members as explained below.

### 5.7.1 Design of Precast U In-Situ Beam

**Stage I: precast U beam**

- Calculate the bending moment (hogging and sagging):

$$\text{Maximum hogging moment} = W_l l^2/2$$

$$\text{Maximum sagging moment} = W_l l_l^2/8$$

Where

$W_l$ = Factored self weight of the beam
$l$ = Centre to centre distance of lifting hook = 0.208L
$l_l$ = length of precast beam
$W_l$ = 1.5 × self weight of beam

Self weight of beam = (width of section × depth of precast bottom) + 2 (depth of precast side wall × thickness of precast side wall) × unit weight of concrete

- Calculate the area of steel reinforcement required:

As per IS 456 (2000)

$$R = \frac{M_u}{bd^2} \tag{5.31}$$

$$\frac{P_t}{100} = \frac{f_{ck}}{2 \times f_y}\left[1 - \sqrt{1 - \frac{4.598 \times R}{f_{ck}}}\right] \tag{5.32}$$

where

$M_u$ = Bending moment in kNm
$d$ = Effective depth in mm
$b$ = Width in mm
$P_t$ = Percentage of steel
$f_{ck}$ = Characteristic compressive strength in N/mm²
$f_y$ = Grade of steel in N/mm²

Effective depth d = Total depth – [(Clear cover + Diameter of main bar + Diameter of stirrup)/2]

b = Thickness of precast side wall
Area of reinforcement $A_{st,req} = P_t \times b \times d/100$
Minimum area of reinforcement $A_{st,min} = 0.85\ bd/f_y$
Area of reinforcement provided $A_{st,s1}$ = maximum of $A_{st,req}$ and $A_{st,min}$
Number of bars required = Area of reinforcement provided/Area of one bar
Provide number of bars more than required

- Calculate shear reinforcement:

Factored shear force $V_u = W_1 l_1/2$
Calculate the area of stirrup legs $A_{sv}$ = Number of legs $\times p \times (\text{diameter of stirrup legs})^2/4$,
Nominal shear stress (due to shear force) $\tau_v = V_u/bd$
Calculate the percentage of reinforcement $P_t = 100 \times A_{st}/(b \times d)$
Design shear stress of concrete $\tau_c$ as per Table 19 from IS 456 (2000)
If $\tau_c > \tau_v$ provide minimum shear reinforcement:
Spacing of stirrups $S_v = (0.87 \times f_y \times A_{sv})/(0.4 \times b)$
If $\tau_c > \tau_v$ provide shear reinforcement as follows:
Spacing of stirrups $S_v = (0.87 \times f_y \times A_{sv} \times d)/V$
Net shear force $V_{us} = V' - \tau_c bd$

### STAGE II: PRECAST U BEAM AND IN-SITU PORTION U BEAM CONSTRUCTION

- Calculate the maximum sagging moment $= W_2 l_1^2/12$:

  where
  - $W_2$ = Total load on the beam
  - $l_1$ = length of precast beam
  - $W_2 = W_1 + 1.5\times$ (Depth of in-situ portion U beam × width of in-situ portion U beam × unit weight of concrete)

- Calculate the area of steel reinforcement required:
  As per IS 456 (2000)

$$R = \frac{M_u}{bd^2} \tag{5.33}$$

$$\frac{P_t}{100} = \frac{f_{ck}}{2 \times f_y}\left[1 - \sqrt{1 - \frac{4.598 \times R}{f_{ck}}}\right] \tag{5.34}$$

where
- $M_u$ = Bending moment in kNm
- d = Effective depth in mm
- b = Width in mm
- $P_t$ = Percentage of steel
- $f_{ck}$ = Characteristic compressive strength in N/mm²
- $f_y$ = Grade of steel in N/mm²

Area of reinforcement $A_{st,req} = P_t \times b \times d/100$

Minimum area of reinforcement $A_{st,min} = 0.85\ bd/f_y$

where

b = Width of the section

d = Total depth – [(Clear cover + Diameter of main bar + Diameter of stirrup)/2]

Area of steel to be provided $A_{st,s2}$ = maximum of $A_{st,req}$ and $A_{st,min}$

- Top reinforcement in in-situ portion

$$\text{Minimum area of reinforcement } A_{st,min} = 0.85\ bd/f_y$$

where

b = Width of in-situ portion

d = Depth of in-situ portion

Number of bars required = Area of reinforcement provided/Area of one bar
Provide the number of bars more than required
Provide minimum shear reinforcement as per Cl.26.5.1.6 IS 456 (2000)

$$\text{Spacing of stirrups } S_v = (0.87 \times f_y \times A_{sv})/(0.4 \times b)(5.35)$$

where

$A_{sv}$ = Area of shear leg = Number of legs × π × (diameter of stirrup legs)$^2$/4

**Stage III: beam-slab construction**

- Calculate maximum sagging moment = $W_3 l_1^2/12$:
  where
  $W_3$ = Factored total load
  $l_1$ = Length of precast beam
  $W_3$ = (Self weight of beam + self weight of slab + construction live load) × 1.5
  Self weight of beam = Width of the section × depth of precast beam × unit weight of concrete
  Self weight of slab = Width of the section × thickness of deck slab × unit weight of concrete
- Calculate the area of steel reinforcement required:
  As per IS 456 (2000):

$$R = \frac{M_u}{bd^2} \tag{5.36}$$

$$\frac{P_t}{100} = \frac{f_{ck}}{2 \times f_y}\left[1 - \sqrt{1 - \frac{4.598 \times R}{f_{ck}}}\right] \tag{5.37}$$

where

$M_u$ = Bending moment in kNm

d = Effective depth in mm

b = Width in mm

$P_t$ = Percentage of steel

$f_{ck}$ = Characteristic compressive strength in N/mm$^2$
$f_y$ = Grade of steel in N/mm$^2$

Area of reinforcement $A_{st,req} = P_t \times b \times d/100$

Minimum area of reinforcement $A_{st,min} = 0.85\ bd/f_y$

where
b = Width of the section
d = Total depth – [(Clear cover + Diameter of main bar + Diameter of stirrup)/2]

Area of steel to be provided $A_{st,s3}$ = Maximum of $A_{st,req}$ and $A_{st,min}$
Area of steel to be provided from stage I, II and III $A_{st,stage}$ = maximum of $A_{st,s1}$ $A_{st,s2}$ and $A_{st,s3}$

### Design of beam (bottom and top reinforcement)

- Calculate the final design moment $M' = M_u + T_u \times (1 + D/b)/1.7$:
  where $M_u$ and $T_u$ are moment and torsion from limit state of collapse (from STAAD analysis)
- Calculate the area of steel reinforcement required:
  As per IS 456 (2000)

$$R = \frac{M_u}{bd^2}$$

$$\frac{P_t}{100} = \frac{f_{ck}}{2 \times f_y}\left[1 - \sqrt{1 - \frac{4.598 \times R}{f_{ck}}}\right]$$

where
$M_u$ = Bending moment in kNm
d = Effective depth in mm
b = Width in mm
$P_t$ = Percentage of steel
$f_{ck}$ = Characteristic compressive strength in N/mm$^2$
$f_y$ = Grade of steel in N/mm$^2$

Area of main reinforcement $A_{st,\ required} = P_t \times b \times d/100$
Minimum area of reinforcement $A_{st,min} = 0.85\ bd/f_y$
Area of steel to be provided $A_{st}$ = maximum of $A_{st,\ required}$ and $A_{st,min}$
Maximum area of reinforcement = 0.04 bD

- Preliminary $A_{st}$ calculation:
  Calculate the area of one bar = $\pi d^2/4$
  Calculate the number of bars required
  Provide number of bars than required

Calculate the area of reinforcement to be provided for resisting moment $A_{st,moment}$ = Area of one bar × Number of bars provided
Area of reinforcement to be provided for precast $A_{st,stage}$
Area of reinforcement to be provided $A_{st,provided}$ = Maximum of $A_{st,stage}$ and $A_{st,moment}$
Area of reinforcement provided is OK when $A_{st,provided} > A_{st}$

- Check the depth of neutral axis:
  Calculate the limiting neutral axis $X_{u,lim}$
  Calculate the actual neutral axis depth $X_{u,actual} = 0.87\, f_y A_{st}/0.36\, f_{ck} b$
  If $X_{u,actual} < X_{u,lim}$ then under reinforced beam whereas $X_{u,actual} < X_{u,lim}$ then the beam is overreinforced and hence revise the section
- Check moment of resistance based on $A_{st}$:

$$\text{Moment of resistance} M_r = 0.87\; f_y A_{st} d \left[ \rightarrow - \frac{A_{st} f_y}{bdf_{ck}} \right] \tag{5.38}$$

  If moment of resistance > Bending moment, design is OK
  If moment of resistance < Bending moment, revise the design
- Check for crack width:
  Calculate the design moment $M' = M_u + T_u \times (1 + D/b)/1.7$
  where $M_u$ and $T_u$ are moment and torsion from limit state of serviceability
  Calculate the stress in steel $f_s = M/A_s Z$

where
M = Bending moment
$A_s$ = Area of steel provided
Z = d − x/3
x is the depth of neutral axis which is given by $bx^2/2 = mA_{st}(d - x)$
m is the modular ratio $m = E_s/E_c$
$E_s$ = 200,000 N/mm$^2$
$E_c = 5000\sqrt{f_{ck}}$
Limiting stress in steel $f_s'$ as per Table 22 IS 456 (2000)
If stress in steel < limiting stress, stress in steel is satisfied
- Check for stress in concrete:

Stress in concrete $f_c = (f_s A_s)/(0.5\, bx)$
Limiting stress in concrete $f_c'$ as per Table 21 IS 456 (2000)
If stress in concrete < limiting stress in concrete, stress in concrete is satisfied
- Check for crack width:

$$W_{cr} = \frac{3a_{cr}\varepsilon_m}{1 + \frac{2(a_{cr} - C_{\min})}{h - x}} \tag{5.39}$$

$$\varepsilon_1 = \frac{f_s}{E_s}\frac{(h - x)}{(d - x)} \tag{5.40}$$

$$\varepsilon_m = \varepsilon_1 - \frac{b(h-x)(a-x)}{3E_s A_s (d-x)} \tag{5.41}$$

$$a_{cr} = \sqrt{\left(\frac{S}{2}\right)^2 + (d_c)^2} - \frac{d_b}{2} \tag{5.42}$$

where

| | |
|---|---|
| S | = Centre to centre spacing of bar |
| $d_c$ | = Effective cover |
| $d_b$ | = Diameter of longitudinal bar |
| x | = Depth of neutral axis |
| $\varepsilon_m = \varepsilon_1$ | = Average strain |
| $E_s$ | = Young modulus of steel |
| $A_s$ | = Area of reinforcement provided |
| b | = Width |
| h | = Total depth |
| $C_{min}$ | = Minimum cover to reinforcement |

- Calculation of side face reinforcement:
  As per IS 456 (2000) Cl.26.5.1.3 when the depth of beam exceeds 750 mm side face reinforcement shall be provided along the two faces
  Minimum area of reinforcement = (0.05/100) × b × D
  Number of bars required = Area of reinforcement/Area of one bar ($\pi d^2/4$)
- Calculation of shear reinforcement
  Calculate shear force
  V = Maximum factored load × Span ($L_y$)
  Final design shear force $V' = V_u + 1.6(T_u/b)$
  Calculate the area of stirrup legs $A_{sv} = \pi \times$ (diameter of stirrup legs)$^2$/4,
  Nominal shear stress (due to shear force) $\tau_v = V'/bd$
  Calculate the percentage of reinforcement $P_t = 100 \times A_{st}/(b \times d)$
  Design shear stress of concrete $\tau_c$ as per Table 19 from IS 456 (2000)
  If $\tau_c > \tau_v$, provide minimum shear reinforcement as follows:
  Spacing of stirrups $S_v = (0.87 \times f_y \times A_{sv})/(0.4 \times b)$
  If $\tau_c < \tau_v$, provide shear reinforcement as follows:
  Spacing of stirrups $S_v = (0.87 \times f_y \times A_{sv} \times d)/V$
  Net shear force $V_{us} = V' - \tau_c bd$

### 5.7.2 Design of In-Situ Beam

- Calculate the bending moment sagging bottom and hogging top manually or using STAAD.Pro software (collapse and serviceability condition). Take corresponding torsion of maximum bending moment.
- Calculate the shear force hogging top (limit state of collapse condition) manually or using STAAD.Pro software.

- Assume the diameter of bar for bottom and top reinforcement, side face reinforcement and stirrups.
- Calculate the effective depth of the section d = Total depth – [(Clear cover + Diameter of main bar + Diameter of stirrup)/2]:
  - Calculate the final design moment $M' = M_u + T_u \times (1 + D/b)/1.7$ (5.43)

where $M_u$ and $T_u$ are moment and torsion from limit state of collapse.

Calculate the limiting moment of resistance of the section

$$M_{u,lim} = 0.36\frac{x_{u,max}}{d}\left(1-0.42\frac{x_{u,max}}{d}\right)bd^2 f_{ck} \quad (5.44)$$

Take $x_{u,max}/d$ from IS 456 (2000) Cl.38.1.
If $M_{u,lim} > M'$, Design the beam as singly reinforced section if not design as doubly reinforced section.

- Calculate the area of steel reinforcement required:

As Per IS 456 (2000)

$$R = \frac{M_u}{bd^2} \quad (5.45)$$

$$\frac{P_t}{100} = \frac{f_{ck}}{2\times f_y}\left[1-\sqrt{1-\frac{4.598\times R}{f_{ck}}}\right] \quad (5.46)$$

where

$M_u$ = Bending moment in kNm
d = Effective depth in mm
b = Width in mm
$P_t$ = Percentage of steel
$f_{ck}$ = Characteristic compressive strength in N/mm$^2$
$f_y$ = Grade of steel in N/mm$^2$

Area of main reinforcement = $P_t \times b \times d/100$
Minimum area of reinforcement $A_{st,min} = 0.85\ bd/f_y$
Maximum area of reinforcement $A_{st,max} = 0.04\ bd$
Calculate the area of one bar = $\pi d^2/4$
Calculate the number of bars required
Provide number of bars than required
Check the depth of neutral axis
Calculate the limiting neutral axis $X_{u,lim}$
Calculate the actual neutral axis depth $X_{u,actual} = 0.87\ f_y A_{st}/0.36\ f_{ck} b$

If $X_{u,actual} < X_{u,lim}$ then under reinforced beam
Check moment of resistance based on $A_{st}$

$$\text{Moment of resistance} M_r = 0.87\, f_y A_{st} d \left[1 - \frac{A_{st} f_y}{bdf_{ck}}\right] \tag{5.47}$$

If moment of resistance > Bending moment, design is OK
If moment of resistance < Bending moment, revise the design

- Check for crack width
  Calculate the design moment $M' = M_u + T_u \times (1 + D/b)/1.7$
  where $M_u$ and $T_u$ are moment and torsion from limit state of serviceability.
  Calculate the stress in steel $f_s = M/A_s Z$

where
M = Bending moment
As = Area of reinforcement Ast
Z = d – x/3
x = depth of neutral axis which is given by bx2/2 = mAst(d – x)
m = modular ratio m = Es/Ec
$E_s$ = 200,000 N/mm2
$E_c$ = 5000 sqrt ($f_{ck}$)

Limiting stress in steel $f_s'$ as per Table 22 IS 456 (2000)
If Stress in steel < limiting stress, stress in steel is satisfied
Check for stress in concrete
Stress in concrete $f_c = (f_s A_s)/(0.5\ bx)$
Limiting stress in concrete $f_c'$ as per Table 21 IS 456 (2000)
If stress in concrete < limiting stress in concrete, stress in concrete is satisfied
Check for crack width

$$W_{cr} = \frac{3a_{cr}\varepsilon_m}{1 + \frac{2(a_{cr} - C_{min})}{h - x}} \tag{5.48}$$

$$\varepsilon_1 = \frac{f_s}{E_s}\frac{(h-x)}{(d-x)} \tag{5.49}$$

$$\varepsilon_m = \varepsilon_1 - \frac{b(h-x)(a-x)}{3E_s A_s (d-x)} \tag{5.50}$$

$$a_{cr} = \sqrt{\left(\frac{S}{2}\right)^2 + (d_c)^2} - \frac{d_b}{2} \tag{5.51}$$

where

$S$ = Centre to centre spacing of bar
$d_c$ = effective cover
$d_b$ = diameter of longitudinal bar
$x$ = depth of neutral axis
$\varepsilon_m - \varepsilon_1$ = average strain
$E_s$ = Young modulus of steel
$A_s$ = Area of reinforcement provided
$b$ = width
$h$ = total depth
$C_{min}$ = minimum cover to reinforcement

### 5.7.3 Design of Rectangular Precast Beam

**Stage I: precast beam**

- Calculate area of reinforcement for hogging and sagging.
- Calculate the bending moment for hogging and sagging.
  Maximum hogging moment = $Wl^2/2$
  Maximum sagging moment = $Wl_1^2/8$

  where
  W = Factored self weight of the beam
  l = Centre to centre distance of lifting hook = 0.208L
  $l_1$ = Length of precast beam

- Calculate the area of steel reinforcement required.
  As Per IS 456 2000

$$R = \frac{M_u}{bd^2} \tag{5.52}$$

$$\frac{P_t}{100} = \frac{f_{ck}}{2 \times f_y}\left[1 - \sqrt{1 - \frac{4.598 \times R}{f_{ck}}}\right] \tag{5.53}$$

where

$M_u$ = Bending moment in kNm
$d$ = Effective depth in mm
$b$ = Width in mm
$P_t$ = Percentage of steel
$f_{ck}$ = Characteristic compressive strength in N/mm$^2$
$f_y$ = Grade of steel in N/mm$^2$

Area of main reinforcement $A_{st,req} = P_t$ x b x d/100
Minimum area of reinforcement $A_{st,min} = 0.85bd/f_y$
Area of main reinforcement provided $A_{st,s1}$ = maximum of $A_{st,req}$ and $A_{st,min}$

- Lifting hook
  Calculate the factored load on a lifting hook

Load on one lifting hook = Factored self weight of beam in kN / Number of lifting hook
Stress in lifting hook = $0.6f_y$
Area of steel required = Factored load x 1000 / stress in lifting hook
Area of one hook = $\pi d^2/4$, where d is the diameter of hook
If area of steel required < area of hook, provided diameter of hook is OK
Actual stress in lifting hook = Factored self weight of beam/Area of one hook
If actual stress in lifting hook < stress in lifting hook, provided diameter of hook is OK

- Developmental length for hook
  Development length of reinforcement $L_d = ø\sigma_s / 4\tau_{bd}$

where
$L_d$ = Development length of reinforcement
$\sigma_s$ = Stress in bar at the section considered at design load = $0.87f_y$
ø = Diameter of reinforcement
$\tau_{bd}$ = Design bond stress as per IS 456 (2000) Cl.26.2.1.1

## DESIGN OF BEAM (BOTTOM AND TOP REINFORCEMENT)

- Calculate the final design moment M' = $M_u$+$T_u$*(1+D/b)/1.7
  where $M_u$ and $T_u$ are moment and torsion from limit state of collapse
- Calculate the area of steel reinforcement required
  As Per IS 456 2000

$$R=\frac{M_u}{bd^2} \tag{5.54}$$

$$\frac{P_t}{100}=\frac{f_{ck}}{2\times f_y}\left[1-\sqrt{1-\frac{4.598\times R}{f_{ck}}}\right] \tag{5.55}$$

where
$M_u$ = Bending moment in kNm
d = Effective depth in mm
b = Width in mm
$P_t$ = Percentage of steel
$f_{ck}$ = Characteristic compressive strength in N/mm²
$f_y$ = Grade of steel in N/mm²

Area of main reinforcement $A_{st,\,required} = P_t$ x b x d/100
Minimum area of reinforcement $A_{st,min} = 0.85bd/f_y$
Area of steel to be provided $A_{st}$ = maximum of $A_{st,\,required}$ and $A_{st,min}$
Maximum area of reinforcement = 0.04Bd

- Preliminary Ast calculation
  Calculate the area of one bar = $\pi d^2 / 4$

Calculate the number of bars required
Provide number of bars than required
Calculate the area of reinforcement to be provided for resisting moment
$A_{st,moment}$ = Area of one bar x Number of bars provided
Area of reinforcement to be provided $A_{st,provided}$ = Maximum of $A_{st,s1}$ and $A_{st,moment}$
Area of reinforcement provided is OK when $A_{st,provided} > A_{st}$

- Check the depth of neutral axis
  Calculate the limiting neutral axis $X_{u,lim}$
  Calculate the actual neutral axis depth $X_{u,\ actual} = 0.87f_yA_{st}/0.36f_{ck}b$
  If $X_{u,actual} < X_{u,lim}$ then under reinforced beam whereas $X_{u,actual} < X_{u,lim}$ then the beam is overreinforced and hence revise the section
- Check moment of resistance based on $A_{st}$

$$\text{Moment of resistance } M_r = 0.87\,f_y A_{st} d\left[1-\frac{A_{st}f_y}{bdf_{ck}}\right] \tag{5.56}$$

If moment of resistance > Bending moment, design is OK
If moment of resistance < Bending moment, revise the design

- Check for crack width

$$\text{Calculate the design moment } M' = M_u + T_u * \left(1+D/b\right)/1.7 \tag{5.57}$$

where $M_u$ and $T_u$ are moment and torsion from limit state of serviceability
Calculate the stress in steel $f_s = M/A_s Z$

where
M = Bending moment
$A_s$ = Area of steel provided
Z = d-x/3
x = Depth of neutral axis which is given by $bx^2/2 = mA_{st}(d-x)$
m = Modular ratio $m = E_s/E_c$
$E_s$ = 200,000 $N/mm^2$
$E_c = 5000\sqrt{f_{ck}}$
Limiting stress in steel fs' as per Table 22 IS 456 (2000)
If Stress in steel < limiting stress, stress in steel is satisfied
Check for stress in concrete
Stress in concrete $f_c = (f_sA_s)/(0.5bx)$
Limiting stress in concrete $f_c$' as per Table 21 IS 456 (2000)
If stress in concrete < limiting stress in concrete, stress in concrete is satisfied
Check for crack width

$$W_{cr} = \frac{3a_{cr}\varepsilon_m}{1+\frac{2\left(a_{cr}-C_{\min}\right)}{h-x}} \tag{5.58}$$

$$\varepsilon_1 = \frac{f_s}{E_s}\frac{(h-x)}{(d-x)} \tag{5.59}$$

$$\varepsilon_m = \varepsilon_1 - \frac{b(h-x)(a-x)}{3E_sA_s(d-x)} \tag{5.60}$$

$$a_{cr} = \sqrt{\left(\frac{S}{2}\right)^2 + (d_c)^2} - \frac{d_b}{2} \tag{5.61}$$

where

$S$ = Centre to centre spacing of bar
$d_c$ = Effective cover
$d_b$ = Diameter of longitudinal bar
x = Depth of neutral axis
$\varepsilon_m = \varepsilon_1$ = Average strain
$E_s$ = Young modulus of steel
$A_s$ = Area of reinforcement provided
b = Width
h = Total depth
$C_{min}$ = Minimum cover to reinforcement

- Calculation of side face reinforcement
  As per IS 456 (2000) Cl.26.5.1.3 when the depth of beam exceeds 750 mm side face reinforcement shall be provided along the two faces
  Minimum area of reinforcement = (0.05/100) x b x D
  Number of bars required = Area of reinforcement / Area of one bar ($\pi d^2/4$)
- Calculation of shear reinforcement
  Calculate Shear force
  V= Maximum Factored load xSpan ($L_y$)
  Final design shear force $V' = V_u + 1.6(T_u/b)$
  Calculate the area of stirrup legs $A_{sv} = \pi$ x (diameter of stirrup legs)$^2$ / 4,
  Nominal shear stress (due to shear force) $\tau_v = V'/bd$
  Calculate the percentage of reinforcement $P_t = 100 \times A_{st}/(b \times d)$
  Design shear stress of concrete $\tau_c$ as per Table 19 from IS 456 (2000)
  If $\tau_c > \tau_v$ provide minimum shear reinforcement
  Spacing of stirrups $S_v = (0.87 \times f_y \times A_{sv})/(0.4 \times b)$
  If $\tau_c < \tau_v$ provide shear reinforcement as follows
  Spacing of stirrups $S_v = (0.87 \times f_y \times A_{sv} \times d)/V$
  Net shear force $V_{us} = V' - \tau_c bd$
- Calculation of development length
  Development length of reinforcement $L_d = \o\sigma_s / 4\tau_{bd}$

  where
  $L_d$ = Development length of reinforcement
  $\sigma_s$ = Stress in bar at the section considered at design load = actual stress in lifting hook

$\phi$ = Diameter of reinforcement
$\tau_{bd}$ = Design bond stress as per IS 456 (2000) Cl.26.2.1.1

## 5.7.4 Design of Slab (Precast + Cast In-Situ)

### Stage I: precast slab

- Calculate the Bending moment (hogging and sagging)
  Maximum hogging moment = $W_1 l^2/2$
  Maximum sagging moment = $W_1 l_1^2/8$
  where
  $W_1$ = Factored self-weight of the slab
  Self-weight of the slab = Overall depth (D)x unit weight of concrete x width
  l = Centre to centre distance of the lifting hook
  $l_1$= Length of precast slab
- Calculate the reinforcement required (hogging top)
  As per IS 456 (2000)
  Calculate the percentage of reinforcement $P_t$

$$\frac{P_t}{100} = \frac{f_{ck}}{2 \times f_y}\left[1 - \sqrt{1 - \frac{4.598 \times R}{f_{ck}}}\right] \tag{5.62}$$

where

$$R = \frac{M_u}{bd^2} \tag{5.63}$$

$M_u$ = Maximum bending moment in kNm
b = Width in m
d = Effective depth in m
To calculate effective depth
d = Total depth (D) – (Clear cover to reinforcement + Diameter of bar + Diameter of stirrups)/ 2
$f_{ck}$ = Characteristic compressive strength of concrete in N/mm$^2$
$f_y$ = Yield strength of steel in N/mm$^2$

Area of steel required $A_{st,req}$= ($P_t$ /100) x (bxd)
Minimum area of reinforcement $A_{st,min}$= 0.12%bd
Provide area of reinforcement maximum of $A_{st,req}$ and $A_{st,min}$
Spacing of reinforcement required = (1000 x area of one bar)/$A_{st}$ provided
Calculate the distribution reinforcement of precast slab
Minimum area of reinforcement $A_{st,min}$= 0.12%bd
Spacing of reinforcement required = (1000 x area of one bar)/$A_{st}$ provided

### Stage II: precast slab and in-situ portion construction

- Calculate the maximum bending moment at top and bottom

Maximum moment bottom = $W_2 l^2/8$
Maximum moment top = $W_2 l^2/12$
where
$W_2$ = Factored load on the slab
l = Length of the slab after placing
$W_2$ = (self weight of the slab + live load) x 1.5
Self-weight of the slab = Overall depth (D)x unit weight of concrete x width

**STAGE III: LIVE LOAD**

Considering crane load is acting on the structure

- Calculate the maximum bending moment at top and bottom
  Maximum moment bottom = $W_3 l^2/24$
  Maximum moment top = $W_3 l^2/12$
  where
  $W_3$ = Total load on the slab
  l = Length of the slab after placing
  $W_3$ = [Self weight of slab + construction live load (maximum of live load or crane load)]
  x1.5 Self-weight of the slab = Overall depth (D)x unit weight of concrete x width
- Crane load calculation
  Maximum corner load and standard pad dimension are taken from crane catalogue
  Maximum area pressure = maximum corner load/(length x width of pad)

  Pressure distribution area on slab
  Length after distribution = length + (2 x (slab thickness + wearing coat))
  Width after distribution = width + + (2 x (slab thickness + wearing coat))
  Dispersed pressure = maximum corner load/(Length after distribution x Width after distribution)
- Calculation of area of reinforcement (top and bottom)
  Calculate the final design moment $M' = M_u + T_u*(1+D/b)/1.7$
  where
  M (bottom reinforcement) = Maximum of moment at bottom due to load $W_3$ and moment at top due to load $W_2$
  M (top reinforcement) = Moment top due to load $W_3$
  As per IS 456 (2000)
  Calculate the percentage of reinforcement $P_t$

$$\frac{P_t}{100} = \frac{f_{ck}}{2 \times f_y}\left[1 - \sqrt{1 - \frac{4.598 \times R}{f_{ck}}}\right] \quad (5.64)$$

where

$$R = \frac{M_u}{bd^2} \qquad (5.65)$$

$M_u$ = Maximum bending moment in kNm
b = Width in m
d = Effective depth in m
To calculate effective depth
d = Total depth (D) – (Clear cover to reinforcement + Diameter of bar + Diameter of stirrups)/ 2
$f_{ck}$ = Characteristic compressive strength of concrete in N/mm$^2$
$f_y$ = Yield strength of steel in N/mm$^2$

Area of steel required $A_{st,req} = (P_t /100)$ x (bxd)
Minimum area of reinforcement $A_{st,min} = 0.12\%$bd
$A_{st,provided}$ = maximum of $A_{st,req}$ and $A_{st,min}$
Maximum area of reinforcement $A_{st,max} = 0.04$bD
Spacing of reinforcement required = (1000 x area of one bar)/$A_{st\ provided}$
Provide spacing less than the required
Area of reinforcement to be provided for resisting moment $A_{st,\ moment}$ = (1000 x area of one bar)/spacing of reinforcement provided
Area of reinforcement to be provided = maximum of $A_{st,provided}$ and $A_{st,\ moment}$

- Check the depth of neutral axis
  Limiting axis neutral depth $X_{u,\ lim} = 0.46$ x d
  Actual neutral axis depth $X_{u,\ actual} = 0.87 f_y A_{st} / 0.36 f_{ck} b$
  If $X_{u,\ actual} < X_{u,\ lim}$ the section is under reinforced whereas $X_{u,\ actual} > X_{u,\ lim}$ the section is over reinforced and hence revise the section
- Check the moment of resistance

$$\text{Moment of resistance of the section } M_r = 0.87\, f_y A_{st} d \left[1 - \frac{A_{st} f_y}{bdf_{ck}}\right] \qquad (5.66)$$

If moment of resistance of the section $M_r$ > Bending M', the section is ok otherwise revise the section

- Calculation of distribution reinforcement
  Minimum area of reinforcement $A_{st,min} = 0.12\%$bd
  Spacing of reinforcement required = (1000 x area of one bar)/$A_{st,provided}$
- Check for crack width
  Calculate the design moment $M' = M_u + T_u*(1+D/b)/1.7$
  where $M_u$ and $T_u$ are moment and torsion from limit state of serviceability
  Calculate the stress in steel $f_s = M/A_s Z$
  where
  M = Bending moment
  $A_s$ = Area of steel provided
  Z = d-x/3

x = Depth of neutral axis which is given by $bx^2/2 = mA_{st}(d-x)$
m = Modular ratio $m = E_s/E_c$
$E_s$ = 200,000 N/mm$^2$
$E_c$ = 5000 sqrt ($f_{ck}$)
Limiting stress in steel $f_s$' as per Table 22 IS 456 (2000)
If Stress in steel < limiting stress, stress in steel is satisfied
Check for stress in concrete
Stress in concrete $f_c = (f_s A_s)/(0.5bx)$
Limiting stress in concrete $f_c$' as per Table 21 IS 456 (2000)
If stress in concrete < limiting stress in concrete, stress in concrete is satisfied
Check for crack width

$$W_{cr} = \frac{3a_{cr}\varepsilon_m}{1+\dfrac{2(a_{cr}-C_{min})}{h-x}} \tag{5.67}$$

$$\varepsilon_1 = \frac{f_s}{E_s}\frac{(h-x)}{(d-x)} \tag{5.68}$$

$$\varepsilon_m = \varepsilon_1 - \frac{b(h-x)}{3E_s A_s(d-x)} \tag{5.69}$$

$$a_{cr} = \sqrt{\left(\frac{S}{2}\right)^2 + (d_c)^2} - \frac{d_b}{2} \tag{5.70}$$

where

| | | |
|---|---|---|
| S | = | Centre to centre spacing of bar |
| $d_c$ | = | Effective cover |
| $d_b$ | = | Diameter of longitudinal bar |
| x | = | Depth of neutral axis |
| $\varepsilon_m = \varepsilon_1$ | = | Average strain |
| $E_s$ | = | Young modulus of steel |
| $A_s$ | = | Area of reinforcement provided |
| b | = | Width |
| h | = | Total depth |
| $C_{min}$ | = | Minimum cover to reinforcement |

- Calculation of shear reinforcement
  Calculate Shear force
  V= Maximum Factored load x Span ($L_y$)
  Final design shear force $V' = V_u + 1.6(T_u/b)$
  Calculate the area of stirrup legs $A_{sv} = \pi \times$ (diameter of stirrup legs)$^2$ / 4,
  Nominal shear stress (due to shear force) $\tau_v = V'/bd$
  Calculate the percentage of reinforcement $P_t = 100 \times A_{st}/(b \times d)$
  Design shear stress of concrete $\tau_c$ as per Table 19 from IS 456 (2000)
  If $\tau_c > \tau_v$ provide minimum shear reinforcement

Spacing of stirrups $S_v = (0.87 \times f_y \times A_{sv})/(0.4 \times b)$
If $\tau_c < \tau_v$ provide shear reinforcement as follows
Spacing of stirrups $S_v = (0.87 \times f_y \times A_{sv} \times d)/V$
Net shear force $V_{us} = V' - \tau_c bd$

- Calculation of development length
  Development length of reinforcement $L_d = \phi\sigma_s / 4\tau_{bd}$

  where
  $L_d$ = Development length of reinforcement
  $\sigma_s$ = Stress in bar at the section considered at design load = $0.87f_y$
  $\phi$ = Diameter of reinforcement
  $\tau_{bd}$ = Design bond stress as per IS 456 (2000) Cl.26.2.1.1

### 5.7.5 Design of Precast Slab

- Check the ratio of Longer span $L_y$ to shorter span $L_x$ if $L_y/L_x < 2$ then the slab is two-way and if $L_y/L_x > 2$ then the slab is one-way
- Calculate the factored load acting on the slab
  Dead load = Overall depth (D)x unit weight of concrete
  Unit weight of concrete = 25 kN/m$^3$
  Factored load = (Dead load + Live load) × 1.5
- Calculate the moment as per edge conditions from ANNEX D IS 456 (2000)

$$M_x = \alpha_x \times w \times L_x^2$$
$$M_y = \alpha_y \times w \times L_x^2$$

  where
  $M_x$ and $M_y$ = Moments on strips of unit width spanning $L_x$ and $L_y$ respectively
  $\alpha_x$ and $\alpha_y$ = Bending moment coefficients as per Table 26 IS 456 (2000)
  w = Total design load per unit area
  $L_x$ and $L_y$ = lengths of the shorter span and longer span respectively
- Check for depth
  $M_u = 0.138 \times f_{ck} \times b \times d_{req}^2$
  If $d_{req} < d$, the depth is adequate
- Calculate the area of reinforcement required
  $M_u = 0.87 \times f_y \times A_{st} \times d\,[1- ((A_{st} \times f_y)/ (b \times d \times f_{ck}))]$as per ANNEX G, G.1.1 from IS 456 (2000)
  Calculate the spacing of bars
  Spacing of bars = $1000 \times a_{st}/A_{st}$
  where
  $a_{st}$ = Area of one bar
  $A_{st}$ = Area of reinforcement provided
- Check for crack width
  Calculate the design moment $M' = M_u + T_u*(1+D/b)/1.7$
  where $M_u$ and $T_u$ are moment and torsion from limit state of serviceability
  Calculate the stress in steel $f_s = M/A_s Z$

where
M = Bending moment
$A_s$ = Area of steel provided
Z = d-x/3
x = Depth of neutral axis which is given by $bx^2/2 = mA_{st}(d-x)$
m = Modular ratio $m = E_s/E_c$
$E_s$ = 200,000 N/mm$^2$
$E_c$ = 5000 sqrt ($f_{ck}$)

Limiting stress in steel $f_s'$ as per Table 22 IS 456 (2000)
If Stress in steel < limiting stress, stress in steel is satisfied
Check for stress in concrete
Stress in concrete $f_c = (f_sA_s)/(0.5bx)$
Limiting stress in concrete $f_c'$ as per Table 21 IS 456 (2000)
If stress in concrete < limiting stress in concrete, stress in concrete is satisfied
Check for crack width

$$W_{cr} = \frac{3a_{cr}\varepsilon_m}{1+\frac{2(a_{cr}-C_{min})}{h-x}} \quad (5.71)$$

$$\varepsilon_1 = \frac{f_s}{E_s}\frac{(h-x)}{(d-x)} \quad (5.72)$$

$$\varepsilon_m = \varepsilon_1 - \frac{b(h-x)(a-x)}{3E_sA_s(d-x)} \quad (5.73)$$

$$a_{cr} = \sqrt{\left(\frac{S}{2}\right)^2 + (d_c)^2} - \frac{d_b}{2} \quad (5.74)$$

where
S = Centre to centre spacing of bar
$d_c$ = Effective cover
$d_b$ = Diameter of longitudinal bar
x = Depth of neutral axis
$\varepsilon_m = \varepsilon_1$ = Average strain
$E_s$ = Young modulus of steel
$A_s$ = Area of reinforcement provided
b = Width
h = Total depth
$C_{min}$ = Minimum cover to reinforcement

- Calculate shear reinforcement
  Calculate Shear force
  V= Factored load × Span ($L_y$)

Final design shear force $V' = V_u + 1.6(T_u/b)$ (5.75)

Calculate the area of stirrup legs $A_{sv} = \pi$ x (diameter of stirrup legs)$^2$ / 4,
Nominal shear stress (due to shear force) $\tau_v = V'/bd$
Calculate the percentage of reinforcement $P_t = 100 \times A_{st}/(b \times d)$
Design shear stress of concrete $\tau_c$ as per Table 19 from IS 456 (2000)
If $\tau_c > \tau_v$ provide minimum shear reinforcement
Spacing of stirrups $S_v = (0.87 \times f_y \times A_{sv})/(0.4 \times b)$
If $\tau_c < \tau_v$ provide shear reinforcement as follows
Spacing of stirrups $S_v = (0.87 \times f_y \times A_{sv} \times d)/V$
Net shear force $V_{us} = V' - \tau_c bd$

## 5.8 PILE LOAD TESTS

### 5.8.1 Introduction

Pile load test method is used for determining the structural capacity of a pile with respect to soil in which it is installed. In-situ test methods are more reliable for computing the capacity of a pile.

### 5.8.2 Necessary Information

As per IS 2911 (Part 4), based on following information, type of test carried out on pile will be proposed.

1. Type of pile
    - Material and reinforcement details of a pile
    - Method of pile installation
    - Pile depth and details of cross-sections
2. Type of test
    - Space availability for the pile load test
    - Soil test should be conducted on the location where the pile is driven for testing
3. Safe load and ultimate load capacity methods are based on
    - Availability of anchors or kentledge for reaction
    - Loading/unloading of piles even when the scour is expected
    - Information about planning and previous similar tests conducted

### 5.8.3 Pile Load Test

As per IS 2911 (Part 4) (2013), for each type of loading (i.e., vertical, lateral and pull out) two tests will be conducted, namely, initial and routine test.

***Initial Test:*** Conducted to achieve one or more of the following objectives:

- To determine the ultimate load capacity and safe load by applying factor of safety
- To provide guideline for setting up the limits of acceptance for routine tests

- To study the effect of piling on the stability of adjacent structures, and assess the suitability of type of piles to be used
- To get an idea of suitability of piling system
- To check on the calculated loads by static and dynamic formulae
  As per IS 2911 (Part 4) (2013), the number of initial tests may be selected as given below depending upon the nature of sub strata, number of piles and past experience at the site.
  a. For small size projects (for piles less than 1000 numbers), a minimum of two tests.
  b. For large size projects (for piles more than 1000 numbers), a minimum of two tests for first 1000 piles and additional one test for every additional 1000 piles and part thereof.
  c. The number of tests for may be increased/decreased depending upon whether the strata is erratic/uniform subjected to minimum of two tests.
  d. The test load provision in the initial load test shall not be less than the estimated safe load multiplied by minimum factor of safety of 2.5.

***Routine Test:*** Conducted to achieve one or more of the following objectives:

- To determine the safe load on the pile
- To check the extent of safety for the specific functional requirement of the pile at working load
- Detection of any unusual performance different to the finding of the initial test
  As per IS 2911 (Part 4) (2013), the number of routine tests shall be 0.5% of the total number of piles, subject to a minimum of one (1) test. The number of tests may be increased up to 2% in particular cases depending upon the nature, type of structure and sub strata condition.

**General procedure for all types of pile load tests:**

- Pile or group of pile is taken for test. Pile load test can be performed either on working pile which forms the foundation of the structure or on a test pile.
- The test load is applied with the help of calibrated jack placed over a rigid circular or square plate which in turn is placed on the head of the pile projecting above ground level.
- The load is applied in equal increments of about one-fifth of the estimated allowable load.
- The settlements are recorded with the help of three dial gauges of sensitivity 0.02 mm, symmetrically arranged over the test plate, and fixed to an independent datum bar.
- The piles are loaded until ultimate load is reached.
- The load is increased to a value 2.5 times the estimated allowable load or to a load which causes a settlement equal to one-tenth of the pile diameter, whichever occurs earlier.
- The results are plotted in the form of load-settlement curve.

- The ultimate load is indicated by load settlement curve approaching vertical.
- If the ultimate load cannot be obtained means, the allowable load is taken as follows:
  a. One-half to one-third of the final load, which causes settlement equal to 10% of pile diameter.
  b. Two-thirds of the final load, which causes a total settlement of 12 mm.
  c. Two-thirds of final load, which causes a net settlement of 6 mm.

### 5.8.4 Types of Pile Load Tests

As per IS 2911 (Part 4) (2013), based on pile type and site condition various types of test can be performed depending upon cost and estimate. The following are the different test methods available:

1. Vertical load test on piles
2. Lateral load test on piles
3. Pull-out test on piles

#### 5.8.4.1 Vertical Load Test on Pile

In this type of test, vertical compressive load is applied to the pile top by using a hydraulic jack against a rolled steel joist or suitable load frame of providing reaction and the settlement is recorded by positioned dial gauges. Various methods of testing the piles are characterized by the duration of force applied to the pile and the strain induced in the pile is recorded. Tests involving large forces applied for longer periods are called static load tests which are used to measure pile load capacity, and low strain energy tests are used to assess pile integrity. In the high strain dynamic load test, though the force is comparable to static test, it is applied over a much shorter period than in a static load test.

#### 5.8.4.2 Static Pile Load Test

At present the most frequently used types of static load testing are

1. Maintained load test (MLT)
2. Constant rate of penetration test (CRP)

This static pile load test will be carried out as stipulated in IS 2911 (Part 4) (2013).

***Pile head*** – The pile head shall be chipped off till sound concrete is met wherever applicable. The reinforcement shall be cut off and head levelled with plaster of Paris. A bearing plate with a hole at centres placed on the pile head for the jack to rest.

***Application of load*** – A series of vertical downward load is applied on the pile as shown in Figure 5.8. Each incremental load shall be carried out after the pile achieving 20% of the safe load. The test load shall be maintained for 24 hours. This type of load application is not suitable for CRP method.

**Reaction** – The reaction may be obtained from the following

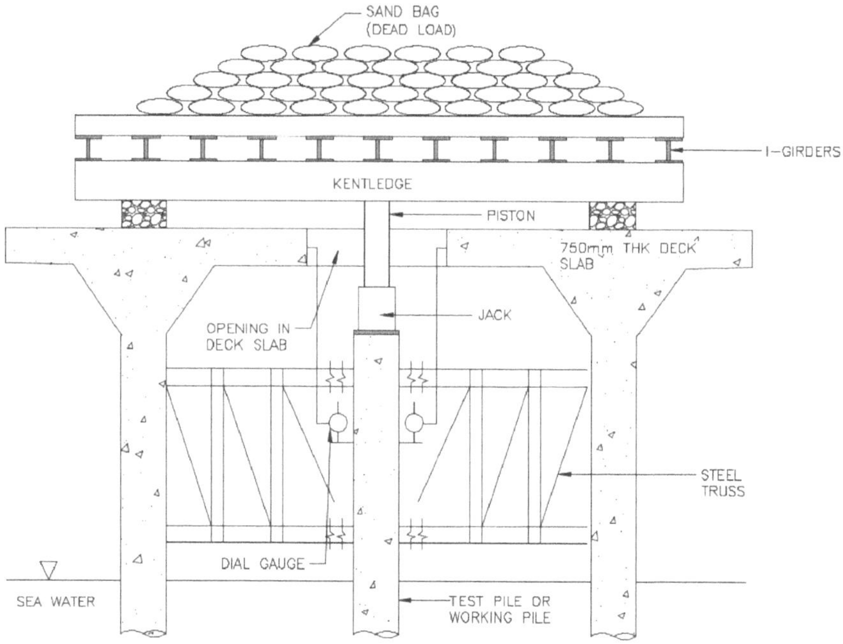


**FIGURE 5.8** Static pile load test.

1. Kentledge shall be designed to get the desired reaction on the piles. The centre of gravity of the kentledge should normally be on the axis of the pile and the load applied by the jack should also be coaxial with this pile.
2. Anchor piles shall be placed at a centre-to-centre distance of test pile not less than 3 times the pile diameter which subject to a minimum distance of 2 m. If anchor piles are working piles, residual uplift should be within the limit.
3. Rock anchors should be at the distance of edge of the piles at rock level being 2 times the test pile shaft diameter or 1.5 m, whichever is greater.

Reaction applied for the test should be 25% more than the final test load proposed to be applied.

**Settlement** – 2 dial gauges for a single pile and 4 dial gauges for a group of piles with 0.01 mm sensitivity shall be used. They shall be positioned at equal distance around the piles on datum bars resting on immovable supports at a distance of 3D (minimum of 1.5 m) where D is the diameter of pile or circumscribing for non-circular piles. This process will not be applicable for the CRP method.

**Initial test** – The safe load on a single pile shall be the least of the following:

1. On two-thirds of the final load, the pile attains a total displacement of 12 mm. In other case based on nature and type of structure safe load should be corresponding to the total displacement.

2. In case of uniform diameter piles, with 50% of the final load pile attains a total displacement equals to 10% of the pile diameter. In case of under reamed piles, total displacement equals to 7.5% of the bulb diameter.

**Routine test –** The maximum settlement at test load should not exceed 12 mm.

#### 5.8.4.3 Maintained Load Test (MLT)

In the MLT, the load is applied to the pile in discrete increments and settlement for each loading is monitored. In each stage, the load is maintained until the displacement of the pile is either 0.1 mm in the first 30 minutes or 0.2 mm in the first 1–2 hours, whichever occurs first. Subsequent load increments are only applied when the above-specified criteria are achieved. If the limit of permissible displacement not exceeded further testing of pile is not required. The test load should be maintained for 24 hours. The MLT method is normally the most suitable in determining the load/settlement performance of a pile under working loads and at 1.5 times working load conditions.

#### 5.8.4.4 Constant Rate of Penetration (CRP) Test

In the CRP test, the load required to cause a pile to penetrate into the ground at a constant rate is monitored until either the maximum specified test load is achieved or 'failure' of the pile occurs. The readings of time, penetration and load is recorded in close intervals. The rate of penetration of friction pile and end-bearing piles should be 0.75 mm and 1.5 mm per minute. The performance of the test takes less than 24 hours, excluding erecting and dismantling the test equipment. A test curve is drawn between load and penetration to determine the ultimate load capacity.

#### 5.8.4.5 Dynamic Pile Load Test

It is a method to determine the bearing capacity of a pile by applying impact load to the pile head. It consists of a high strain dynamic test and a low strain dynamic test.

#### 5.8.4.6 High Strain Dynamic Test

This method is only applicable to driving piles and bored piles. In this method hammer is dropped from a certain height (0.5–3 m) on the pile head and monitoring the response of pile at each stage. Hammer weight shall be based on 1%–2% of the test load or 7%–10% of the dead weight of the pile, whichever is higher. Sensors are used to detect the bending of piles while driving and measured in the pile driving analyser.

Following are the outputs obtained by this test.

- Pile load capacity
- Stress developed in the pile
- Displacement of pile during driving
- Depth of pile installation
- Blow counts required for driving operation

The use of dynamic testing after calibration with a particular geological profile will allow more comprehensive testing at a low cost in comparison to static testing. Typically, a dynamic test will take about 15 minutes to perform on a precast concrete pile using the piling rig hammer.

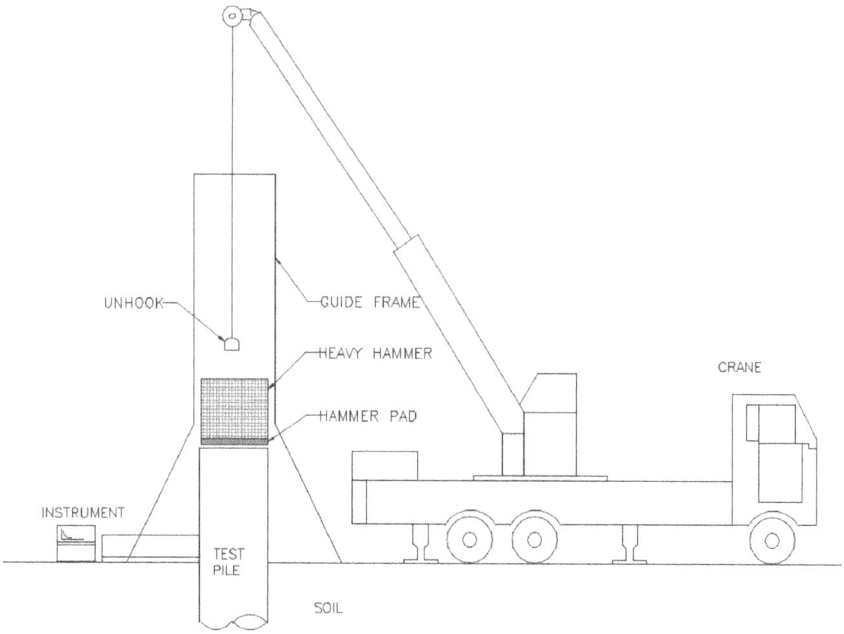


**FIGURE 5.9** High strain dynamic test.

### 5.8.4.7 Low Strain Dynamic Test

It is a simple and effective method. The transient dynamic response (TDR) method is used to measure the pile length/depth, pile head stiffness and flexibility of the pile shaft. This method is carried out by a handheld hammer to produce an impact on the pile head. An accelerometer measures the response of the pile to the impact which is placed on top of the pile head. It is also known as the pile integrity test.

### 5.8.4.8 Lateral Load Test

This test will be carried out as stipulated in IS 2911 (Part 4) (2013).

Based on the position of the gauge, the lateral load test method differs.

**Method 1:** A hydraulic jack with a measuring gauge should be placed horizontally between two piles or pile groups under test as shown in Figure 5.12. The full load imposed by the jack shall be taken as the lateral resistance of each pile or group. The loading should be increased about 20% of the estimated safe load and subsequent increments shall be applied after the rate of displacement is nearer to 0.1 mm per 30 minutes. If the cut-off level is unapproachable, two dial gauges of 0.01 mm sensitivity spaced at 30 cm horizontally, one above the other, will measure the displacement. If the cut-off level is approachable, one dial gauge placed diametrically opposite to the jack shall measure the displacement. Whenever the axes of the dial gauge and jack are not coordinating a line, then two dial gauges may be kept at a distance of 30 cm at a suitable height and the displacement interpolated at load point from similar triangles. The displacement should be at the cut-off level of the pile.

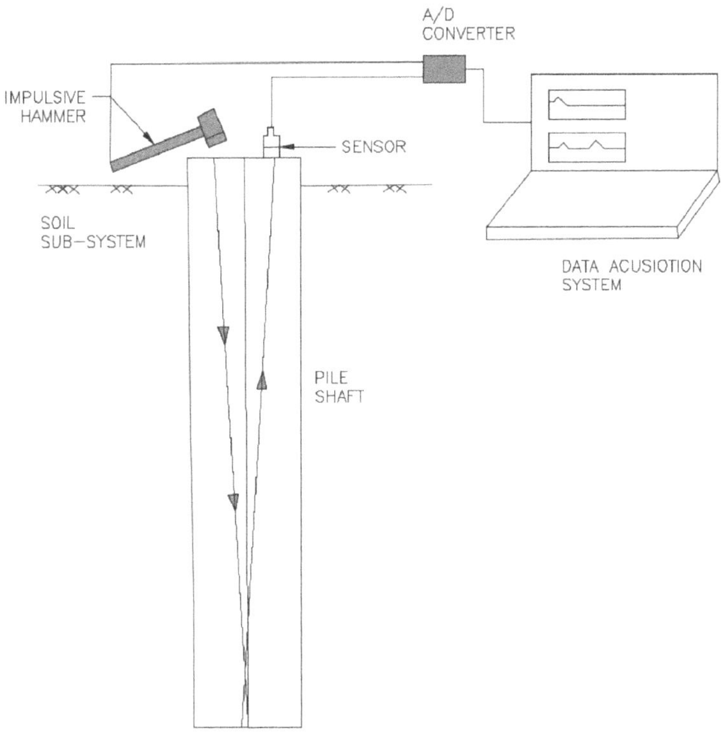


**FIGURE 5.10** Low strain dynamic test.

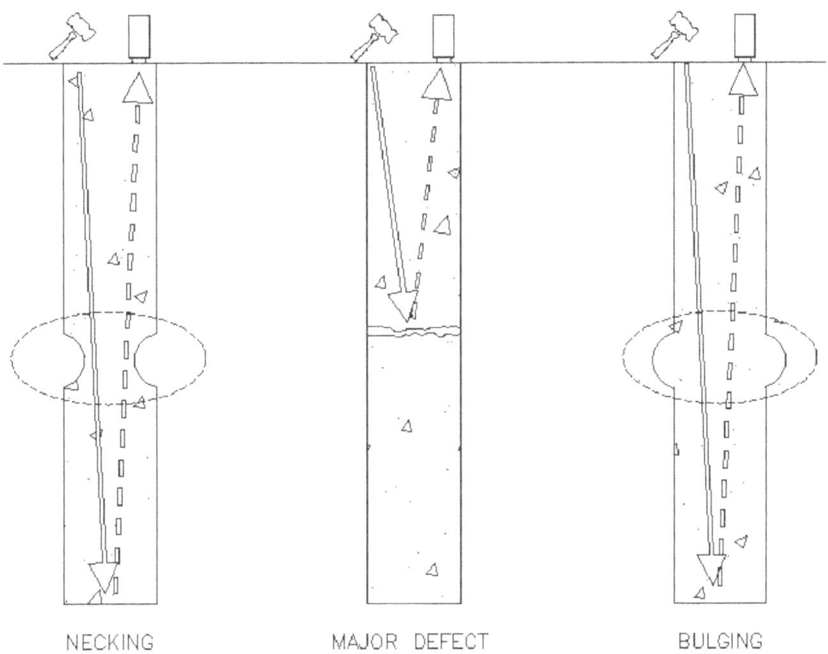


**FIGURE 5.11** Impulse path for various defects of low strain dynamic test.

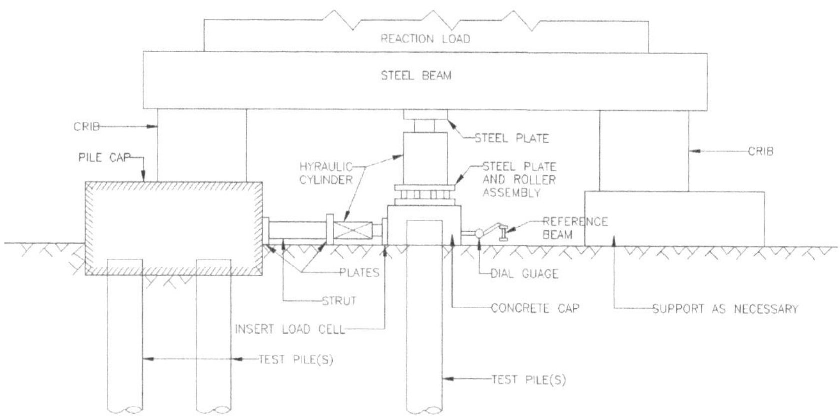


**FIGURE 5.12** Lateral load test on pile.

**Method 2:** This method is performed by placing a dial gauge on the pile surface and a glass piece of 20–30 mm square on the sides of the pile. The tips of the dial gauge shall rest on the central portion of the glass plate.

**Initial test** – The safe load on the pile shall be the least of the following:

1. 50% of the final load, at which the total displacement increases to 12 mm.
2. Final load at which the total displacement corresponds to 5 mm.
3. Load corresponding to any other specified displacement as per performance requirements.

**Routine test** – The test load should be equal to the working load and the deflection should not exceed 5 mm.

### 5.8.4.9 Pull-Out Test on Pile

This test will be carried out as stipulated in IS 2911 (Part IV) (2013).

A suitable set-up shall be designed to provide an uplift force to the piles by hydraulic jack with gauge. The hydraulic jack is made to rest on rolled steel joist(s) resting on two supports on the ground. A lifting frame is attached to the top of the test pile. When the jack is operated, the jack respond against a frame, and the pile gets pulled up, and the reaction is transferred to the ground through the supports which are at least 2.5D away from the test pile (where D is pile diameter). The framework can be attached to the pile top with the reinforcement bars, which may be threaded or welded. As an alternative, sometimes central rod is embedded centrally in a pile to take the pile load for a length equal to the bond length. The number of rods has to be increased for taking larger loads. For routine tests, the framework is normally attached to the reinforcing bars, but a central rod may also be used in case the upper portion of the pile is required to be built up.

The test pile should have adequate reinforcement to withstand pulling. In order to tolerate the neck tension, additional reinforcement may be provided to the testing

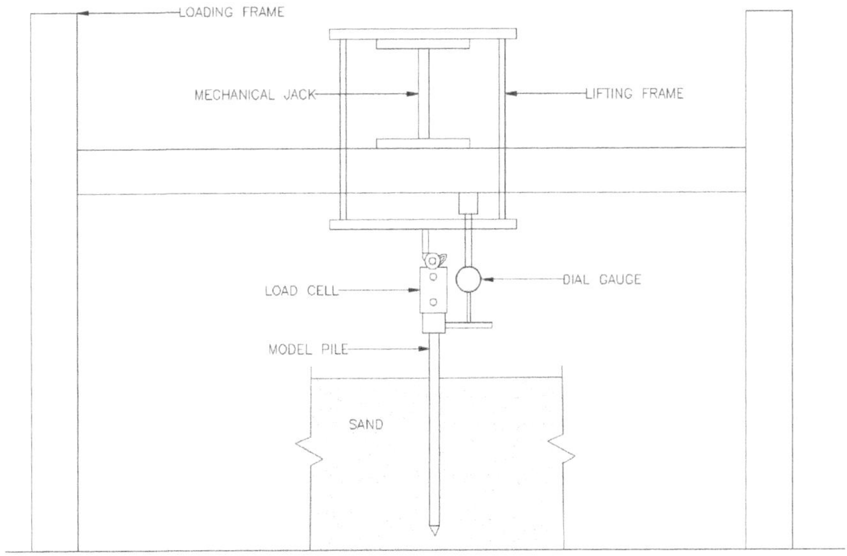


**FIGURE 5.13** Model pile pull-out test.

piles. The load increments and the resulting displacements shall be as per the case of a vertical load test.

The safe uplift load shall be taken as the least of the following:

1. For piles ≤ 600 mm φ, two-thirds of the load at which the total displacement is 12 mm or the load corresponding to a specified permissible uplift.
2. For piles > 600 mm φ, two-thirds of the load at which the total displacement is 18 mm or maximum of 2% of pile diameter or the load corresponding to a specified permissible uplift.
3. Half of the load at which the load-displacement curve shows a clear break.

**Initial test –** It is carried out up to 2.5 times the estimated safe uplift load or until the load-displacement curve shows a clear break.

**Routine test –** It is carried out by 1.5 times the working load or 12 mm total displacement for piles ≤ 600 mm φ or 18 mm total displacement for piles > 600 mm φ, whichever is earlier.

### 5.8.5 Pile Load Test – Field Testing Pictures

#### Examples

1. For the given plan and cross-section, design the following:
   a. Anchored diaphragm wall
   b. Dead man wall

Grade of concrete is M30 and grade of steel is Fe500.

**FIGURE 5.14** Low and high strain dynamic test.

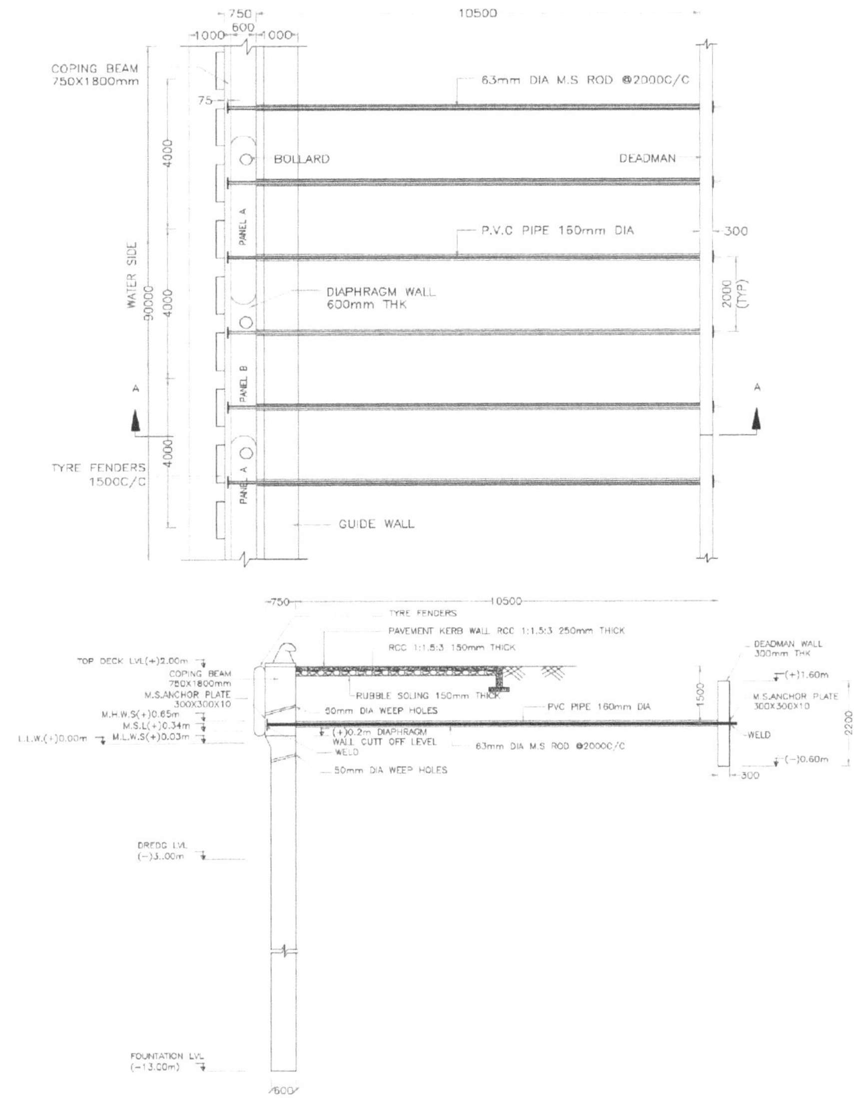


**FIGURE 5.15** Plan and cross-section of anchored diaphragm wall.

### *Given data*

Characteristic compressive strength of concrete $f_{ck}$ = 30 N/mm$^2$
Characteristic strength of steel $f_y$ = 500 N/mm$^2$
Young's modulus of steel = 200,000 N/mm$^2$

### *Solution*

1. Anchored diaphragm wall

   Clear cover to reinforcement = 75 mm (splash zone and immersed in sea water)

Over all depth of wall D = 600 mm
Height of wall H = 13,200 mm (cutoff level to FL)
Tension reinforcement
Diameter of the bar in layer 1 = 20 mm
Distribution reinforcement diameter bar = 16 mm
Compression reinforcement
Diameter of the bar in layer 1 = 16 mm
Distribution reinforcement diameter bar = 16 mm
Diameter of stirrups = 10 mm
Width of the section b = 2000 mm
Effective depth of the section at tension = 600 – (75 + 20 + 12/2) = 499 mm
Effective depth of the section at tension = 600 – (75 + 16 + 12/2) = 503 mm
STAAD Data – Collapse condition
Design bending moment (tension) = 528.33 kNm
Design bending moment (compression) = 283.42 kNm
Design shear force = 384. 49 kNm
STAAD data – Serviceability condition
Design bending moment (tension) = 442.67 kNm
Design bending moment (compression) = 230.78 kNm

---

**DESIGN – WALL TENSION**

| | | | |
|---|---|---|---|
| Moment | M | = | 528.33 kNm |
| As per IS 456 & SP 16 | | | |
| | $R = \frac{Mu}{bd^2}$ | | |
| | | = | $(528.33 \times 10^6)/(2000 \times 499^2)$ |
| | R | = | 1.06 |
| | $\frac{P_t}{100} = \frac{fck}{2 \times fy} 1 - \sqrt{\frac{1 - 4.598 \times R}{fck}}$ | | |
| | $P_t$ | = | 0.25% |
| Area of steel required | $A_{st,reqd}$ | = | 2495 mm$^2$ |
| Minimum area of reinforcement | $A_{st,min}$ | = | (0.15%bD) |
| | | = | 0.15/100 x 2000 x 600 |
| | | = | 1800 mm$^2$ |
| Area of steel to be provided | $A_{st}$ | = | 2495 mm$^2$ |
| (Maximum of $A_{st,reqd}$ and $A_{st,min}$) | | | |
| Maximum area of reinforcement | $A_{st,max}$ | = | 0.04bD |
| (as per IS 456 Cl 26.5.2.1) | | | 0.04 x 2000 x 600 |
| | | = | 48000 mm$^2$ |
| Preliminary Ast calculation | | | |
| Diameter of bar | | = | 20 mm |
| Area of bar | | = | 314.16 mm$^2$ |
| Spacing of reinforcement required | | = | (2000 x314.16) / 2495) |

(Continued)

| | | | |
|---|---|---|---|
| | | = | 251.83 mm |
| Spacing of reinforcement provided | S | = | 150 mm |
| Area of reinforcement to be provided | $A_{st,prov}$ | = | 4188.80 mm$^2$ |
| Area of reinforcement provided is Ok | | | |
| Area of reinforcement to be provided: 4188.8 sq mm | | | |
| Percentage of reinforcement provided = (100 x 4188.8)/(2000 x 499) | | = | 0.42 |
| Layer 1 | | | |
| Area of reinforcement | $A_{st,1}$ | = | 4188.8 mm$^2$ |
| Diameter of bar | | = | 20 mm |
| Area of one bar | | = | 314.16 mm$^2$ |
| Spacing of reinforcement required | | = | (2000 x 314.16) / 4188.8 |
| | | = | 150 mm |
| Spacing of reinforcement provided | | = | 150 mm |
| Area of reinforcement provided | $A_{st,1prov}$ | = | 4188.8 mm$^2$ |
| Area of reinforcement is satisfied | | | |
| Provide Y – 20 @ 150 mm c/c in first layer | | | |
| Total area of reinforcement provided | $A_{st}$ | = | 4188.8 mm$^2$ |
| Horizontal Reinforcement | | | |
| Area of reinforcement | | = | (0.2%bD) |
| (as per IS 456 Cl 26.5.2.1) | | | (0.2/100) x 2000 x 600 |
| | | = | 2400 mm$^2$ |
| Diameter of reinforcement | | = | 16 mm |
| Area of one bar | | = | 201.06 mm$^2$ |
| Required spacing | | = | 2000 x (201.06 / 2400) |
| | | = | 167.55 mm |
| Provided spacing of reinforcement | | say | 150 mm |
| Provide Y – 16 @ 150 mm c/c | | | |

**WALL TENSION – CRACK WIDTH CALCULATION (Doubly reinforced section)**

| | | | |
|---|---|---|---|
| Moment | M | = | 442.67 kNm |
| Torsion | T | = | 0 |
| Design moment | M' | = | $M_u+T_u*(1+D/b)/1.7$ |
| | | = | 442.67 + 0 |
| | M' | = | 442.67 kNm |

Determination of depth of Neutral Axis (x)

$bx^2/2 + (1.5m-1)A_{sc}(x-d') = mA_{st}(d-x)$

where,

| | | | |
|---|---|---|---|
| Effective clear cover | d' | = | 101 mm |

*(Continued)*

(Continued)

| | | | |
|---|---|---|---|
| | b | = | 2000 mm |
| | $A_{sc}$ | = | 2680.8 $mm^2$ |
| | $A_{st}$ | = | 4188.8 $mm^2$ |
| | m | = | $E_s/E_c$ |
| | $E_s$ | = | 200,000 $N/mm^2$ |
| | $E_c$ | = | 5000 sqrt ($f_{ck}$) |
| | | = | 27386.13 $N/mm^2$ |
| Thus, | m | = | 7.3 |
| Rewriting the equation, | | | |
| 2000x(x^2)/2 + (1.5x7.3–1)×2680.8(x-101= 7.3x4188.8(499- x) | | | |
| Hence, | | | |
| 1000(x^2) +[(1.5×7.3–1)2680.8+7.3×4188.8](x)-[7.3×4188.8×499+(1.5×7.3–1)2680.8101] =0 | | | |
| Depth of neutral axis | x | = | 108.39 mm |
| Moment of inertia at NA | | $=bx^3/3 + (1.5m-1)A_{sc}(x-d')^2 + mA_{st}(d-x)^2$ | |
| | $I_{NA}$ | = | 5515906343 $mm^3$ |
| Lever arm (d – x/3) | Z | = | 462.87 mm |
| Stress in steel | $f_s$ | = | $M/A_s Z$ |
| | | = | $(442.67 \times 10^6)/(4188.8 \times 462.87)$ |
| | | = | 228.31 $N/mm^2$ |
| Limiting stress in steel (as per Table 22 IS 456) | $f_s'$ | $= 0.55f_y =$ | 275 $N/mm^2$ |
| Stress in steel is satisfied | | | |
| Stress in concrete | $f_c$ | = | $M_x/I_{NA}$ |
| | | = | 8.70 $N/mm^2$ |
| Limiting stress in concrete (as per Table 21 IS 456) | $f_c'$ | = | 10 $N/mm^2$ |
| Stress in concrete is satisfied | | | |
| Strain | | $\varepsilon_1 = \frac{f_s}{E_s}\frac{(h-x)}{(d-x)}$ | |
| | e1 | = | 1.44E-03 $N/mm^2$ |
| Average steel strain at the level considered | | $\varepsilon_m = \varepsilon_1 - \frac{b(h-x)(a-x)}{3E_sA_s(d-x)}$ | |
| | em | = | 9.44E-04 $N/mm^2$ |
| Determination of $a_{cr}$ | | | |
| At the point directly below the midpoint between main rods | $a_{cr1}$ | $=[(150/2)^2 + (600–499)^2]^{1/2}–20/2 = 115.8$ mm | |
| At the point directly below the main rods | $a_{cr2}$ | = | (600–499) – 20/2 = 91 mm |

(Continued)

| | | | |
|---|---|---|---|
| At the corner point of the section | $a_{cr3}$ | $= [(75)^2 + (600–499)^2]^{1/2}–20/2 = 115.8$ mm | |
| Crack width | | $W_{cr} = \dfrac{3a_{cr}\varepsilon_m}{1+\dfrac{2(a_{cr} - C_{min})}{h-x}}$ | |
| | $W_{cr1}$ | = | 0.281 mm |
| | $W_{cr2}$ | = | 0.242 mm |
| | $W_{cr3}$ | = | 0.281 mm |
| Maximum crack width | $W_{cr}$ | = | 0.281 mm |
| Limiting crack width | | = | 0.3 mm |
| Crack width is satisfied | | | |
| *DESIGN – WALL COMPRESSION* | | | |
| Moment | M | = | 283.42 kNm |
| As per IS 456 & SP 16 | | | |
| | | $R = \dfrac{Mu}{bd^2}$ | (5.76) |
| | | $R = (283.42 \times 10^6)/(2000 \times 503^2)$ | |
| | | = | 0.56 |
| | | $\dfrac{P_t}{100} = \dfrac{fck}{2\times fy} 1 - \sqrt{\dfrac{1-4.598\times R}{fck}}$ | (5.77) |
| | $P_t$ | = | 0.13% |
| Area of steel required | $A_{st,reqd}$ | = | 1307.8 mm² |
| Minimum area of reinforcement | $A_{st,min}$ | = | (0.15%bD) |
| (as per IS 456 Cl 26.5.2.1) | | = 0.15/100 x 2000 x 600 | |
| | | = | 1800 mm² |
| Area of steel to be provided (Maximum of $A_{st,reqd}$ and $A_{st,min}$) | $A_{st}$ | = | 1800 mm² |
| Maximum area of reinforcement | $A_{st,max}$ | = | 0.04bD |
| | | = | 0.04 x 2000 x 600 |
| | | = | 48000 mm² |
| Preliminary $A_{st}$ calculation | | | |
| Diameter of bar | | = | 16 mm |
| Area of bar | | = | 201.06 mm² |
| Spacing of reinforcement required | | = (2000 x 201.06) / 1800 | |
| | | = | 223.4 mm |
| Spacing of reinforcement provided | S | = | 150 mm |
| Area of reinforcement to be provided | $A_{st,prov}$ | = | 2680.8 mm² |

*(Continued)*

(Continued)

| | | | |
|---|---|---|---|
| Area of reinforcement provided is Ok | | | |
| Area of reinforcement to be provided: 2680.8 sq mm | | | |
| Percentage of reinforcement provided = (100 x 2680.8)/(2000 x 503) = 0.27 | | | |
| Layer 1 | | | |
| Area of reinforcement | $A_{st,1}$ | = | 2680.8 $mm^2$ |
| Diameter of bar | | = | 16 mm |
| Area of one bar | | = | 201.06 $mm^2$ |
| Spacing of reinforcement required | | = (2000 x 201.06) / 2680.8 | |
| | | = | 150 mm |
| Spacing of reinforcement provided | | = | 150 mm |
| Area of reinforcement provided | $A_{st,1prov}$ | =(2000 x 201.06) /150 | |
| | | = | 2680.8 $mm^2$ |
| Area of reinforcement is satisfied | | | |
| Provide Y – 16 @ 150 mm c/c in first layer | | | |
| Tota area of reinforcement provided | $A_{st}$ | = | 2680.8 $mm^2$ |
| Horizontal Reinforcement | | | |
| Area of reinforcement | | = | (0.2%bD) |
| (as per IS 456 Cl 26.5.2.1) | | = (0.2/100) x 2000 x 600 | |
| | | = | 2400 $mm^2$ |
| Diameter of reinforcement | | = | 16 mm |
| Area of one bar | | = | 201.06 $mm^2$ |
| Required spacing | = 2000 x (201.06/2400) | | |
| | | = | 167.55 mm |
| Provided spacing of reinforcement | | say | 150 mm |
| Provide Y – 16 @ 150 mm c/c | | | |

**WALL COMPRESSION – CRACK WIDTH CALCULATION**

| | | | |
|---|---|---|---|
| Moment | | = | 230.78 kNm |

Determination of depth of Neutral Axis (x)

$bx^2/2 = mA_{st}(d-x)$

where,

| | | |
|---|---|---|
| b | = | 2000 mm |
| $A_{st}$ | = | 2680.8 $mm^2$ |
| m | = | $E_s/E_c$ |
| $E_s$ | = | 200,000 N/$mm^2$ |
| $E_c$ | = | 5000 sqrt ($f_{ck}$) |
| | = | 27386.13 N/$mm^2$ |

(Continued)

| | | | | |
|---|---|---|---|---|
| Thus, | m | = | 7.3 | |
| Rewriting the equation, | 2000x(x^2)/2 = 7.3x2680.8(503- x) | | | |
| Hence, | 1000(x^2) +19569.84(x)-9843629.52 = 0 | | | |
| Depth of neutral axis | x | = | 89.91 mm | |
| Lever arm (d – x/3) | Z | = | 473.03 mm | |
| Stress in steel | $f_s$ | = | $M/A_s Z$ | |
| | | $= (230.78 \times 10^6)/(2680.8 \times 473.03)$ | | |
| | | = | 181.99 N/mm$^2$ | |
| Limiting stress in steel (as per Table 22 IS 456) | $f_s'$ | $= 0.55f_y =$ | 275 N/mm$^2$ | |
| Stress in steel is satisfied | | | | |
| Stress in concrete | $f_c$ | = | $(f_s A_s)/(0.5bx)$ | |
| | | = | $(181.99 \times 2680.8)/(0.5 \times 2000 \times 89.91)$ | |
| | | = | 5.43 N/mm$^2$ | |
| Limiting stress in concrete (as per Table 21 IS 456) | $f_c'$ | = | 10 N/mm$^2$ | |
| Stress in concrete is satisfied | | | | |
| Strain | | $\varepsilon_1 = \frac{f_s}{E_s}\frac{(h-x)}{(d-x)}$ | | (5.78) |
| | e1 | = | 1.13E-03 N/mm$^2$ | |
| Average steel strain at the level considered | | $\varepsilon_m = \varepsilon_1 - \frac{b(h-x)(a-x)}{3E_s A_s (d-x)}$ | | (5.79) |
| | em | = | 3.44E-04 N/mm$^2$ | |
| Determination of $a_{cr}$ | | | | |
| at the point directly below the midpoint between main rods | $a_{cr1}$ | $= [(150/2)^2 + (600–503)^2]^{1/2} - 16/2$ =114.61 mm | | |
| at the point directly below the main rods | $a_{cr2}$ | = (600–503) – 16/2 = 89 mm | | |
| at the corner point of the section | $a_{cr3}$ | $= [(75)^2 + (600–503)^2]^{1/2} - 16/2$ = 114.61 mm | | |
| Crack width | | $W_{cr} = \frac{3a_{cr} \in_m}{1+2(a_{cr}-C_{min})/(D-x)}$ | h – total depth<br>$C_{min}$ – Clear cover | (5.80) |
| | $W_{cr1}$ | = | 0.102 mm | |
| | $W_{cr2}$ | = | 0.087 mm | |
| | $W_{cr3}$ | = | 0.102 mm | |
| Maximum crack width | $W_{cr}$ | = | 0.102 mm | |
| Limiting crack width | | = | 0.2 mm | |
| Crack width is satisfied | | | | |

*(Continued)*

(Continued)

| | | | | |
|---|---|---|---|---|
| **DESIGN – SHEAR** | | | | |
| Shear force | V | = | 384.49 kN | |
| Diameter of shear reinforcement | | = | 12 mm | |
| Yield strength of shear link | $f_{yv}$ | = | 415 N/mm$^2$ | |
| Number of clips per meter | | = | 8 | |
| Area of shear clips | $A_{sv}$ | = | 904.78 mm$^2$ | |
| | $H_w/L_w$ | = | 6.6 | |
| Calculation for shear reinforcement | | | | |
| Nominal shear stress (Due to shear force) | $\tau_{vw}$ | = | V/0.8td | |
| | | = (384.49 x 10$^3$)/(0.8 x 2000 x 499) | | |
| | | = | 0.48 N/mm$^2$ | |
| Percentage of reinforcement | | = | 0.27 | |
| (Minimum of compression and tension) | | | | |
| Maximum shear stress of concrete | $\tau_{cmax}$ | = | $0.17\times f_{ck}$ | |
| From IS 456 (2000) Cl.32.4.2 | | | | |
| Limit state method maximum shear stress of concrete | | = | 5.1 N/mm$^2$ | |
| Hence safe | | | | |
| Design shear strength of concrete ($H_w/L_w \geq 1$) From IS 456 (2000) Cl.32.4.2 | $\tau_{cw}$ | = | $K_2\sqrt{fck}\left(\dfrac{\dfrac{H_w}{L_W}+1}{\dfrac{H_w}{L_W}+1}\right)$ | (5.81) |
| | | = | 0.33 N/mm$^2$ | |
| Design shear strength of concrete | $\tau_{cw}$ | = | $K_3\sqrt{f_{ck}}$ | |
| | | = | 0.82 N/mm$^2$ | |
| | | = | 0.82 N/mm$^2$ | |
| | $\tau_{cw}$ | = | 0.82 N/mm$^2$ | |
| From Table 19 – IS 456 (2000) | $\tau_{cmax}$ | = | 3.7 N/mm$^2$ | |
| | $\tau_{c,max} > \tau_{cw}$ | | $>\tau_{vw}$ | |
| Provide minimum shear reinforcement | | | | |
| Spacing of stirrups (Minimum shear reinforcement) $S_v$ | | = | (0.87 x $f_y$ x $A_{sv}$)/(0.4*b) | |

(Continued)

| | | |
|---|---|---|
| As per IS 456 (2000) Cl.26.5.1.6 | = | (0.87 x 415 x 904.78)/(0.4 x 2000) |
| | = | 408.34 mm |
| $S_v$ | = | 150 mm |

Provide 12 mm dia bars @ 150 mm c/c (8 clips per meter)

***Deadman wall***

Clear cover to reinforcement = 50 mm (Below ground level)
Overall depth of the wall (thickness) D = 300 mm
Height of the wall H = 2200 mm
Tension reinforcement
Diameter of the bar in layer 1 = 12 mm

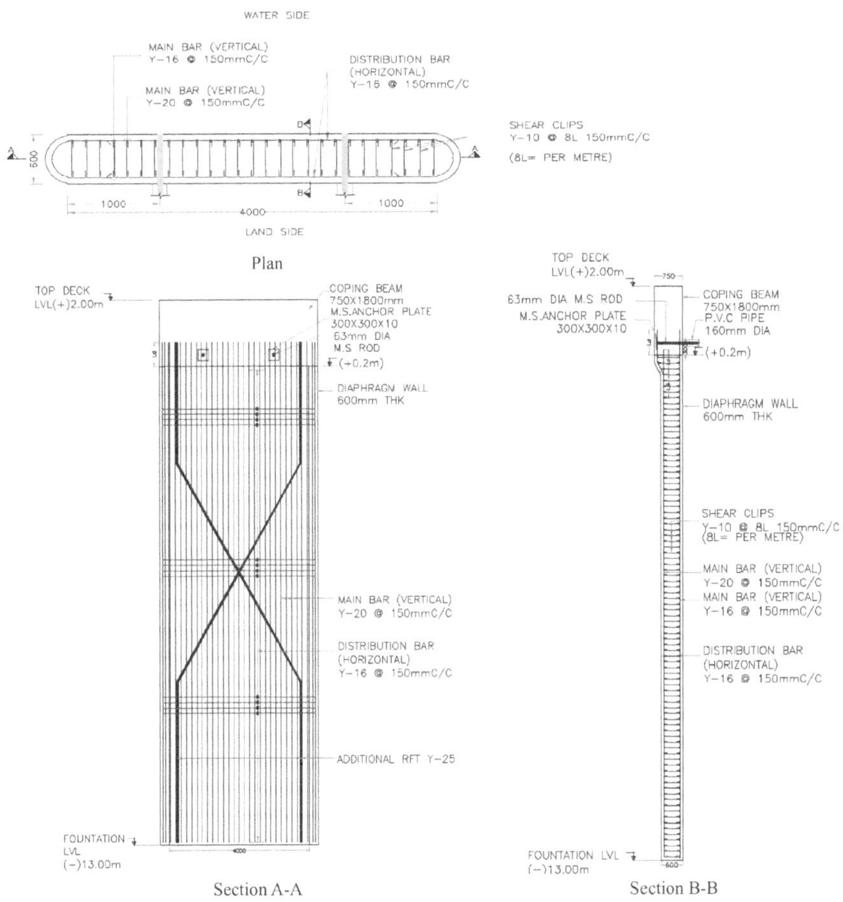


**FIGURE 5.16** Reinforcement details of anchored diaphragm wall.

Distribution reinforcement diameter bar = 12 mm
Compression reinforcement
Diameter of the bar in layer 1 = 12 mm
Distribution reinforcement diameter bar = 12 mm
Diameter of stirrups = 10 mm
Width of the section b = 2000 mm
Effective depth of the section at tension = 300 – (50 + 12 + 10/2) = 233 mm
Effective depth of the section at tension = 300 – (50 + 12 + 10/2) = 233 mm
STAAD Data – Collapse condition
Design bending moment (tension) = 34.987 kNm
Design bending moment (compression) = 0
Design Shear Force = 47.982 kNm
STAAD data – Serviceability condition
Design bending moment (tension) = 28.138 kNm
Design bending moment (compression) = 0 kNm

---

**DESIGN – WALL TENSION**

| | | | | |
|---|---|---|---|---|
| Moment | M | = | 34.897 kNm | |
| As per IS 456 & SP 16 | | | | |
| | $R = \frac{Mu}{bd^2}$ | | | (5.82) |
| | $= (34.897 \times 10^6)/(2000 \times 233^2)$ | | | |
| | R | = | 0.32 | |
| | $\frac{P_t}{100} = \frac{fck}{2 \times fy} 1 - \sqrt{\frac{1 - 4.598 \times R}{fck}}$ | | | (5.83) |
| | $P_t$ | = | 0.07% | |
| Area of steel required | $A_{st,reqd}$ | = | (0.07/100) x (2000 x 233) | |
| | | = | 326.2 mm$^2$ | |
| Minimum area of reinforcement | $A_{st,min}$ | = | (0.15%bD) | IS 456 Cl 26.5.2.1 |
| | | = | (0.15/100) x (2000 x 300) | |
| | | = | 900 mm$^2$ | |
| Area of steel to be provided | $A_{st}$ | = | 900 mm$^2$ | |
| Maximum area of reinforcement | $A_{st,max}$ | = | 0.04bD | |
| | | = | 0.04 x 2000 x 300 | |
| | | = | 24000 mm$^2$ | |
| Preliminary $A_{st}$ calculation | | | | |
| Diameter of bar | | = | 12 mm | |
| Area of bar | | = | 113.1 mm$^2$ | |

(Continued)

| | | | | |
|---|---|---|---|---|
| Spacing of reinforcement required | | = | (2000 x 113.1) / 900 | |
| | | = | 251.333 mm | |
| Spacing of reinforcement provided | S | = | 150 mm | |
| Area of reinforcement to be provided | $A_{st,prov}$ | = | 1508 mm$^2$ | |
| Area of reinforcement provided is Ok | | | | |
| Area of reinforcement to be provided: 1508 Sq.mm | | | | |
| Percentage of reinforcement provided | $P_t$ | = | (100 x 1508)/ (2000 x 233) | |
| | | = | 0.32 | |
| Layer 1 | | | | |
| Area of reinforcement | $A_{st,1}$ | = | 1508 mm$^2$ | |
| Diameter of bar | | = | 12 mm | |
| Area of one bar | | = | 113.1 mm$^2$ | |
| Spacing of reinforcement required | | = | (2000 x 113.1) / 1508 | |
| | | | 150 mm | |
| Spacing of reinforcement provided | | = | 150 mm | |
| Area of reinforcement provided | $A_{st,1prov}$ | = | 1508 mm$^2$ | |
| Area of reinforcement is satisfied | | | | |
| Provide Y – 12 @ 150 mm c/c in first layer | | | | |
| Total area of reinforcement provided | $A_{st}$ | = | 1508 mm$^2$ | |
| Horizontal Reinforcement | | | | |
| Area of reinforcement | | = | (0.2%bD) | IS 456 Cl 26.5.2.1 |
| | | = | (0.2 /100) x 2000 x 300 | |
| | | = | 1200 mm$^2$ | |
| Diameter of reinforcement | | = | 12 mm | |
| Area of one bar | | = | 113.10 mm$^2$ | |
| Required spacing | | = | 2000 x (113.10 / 1200) | |
| | | = | 188.50 mm | |
| Provided spacing of reinforcement | | say | 150 mm | |
| Provide Y – 12 @ 150 mm c/c | | | | |

*(Continued)*

(Continued)

| | | | |
|---|---|---|---|
| **WALL TENSION – CRACK WIDTH CALCULATION** | | | |
| Moment | | = | 28.138 kNm |
| Determination of depth of Neutral Axis (x) | | | |
| $bx^2/2 = mA_{st}(d\text{-}x)$ | | | |
| where, | | | |
| | b | = | 2000 mm |
| | $A_{st}$ | = | 1508 mm² |
| | m | = | $E_s/E_c$ |
| | $E_s$ | = | 200,000 N/mm² |
| | $E_c$ | = | 5000 sqrt ($f_{ck}$) |
| | | = | 27386.13 N/mm² |
| Thus, | m | = | 7.3 |
| Rewriting the equation, | 2000×(x^2)/2 = 7.3×1508(233- x) | | |
| Hence, | 1000(x^2) +11008.4(x)-2564957.2 = 0 | | |
| Depth of neutral axis | x | = | 45.44 mm |
| Lever arm (d – x/3) | Z | = | 217.85 mm |
| Stress in steel | $f_s$ | = | $M/A_s Z$ |
| | | = | $(28.138\times 10^6)/(1508 \times 217.85)$ |
| | | = | 85.65 N/mm² |
| Limiting stress in steel | $f_s$‘ | = | $0.55f_y$ (as per Table 22 IS 456 (2000)) |
| | | = | 275 N/mm² |
| Stress in steel is satisfied | | | |
| Stress in concrete | $f_c$ | = | $(f_s A_s)/(0.5bx)$ |
| | | = | $(85.65 \times 1508)/(0.5 \times 2000 \times 45.44)$ |
| | | = | 2.84 N/mm² |
| Limiting stress in concrete | $f_c$‘ | = | 10 N/mm² (as per Table 21 IS 456 (2000)) |
| Stress in concrete is satisfied | | | |
| Strain | | | $\varepsilon_1 = \dfrac{f_s}{E_s}\dfrac{(h-x)}{(d-x)}$ (5.84) |
| | $e_1$ | = | 5.81E-04 N/mm² |
| Average steel strain at the level considered | | | $\varepsilon_m = \varepsilon_1 - \dfrac{b(h-x)(a-x)}{3E_s A_s (d-x)}$ (5.85) |

(Continued)

| | | | |
|---|---|---|---|
| | $e_m$ | = | -1.82E-04 N/mm$^2$ |
| Determination of $a_{cr}$ | | | |
| At the point directly below the midpoint between main rods | $a_{cr1}$ | | = [(150/2)$^2$ + (300–233)$^2$]$^{1/2}$– 12/2= 94.57 mm |
| At the point directly below the main rods | $a_{cr2}$ | = | (300–233) – 12/2 = 61 mm |
| At the corner point of the section | $a_{cr3}$ | = | [(50)$^2$ + (300–233)$^2$]$^{1/2}$–12/2 = 77.6 mm |
| Crack width | | | $W_{cr} = \frac{3a_{cr}\varepsilon_m}{1+\frac{2(a_{cr} - C_{min})}{h-x}}$ (5.86) |
| | $W_{cr1}$ | = | –0.038 mm |
| | $W_{cr2}$ | = | –0.031 mm |
| | $W_{cr3}$ | = | –0.035 mm |
| Maximum crack width | $W_{cr}$ | = | –0.031 mm |
| Limiting crack width | | = | 0.3 mm |
| Crack width is satisfied | | | |
| **DESIGN – WALL COMPRESSION** | | | |
| Moment | M | = | 0 kNm |
| As per IS 456 & SP 16 | | | |
| | $R = \frac{Mu}{bd^2}$ | | (5.87) |
| | R | = | 0 |
| | $\frac{P_t}{100} = \frac{fck}{2 \times fy} 1 - \sqrt{\frac{1-4.598 \times R}{fck}}$ | | (5.88) |
| | $P_t$ | = | 0% |
| Area of steel required | $A_{st,reqd}$ | = | 0 mm$^2$ |
| Minimum area of reinforcement | $A_{st,min}$ | = | (0.15%bD) |
| | | = | (0.15/100) × 2000 × 300 |
| | | = | 900 mm$^2$ |
| Area of steel to be provided | $A_{st}$ | = | 900 mm$^2$ |
| Maximum area of reinforcement | $A_{st,max}$ | = | 0.04bD |
| | | = | 0.04 × 2000 × 300 |
| | | = | 24000 mm$^2$ |

*(Continued)*

(Continued)

| | | | |
|---|---|---|---|
| Preliminary $A_{st}$ calculation | | | |
| Diameter of bar | | = | 12 mm |
| Area of bar | | = | 113.1 mm$^2$ |
| Spacing of reinforcement required | | = | (2000 × 113.1)/900 |
| | | = | 251.333 mm |
| Spacing of reinforcement provided | S | = | 150 mm |
| Area of reinforcement to be provided | $A_{st,prov}$ | = | 1508 mm$^2$ |

Percentage of reinforcement provided $P_t$= (100 × 1508)/ (2000× 233) = 0.32

Area of reinforcement provided is Ok

Area of reinforcement to be provided: 1508 Sq.mm

| | | | |
|---|---|---|---|
| Check: depth of neutral axis | | | |
| Limiting neutral axis depth | $X_{u,lim}$ | = | 107.18 mm |
| Actual neutral axis depth | $X_{u,actual}$ | = | $0.87f_yA_{st}/0.36f_{ck}b$ |
| | = (0.87× 500 × 1508)/(0.36 × 30 × 2000) | | |
| | $X_{u,actual}$ | = | 30.37 mm |

$X_{u,actual} < X_{u,lim}$, Hence, under reinforced beam

| | | | |
|---|---|---|---|
| Check: Moment of Resistance | | | |
| Moment of resistance of the section | $M_r$ | = | $= 0.87 f_y A_{st} d 1 - \frac{A_{st} f_y}{bdf_{ck}}$ |
| | $M_r$ | = | 144.6 kNm |
| Bending moment | M' | = | 0 |

Moment of Resistance of the section > Bending moment, Hence Ok

| | | | |
|---|---|---|---|
| Layer 1 | | | |
| Area of reinforcement | $A_{st,1}$ | = | 1508 mm$^2$ |
| Diameter of bar | | = | 12 mm |
| Area of one bar | | = | 113.1 mm$^2$ |
| Spacing of reinforcement required | | = | (2000 × 113.1) / 1508 |
| | | = | 150 mm |
| Spacing of reinforcement provided | | = | 150 mm |
| Area of reinforcement provided | $A_{st,1prov}$ | = | 1508 mm$^2$ |

Area of reinforcement is satisfied

(Continued)

| | | | |
|---|---|---|---|
| Provide Y – 12 @ 150 mm c/c in first layer | | | |
| Total area of reinforcement provided | $A_{st}$ | = | 1500 mm$^2$ |
| Horizontal Reinforcement | | | |
| Area of reinforcement | | = | (0.2%bD) |
| | | = | (0.2/100) x 2000 x 300 |
| | | = | 1200 mm$^2$ |
| Diameter of reinforcement | | = | 12 mm |
| Area of one bar | | = | 113.10 mm$^2$ |
| Required r spacing | | = | 2000 × (113.10 / 1200) |
| | | = | 188.50 mm |
| Provided spacing of reinforcement | | say | 150 mm |
| Provide Y – 12 @ 150 mm C/C | | | |
| **WALL COMPRESSION – CRACK WIDTH CALCULATION** | | | |
| Moment | | = | 0 kNm |
| Determination of depth of Neutral Axis (x) | | | |
| $bx^2/2 = mA_{st}(d\text{-}x)$ | | | |
| where, | | | |
| | b | = | 2000 mm |
| | $A_{st}$ | = | 1508 mm$^2$ |
| | m | = | $E_s/E_c$ |
| | $E_s$ | = | 200,000 N/mm$^2$ |
| | $E_c$ | = | 5000 sqrt ($f_{ck}$) |
| | | = | 27386.13 N/mm$^2$ |
| Thus, | m | = | 7.3 |
| Rewriting the equation, | 2000×(x^2)/2 = 7.3×1508(233- x) | | |
| Hence, | 1000(x^2) +11008.4(x)-2564957.2 = 0 | | |
| Depth of neutral axis | x | = | 45.44 mm |
| Lever arm (d – x/3) | Z | = | 217.85 mm |
| Stress in steel | $f_s$ | = | M/As Z |
| | | = | 0 |
| Limiting stress in steel | $f_s$' | = | 275 N/mm$^2$ |
| Stress in steel is satisfied | | | |

*(Continued)*

(Continued)

| | | | | |
|---|---|---|---|---|
| Stress in concrete | $f_c$ | = | $(f_sA_s)/(0.5bx)$ | |
| | | = | 0 | |
| Limiting stress in concrete | $f_c$‘ | = | 10 N/mm$^2$ | |
| Stress in concrete is satisfied | | | | |
| Strain | | $\varepsilon_1 = \frac{f_s}{E_s}\frac{(h-x)}{(d-x)}$ | | (5.89) |
| | $e_1$ | = | 0.00E+00 N/mm$^2$ | |
| Average steel strain at the level considered | | $\varepsilon_m = \varepsilon_1 - \frac{b(h-x)(a-x)}{3E_sA_s(d-x)}$ | | (5.90) |
| | $e_m$ | = | −7.64E-04 N/mm$^2$ | |
| Determination of $a_{cr}$ | | | | |
| At the point directly below the midpoint between main rods | $a_{cr1}$= [(150/2)$^2$ + (300–233)$^2$]$^{1/2-}$12/2= 94.57 mm | | | |
| At the point directly below the main rods | $a_{cr2}$ | = | (300–233) – 12/2 = 61 mm | |
| At the corner point of the section | $a_{cr3}$ | = [(50)$^2$ + (300–233)$^2$]$^{1/2-}$12/2 = 77.6 mm | | |
| Crack width | $W_{cr} = \frac{3a_{cr}\in_m}{1+2(a_{cr}-C_{min})/(D-x)}$ | | (5.91) | |
| | $W_{cr1}$ | = | −0.16 mm | |
| | $W_{cr2}$ | = | −0.129 mm | |
| | $W_{cr3}$ | = | −0.146 mm | |
| Maximum crack width | $W_{cr}$ | = | −0.129 mm | |
| Limiting crack width | | = | 0.2 mm | |
| Crack width is satisfied | | | | |

**DESIGN – SHEAR**

| | | | |
|---|---|---|---|
| Shear force | V | = | 47.892 kN |
| Diameter of shear reinforcement | | = | 10 mm |
| Yield strength of shear link | $f_{yv}$ | = | 415 N/mm$^2$ |
| Number of clips per meter | | = | 8 |
| Area of shear clips | $A_{sv}$ | = | 628.32 mm$^2$ |
| $H_w/L_w$ | | = | 1.1 |

(Continued)

| | | | |
|---|---|---|---|
| Calculation for shear reinforcement | | | |
| Nominal shear stress (Due to shear force) (as per cl 32.4.2 IS 456 (2000)) | $\tau_{vw}$ | = | V/0.8td |
| | | = (47.892 x 1000)/ (0.8 x 2000 x 233) | |
| | | = | 0.13 N/mm$^2$ |
| Percentage of reinforcement | | = | 0.32 |
| Maximum shear stress of concrete (as per cl 32.4.2 IS 456 (2000)) | $\tau_{cmax}$= | $0.17\times f_{ck}$ | |
| Limit state method maximum shear stress of concrete= | | 5.1 N/mm$^2$ | |
| Hence safe | | | |
| Design shear strength of concrete ($H_w/L_w$ ≥1) (as per cl 32.4.3 IS 456 (2000)) | $\tau_{cw}$ | = | |
| | | = | 5.18 N/mm$^2$ |
| Design shear strength of concrete (as per cl 32.4.3 IS 456 (2000)) | $\tau_{cw}$ | = | $K_3\sqrt{f_{ck}}$ |
| | | = | 0.15 x $(30)^{1/2}$ |
| | | = | 0.82 N/mm$^2$ |
| | | = | 0.82 N/mm$^2$ |
| | $\tau_{cw}$ | = | 5.18 N/mm$^2$ |
| As per Table 19 IS 456 (2000) $\tau_{cmax}$ | | = | 3.7 N/mm$^2$ |
| $\tau_{c,max}$ | < | $\tau_{cw}$ | $>\tau_{vw}$ |
| Provide minimum shear reinforcement | | | |
| Spacing of stirrups (Minimum shear reinforcement) $S_v$ | | $= (0.87 \times f_y \times A_{sv})/(0.4\times b)$ | |
| (as per IS 456 (2000) Cl.26.5.1.6) | | | |
| $= (0.87 \times 415 \times 628.32)/(0.4 \times 2000)$ | | | |
| | | = | 283.57 mm |
| | $S_v$ | = | 150 mm |
| Provide 10 mm diameter bars @ 150 mm C/C (8 clips per meter) | | | |

**Results:**

It is essential to adapt infrastructure to evolving industry needs for sustainable development, ensuring continued growth and success in maritime operations. The vessels of size 200,000 DWT is placed after excavating the soil to a depth of approximately 2 m. To attain the necessary dredge depth, it is advised to position it roughly 9 m away from the face of the berth, considering a slope ratio of 1:4.5.

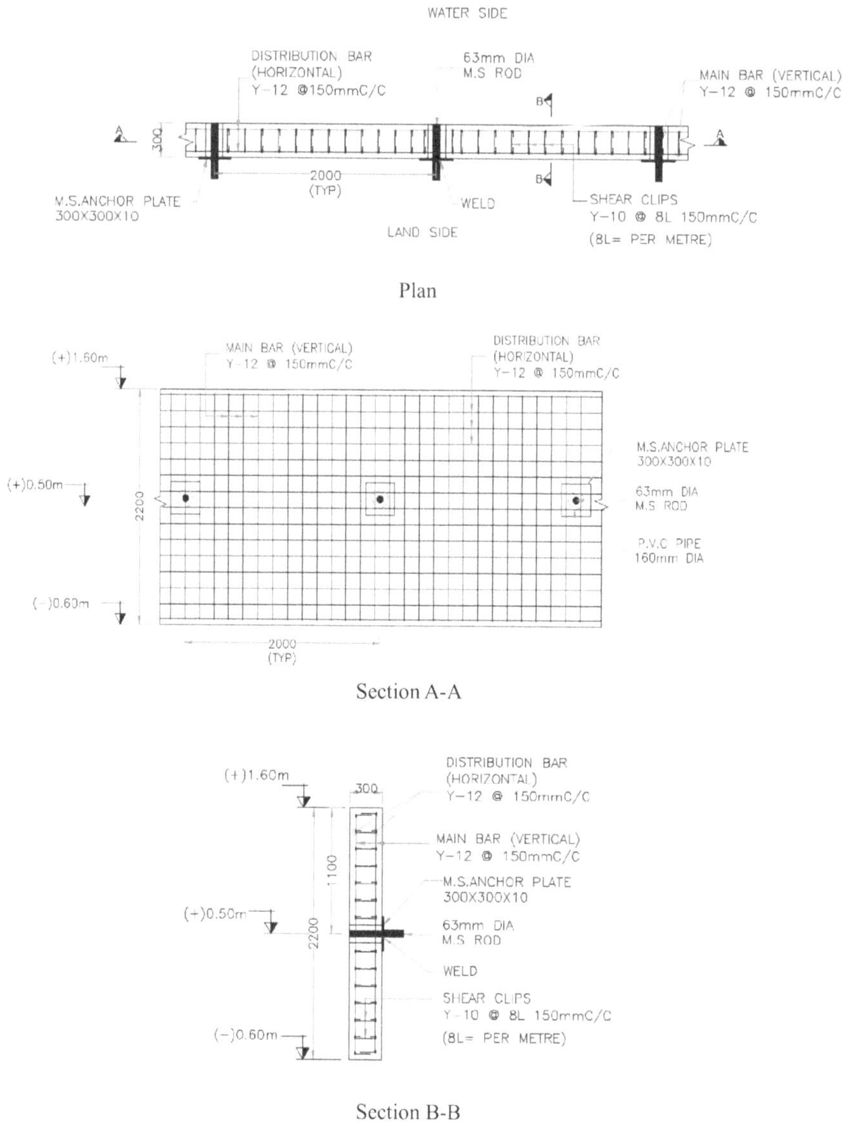


**FIGURE 5.17** Reinforcement details of dead man wall.

A pneumatic fender of size 3300 mm x 6500 mm with initial pressure of 7.1 psi is recommended; by replacing the existing fenders then only, it is possible to berth the 200,000 DWT vessel, since the existing fenders in one crib may not be in contact with the vessel during inclined berthing. Pneumatic fenders offer a linear energy-reaction relationship, ensuring higher capacity than required to minimize reaction and enhance crib safety. Pneumatic fenders are preferable, considering the likelihood of wave oscillations at this location.

# Annexure I – Case Study

## STABILITY OF IRON ORE BERTH TO HANDLE DEEP DRAFT VESSELS

### Problem Definition

The existing berth has an overall size of 270 m × 30.40 m supported on eight numbers of cribs of 18 m × 27.30 m with founding level of −20.25 m. The top level of the crib is +2.50 m, and the deck top level is +4.40 m. The cribs are resting on fills of depth 4 m consisting of (i) fine fill, (ii) coarse fill, and (iii) very coarse fill. Each crib has a fender wall on both sides from the deck top up to 0.00 m CD and is fitted with four numbers of vertical super arch fenders SA 800H/R1.

This iron ore berth is proposed to handle 200,000 DWT vessels. To handle 200,000 DWT vessels, it is recommended to increase the dredge level up to −20.0 m CD. To handle 200,000 DWT an additional mooring dolphin of size 17 m × 17 m with 1200 mm diameter piles with a 4 m thick deck slab was constructed on the east side of the ore berth. For bigger vessel, a new fender system is needed.

### Method of Analysis

Berthing loads are calculated in accordance with IS 4651 (Part 3) (2020). The fender system is designed to accommodate the design vessel of 200,000 DWT. The ultimate berthing energy is found to be 158 Tm. Based on this, the size of the pneumatic fender is finalized. The Sainflou method for wave force due to the Clapotis formation is used for wave force calculation. The wave force was 294 T. A uniform live load of 50 kN/m$^2$ is considered above the berth. The slope from −18.0 m (existing dredge level) to −20.0 m (proposed dredge level) is considered to be 1:4.5, 9 m away from the face of the berth. As per PIANC-2014 guidelines, a net under keel clearance (UKC net) of 0.5 m is sufficient. The properties of the soil layers modelled in PLAXIS 2D are given in the below table. The Mohr-Coulomb model is used for the material model.

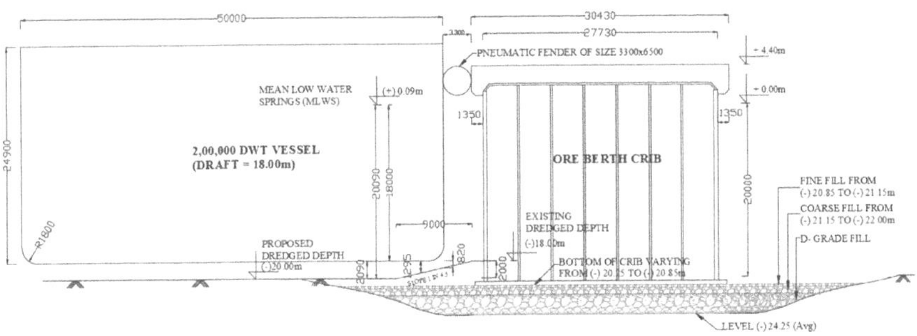


**FIGURE 5.18** Arrangement of 200,000 DWT vessel at ore berth (draft 18.00 m).

**TABLE 5.19**
**Soil Parameters**

| **Soil level** | **−18.0 to −23.2 m** | **−23.2 to −33.0 m** | **−20.85 to −21.15 m** | **−21.15 to −22.0 m** | **−22.0 to −24.25 m** | **+3.40 to −20.39 m** |
|---|---|---|---|---|---|---|
| Material name | Stiff sandy clay | Stiff sandy clay | Fine fill | Coarse fill | Very Coarse fill | Fill up soil |
| Material type | Undrained | Undrained | Drained | Drained | Drained | Drained |
| $\gamma_{sat}$ (kN/m3) | 18 | 18 | 18 | 18 | 18 | 18 |
| $\gamma_{unsat}$ (kN/m3) | 17 | 17 | 17 | 17 | 17 | 17 |
| E (kN/m2) | 10,000 | 12,160 | 120,000 | 135,000 | 150,000 | 100,000 |
| C (kN/m2) | 20 | 110 | 1 | 1 | 1 | 1 |
| Poisson's ratio | 0.25 | 0.25 | 0.3 | 0.3 | 0.3 | 0.3 |
| $\phi$ (degrees) | 10 | 10 | 30 | 30 | 30 | 30 |

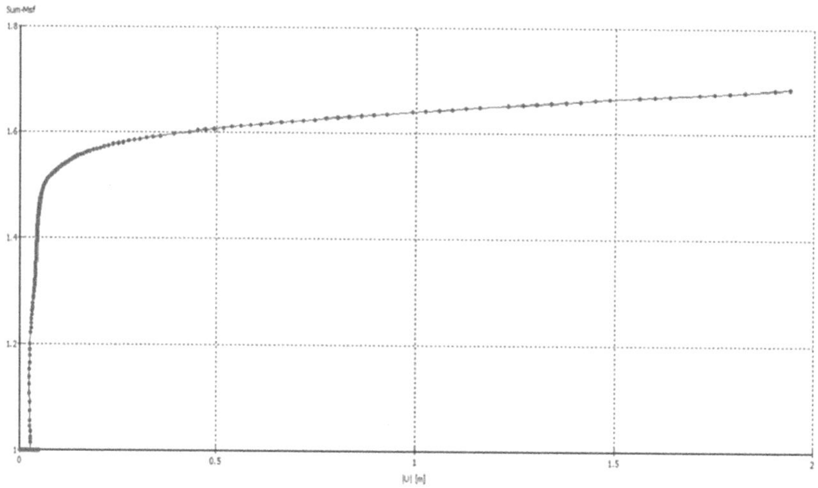


**FIGURE 5.19** Stability factor from PLAXIS (1.6846).

To determine the safe slope for the soil to be dredged, stability factor has to be checked. The stability factor for the dredge depth of –20.0 m is shown in Figure 5.19. The effective stress along the base of the berth is shown in Figure 5.20. The settlement and stress from the analysis are found to be within limits.

The allowable stability factor for a gravity structure in PLAXIS is 1.5. From the analysis, it can be seen that the stability factor is 1.684. The effective upward force is less than the self-weight of the structure. Hence it is safe. The settlement is found to be within the allowable limit. Stability against sliding and overturning for the combination of forces is found to be safe.

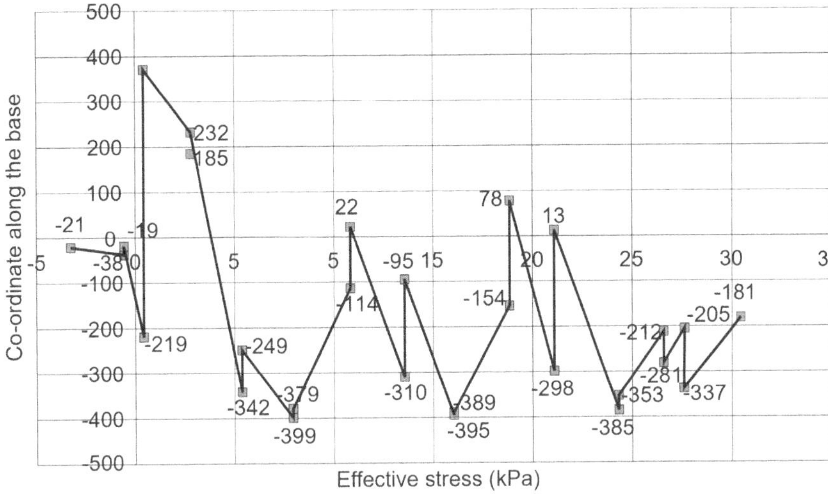


**FIGURE 5.20** Effective stress along the base.

The founding levels of the cribs are –20.25 and –20.85 m. Hence, dredging up to –20.0 m is possible only after giving a minimum clearance from the face of the berth. To find out the minimum clearance required from the face of the berth, a PLAXIS 2D analysis was carried out.

# Annexure II – Pile Foundation Capacity

**EXAMPLE:**

## 1. Pile in Sand Foundation Capacity Calculation

| | | | | | |
|---|---|---|---|---|---|
| **Loose sand – GRID D29 – –35.14** | | | | | |
| Bore Hole No. | BH | = | **MBH 04 BJ – GRID D29** | | |
| Dredge level/Bed Level | | = | –13.08 | m from CD | |
| Outer Diameter of pile | Do | = | 1200 | mm | |
| Thickness of Pile | t | = | 25.75 | mm | |
| Diameter of pile | Di | = | 1148.5 | mm | |
| Factor of Safety for Rock | FOS | = | 1.5 | | |
| **Skin Frictional Resistance** | Qsf | = | **$Qsf_1+Qsf_2+...+Qsf_n$** | | |
| **Layer – 1 (Outer skin friction)** | | | | | |
| Soil Strata Level | | = | –13.08 | to | –17.96 |
| Soil Strata type | | = | Very Loose Sand | | |
| Length of segment | L | = | 4.88 | m | |
| Skin friction capacity | **$Qsf_1$** | = | **Kpotanδ** | | |
| Density of soil | γ | = | 15.400 | $kN/m^3$ | |
| Submerged weight of soil | γ’ | = | 5.150 | $kN/m^3$ | |
| Coefficient of Earth Pressure | K | = | 1.0 | | |
| Effective Overburden Pressure | Pdi | = | $(\gamma_1 h_1)/2$ | | |
| | | = | (5.15x 4.88/2) | | |
| | | = | 12.57 | $kN/m^2$ | |
| Angle of wall friction b/w. pile & soil from API code | δ | = | 15.00 | degree | Table 6.4.3–1 |
| Surface area of pile stem | Asi | = | pDL | | |
| | | = | 3.14x1.2x4.88 | | |
| | | = | 18.39 | $m^2$ | |
| Skin friction capacity of this layer | $Qsf_1$ | = | 1x12.566xtan15x18.39 | | |
| | **$Qsf_1$** | = | **61.92** | **kN** | |
| | | = | 1x12.566xtan15 | | |
| Skin friction as per API Code | | = | **3.37** | **kPa** | |
| Limiting Skin friction from API Code for very loose sand | | = | **47.80** | **kPa** | |

(Continued)

| | | | | | |
|---|---|---|---|---|---|
| Soil Strata Level | | = | −17.96 | to | −32.69 |
| Soil Strata type | | = | Loose | | |
| Length of segment | L | = | 14.73 | m | |
| Skin friction capacity | **$Qsf_2$** | = | **Kpotanδ** | | |
| | | | | | |
| Density of soil | γ | = | 15.400 | $kN/m^3$ | |
| Submerged weight of soil | γ' | = | 5.150 | $kN/m^3$ | |
| Coefficient of Earth Pressure | K | = | 1.0 | | |
| Effective Overburden Pressure | Pdi | = | $(\gamma_1 h_1) +((\gamma 2h2)/2)$ | | |
| | | = | (5.15x4.88) +(5.15x14.73/2) | | |
| | | = | 63.06 | $kN/m^2$ | |
| Angle of wall friction b/w. pile & soil from API code | δ | = | 20.00 | degree | Table 6.4.3–1 |
| Surface area of pile stem | Asi | = | pDL | | |
| | | = | 3.14x1.2x14.73 | | |
| | | = | 55.5 | $m^2$ | |
| Skin friction capacity of this layer | $Qsf_2$ | = | 1x63.06175xtan20x55.5 | | |
| | **$Qsf_2$** | **=** | **1273.87** | **kN** | |
| Skin friction as per API Code | | **=** | **22.95** | **kPa** | |
| Limiting Skin friction from API Code for very loose sand | | **=** | **47.80** | **kPa** | |
| **Layer – 3 (Outer skin friction)** | | | | | |
| Soil Strata Level | | = | −32.69 | to | −33.08 |
| Soil Strata type | | = | Medium sand | | |
| Length of segment | L | = | 0.39 | m | |
| Skin friction capacity | **$Qsf_3$** | = | **Kpotanδ** | | |
| | | | | | |
| Density of soil | γ | = | 15.400 | $kN/m^3$ | |
| Submerged weight of soil | γ' | = | 5.150 | $kN/m^3$ | |
| Coefficient of Earth Pressure | K | = | 1.0 | | |
| Effective Overburden Pressure | Pdi | = | $(\gamma_1 h_1)+(\gamma 2h2)+((\gamma 3h3)/2)$ | | |
| | | = | (5.15x4.88)+(5.15x14.73)+(5.15x0.39/2) | | |
| | | = | 102.00 | $kN/m^2$ | |
| Angle of wall friction b/w. pile & soil from API code | δ | = | 25.00 | degree | |
| Surface area of pile stem | Asi | = | pDL | | |
| | | = | 3.14x1.2x0.39 | | |
| | | = | 1.47 | $m^2$ | |

*(Continued)*

(Continued)

| | | | | | |
|---|---|---|---|---|---|
| Skin friction capacity of this layer | $Qsf_3$ | = | 1x101.99575xtan25x1.47 | | |
| | **Qsf3** | = | **69.92** | **kN** | |
| Skin friction as per API Code | | = | **47.56** | **kPa** | |
| Limiting Skin friction from API Code for Loose sand | | = | **67.00** | **kPa** | |
| **A** | **Qsf** | | **1405.71** | **kN** | |
| **Layer – 1 (Inner Skin friction)** | | | | | |
| Soil Strata Level | | = | −13.08 | to | −17.96 |
| Soil Strata type | | = | Very Loose Sand | | |
| Length of segment | L | = | 4.88 | m | |
| Skin friction capacity | $\mathbf{Qisf_1}$ | = | **KPditanδ** | | |
| Density of soil | γ | = | 15.400 | $kN/m^3$ | |
| Submerged weight of soil | γ’ | = | 5.150 | $kN/m^3$ | |
| Coefficient of Earth Pressure | K | = | 1.000 | | |
| Effective Overburden Pressure | Pdi | = | $(\gamma_1 h_1)/2$ | | |
| | | = | (5.15x 4.88/2) | | |
| | | = | 12.57 | $kN/m^2$ | |
| Angle of wall friction b/w. pile & soil from API code | δ | = | 15.00 | degree | |
| Surface area of pile stem | Asi | = | pDL | | |
| | | = | 3.14x1.1485x4.88 | | |
| | | = | 17.6 | $m^2$ | |
| | $Qisf_1$ | = | 1x12.566xtan15x17.6 | | |
| | $\mathbf{Qisf_1}$ | = | **59.26** | **kN** | |
| **Layer – 2 (Inner Skin friction)** | | | | | |
| Soil Strata Level | | = | −17.96 | to | −32.69 |
| Soil Strata type | | = | Loose sand | | |
| Length of segment | L | = | 14.73 | m | |
| Skin friction capacity | $\mathbf{Qisf_2}$ | = | **KPditanδ** | | |
| Density of soil | γ | = | 15.400 | $kN/m^3$ | |
| Submerged weight of soil | γ’ | = | 5.150 | $kN/m^3$ | |
| Coefficient of Earth Pressure | K | = | 1.000 | | |
| Effective Overburden Pressure | Pdi | = | $(\gamma_1 h_1)+((\gamma 2h2)/2)$ | | |
| | | = | (5.15x4.88)+(5.15x14.73/2) | | |
| | | = | 63.06 | $kN/m^2$ | |
| Angle of wall friction b/w. pile & soil from API code | δ | = | 20.00 | degree | |
| Surface area of pile stem | Asi | = | pDL | | |

(Continued)

| | | | | |
|---|---|---|---|---|
| | | = | 3.14x1.1485x14.73 | | |
| | | = | 53.12 | $m^2$ | |
| | $Qisf_2$ | = | 1x63.06175xtan20x53.12 | | |
| | **$Qisf_2$** | = | **1219.24** | **kN** | |
| **Layer – 3 (Inner Skin friction)** | | | | | |
| Soil Strata Level | | = | −32.69 | to | −33.08 |
| Soil Strata type | | = | Medium sand | | |
| Length of segment | L | = | 0.39 | m | |
| Skin friction capacity | **$Qisf_3$** | = | **KPditanδ** | | |
| Density of soil | γ | = | 15.400 | $kN/m^3$ | |
| Submerged weight of soil | γ' | = | 5.150 | $kN/m^3$ | |
| Coefficient of Earth Pressure | K | = | 1.000 | | |
| Effective Overburden Pressure | Pdi | = | $(\gamma_1 h_1)+(\gamma 2h2)+((\gamma 3h3)/2)$ | | |
| | | = | (5.15x4.88)+(5.15x14.73)+(5.15x0.39/2) | | |
| | | = | 102.00 | $kN/m^2$ | |
| Angle of wall friction b/w. pile & soil from API code | δ | = | 25.00 | degree | |
| Surface area of pile stem | Asi | = | pDL | | |
| | | = | 3.14x1.1485x0.39 | | |
| | | = | 1.41 | $m^2$ | |
| | **$Qisf_3$** | = | 1x101.99575xtan25x1.41 | | |
| | **$Qisf_3$** | = | **67.06** | **kN** | |
| **Inner skin friction Resistance (Sand) B** | **Qisf** | = | **1345.56** | **kN** | |
| **Ultimate End Bearing Resistance (Dense Sand)** | | | | | |
| **Ultimate End Bearing Resistance** | **Qeb** | = | **(PoNq)** | | |
| Diameter of pile | D | = | 1.20 | m | |
| Cross-sectional area of pile | $A_p$ | = | | | |
| | | = | 1.13 | $m^2$ | |
| Effective unit weight of soil at pile tip | γ | = | 5.150 | $kN/m^3$ | |
| Effective Overburden Pressure at pile tip | $P_o$ | = | | | |
| | | = | 103.00 | m | |
| Bearing Capacity factor | $N_q$ | = | 40.00 | | |
| **Ultimate End Bearing Resistance C** | **Qeb** | = | **4659.61** | | |
| Calculated End Bearing as per API Code | | = | **4120.00** | **kPa** | |

*(Continued)*

(Continued)

| | | | | |
|---|---|---|---|---|
| Limiting End Bearing from API Code for Dense sand | | **=** | **9600.00** | **kPa** |
| **Annular End bearing D** | **Qaeb** | = | **(3.14*(1.2^2–1.1485^2)/4)*Po Nq** | |
| | | = | **391.37** | **kN** |
| Allowable Outer skin friction Resistance (SAND) | | = | Qsf/1.5 | |
| Allowable Outer skin friction Resistance (soil) | | = | **937.14** | |
| Allowable inner skin friction Resistance (Sand) | | | Qisf/1.5 | |
| Allowable inner skin friction Resistance (Sand) | | = | **897.04** | |
| Allowable End Bearing Resistance | | = | Qeb/1.5 | |
| Allowable End Bearing Resistance | | = | **3106.41** | |
| Allowable Annular End Bearing Resistance | | | Qaeb/1.5 | |
| Allowable Annular End Bearing Resistance | | = | **260.91** | |
| The ultimate load capacity of pile (A+C) | | = | 6065.32 | kN |
| The Allowable load capacity of pile (A+C)/1.5 | | = | 4043.55 | kN |
| Ultimate Outer skin friction + Inner Skin friction + Annular end bearing (A+B+D) | | = | 3142.64 | kN |
| Allowable Outer Skin friction +Allowable inner skin friction +Allowable annular end bearing | | = | 2095.09 | kN |
| Allowable capacity considering 80% efficiency for not having minimum spacing | | = | 1676.07 | kN |
| Axial load of pile from STAAD- A row pile in Berthing Jetty | | = | 1670.00 | kN |

**Allowable load capacity of pile is greater than Axial load of pile from STAAD- D row pile in Berthing Jetty**

| | | | |
|---|---|---|---|
| **1676.07** | > | **1670.00** | **kN** |

**Pile is safe for founding level –33.08 m CD**

**As per APR RP2A,**

**Limiting Skin Friction**

(Continued)

| | | | |
|---|---|---|---|
| Very loose Sand | = | 47.80 | kPa |
| Loose Sand | = | 67.00 | kPa |
| Medium Sand | = | 81.30 | kPa |
| Dense Sand | = | 95.70 | kPa |
| Very Dense Sand | = | 114.80 | kPa |
| **As per APR RP2A, limiting end bearing resistance** | | | |
| Very loose Sand | = | 1.90 | MPa |
| Loose Sand | = | 2.90 | MPa |
| Medium Sand | = | 4.80 | MPa |
| Dense Sand | = | 9.60 | MPa |
| Very Dense Sand | = | 12.00 | MPa |

## 2. Pile in Rock Foundation Capacity Calculation

| | | | | | | |
|---|---|---|---|---|---|---|
| Bore Hole No. | BH | = | **4** | | | |
| Dredge level/Bed Level | | = | −13 | m from CD | | |
| Founding level | | = | −35 | m from CD | | |
| Diameter of pile | D | = | 1300 | mm | | |
| Factor of Safety (Soil) | FOS | = | 2.5 | | | |
| Factor of Safety (Rock) | FOS | = | 3 | | | |
| **Level – 1** | | | | | | |
| Soil Strata Level | | = | −13.00 | to | −16.80 | m |
| Soil Strata type | | = | **Greyish black to Brown marine clay with shells** | | | |
| Length of segment | L | = | 3.8 | m | | |
| Skin friction capacity | **$Qsf_2$** | = | **$\alpha CusAs$** | | | |
| | N | = | **3.00** | | | |
| Density of soil | $\gamma$ | = | 17.500 | kN/m$^3$ | | |
| Submerged weight of soil | $\gamma'$ | = | 7.250 | kN/m$^3$ | | |
| Average cohesion/shear strength | Cus | = | **0.018** | MPa | | |
| | | | 18.00 | kN/m$^2$ | | |
| Reduction factor | $\alpha$ | = | **0.30** | | | |
| Surface area of pile stem | $A_s$ | | $\pi DL$ | | | |
| | | = | 15.52 | m2 | | |
| | **$Qsf_1$** | = | 83.81 | **kN** | | |
| **Level – 2** | | | | | | |
| Soil Strata Level | | = | −16.80 | to | −24.20 | m |
| Soil Strata type | | = | **Stiff brown to yellowish brown sandy clay** | | | |
| Length of segment | L | = | 7.4 | m | | |
| Skin friction capacity | **$Qsf_2$** | = | **$\alpha CusAs$** | | | |

(Continued)

(Continued)

| | | | | | | |
|---|---|---|---|---|---|---|
| | N | = | **35.00** | | | |
| Density of soil | $\gamma$ | = | 22.000 | kN/m³ | | |
| Submerged weight of soil | $\gamma'$ | = | 11.750 | kN/m³ | | |
| Average cohesion/shear strength | Cus | = | **0.210** | MPa | | |
| | | | 210.00 | kN/m² | | |
| Reduction factor | $\alpha$ | = | **0.30** | | | |
| Surface area of pile stem | $A_s$ | | $\pi DL$ | | | |
| | | = | 30.22 | m2 | | |
| | **$Qsf_2$** | = | 1903.99 | **kN** | | |
| **Level – 3** | | | | | | |
| Soil Strata Level | | = | –24.20 | to | –31.40 | m |
| Soil Strata type | | = | **Silty fine sand with clay lumps** | | | |
| Length of segment | L | = | 7.2 | m | | |
| Skin friction capacity | **$Qsf_3$** | = | **KPditanδAsi** | | | |
| | N | = | **35.00** | | | |
| Density of soil | $\gamma$ | = | 18.000 | kN/m³ | | |
| Submerged weight of soil | $\gamma'$ | = | 7.750 | kN/m³ | | |
| Coefficient of Earth Pressure | K | = | 1.100 | | | |
| Effective Overburden Pressure | Pdi | = | $(\gamma_1 h_1) + (\gamma_2 h_2)/2$ | | | |
| | | = | (7.25 x3.8) + (7.75x 7.2/2) | | | |
| | | = | 55.45 | kN/m² | | |
| Angle of internal friction | $\varphi$ | = | 32.00 | | | |
| Angle of wall friction b/w. pile & soil | $\delta$ | = | 32.00 | degree | | |
| Surface area of pile stem | Asi | = | pDL | | | |
| | | = | 3.14x1.3x7.2 | | | |
| | | = | 29.39 | m2 | | |
| | $Qsf_3$ | = | 1.1x55.45xtan32x 29.39 | | | |
| | **$Qsf_3$** | = | **1120.17** | **kN** | | |
| **Level – 4** | | | | | | |
| Soil Strata Level | | = | –31.40 | to | –33.10 | m |
| Soil Strata type | | = | **Soft Rock** | | | |
| Length of segment | L | = | 1.7 | m | | |
| Friction capacity | **$Qsf_4$** | = | **αCusAs** | | | |
| N Value | N | = | >100 | | | |
| Density of soil | $\gamma$ | = | 22.000 | kN/m³ | | |

(Continued)

| | | | | | | |
|---|---|---|---|---|---|---|
| Submerged weight of soil | γ' | = | 11.750 | $kN/m^3$ | | |
| Reduction factor | α | = | 0.9 | | | |
| Average shear strength | Cus | | 600 | $kN/m^2$ | | |
| Surface area of pile stem | As | = | 3.14x1.3x1.7 | | | |
| | | = | pDL | | | |
| | | = | 6.94 | m2 | | |
| | **$Qsf_4$** | **=** | **3749.18** | **kN** | | |
| **Level – 5** | | | | | | |
| Soil Strata Level | | = | –33.10 | to | –35.00 | m |
| Soil Strata type | | = | **Rock** | | | |
| Length of segment | L | = | 1.9 | m | | |
| Friction capacity | **$Qsf_5$** | **=** | **αCusAs** | | | |
| N Value | N | = | >100 | | | |
| Density of soil | γ | = | 28.000 | $kN/m^3$ | | |
| Submerged weight of soil | γ' | = | 17.750 | $kN/m^3$ | | |
| Reduction factor | α | = | 0.9 | | | |
| Average shear strength | Cus | | 1200 | $kN/m^2$ | | |
| Surface area of pile stem | As | = | 3.14x1.3x1.9 | | | |
| | | = | pDL | | | |
| | | = | 7.76 | m2 | | |
| | **Qsf5** | **=** | **8380.51** | **kN** | | |
| Ultimate skin friction Resistance (soil) | Qsf | = | $Qsf_1+Qsf_{2+}Qsf_3$ | | | |
| **Ultimate skin friction Resistance (soil)** | **Qsf** | **=** | **3107.97** | **kN** | | |
| Ultimate skin friction Resistance (Rock) | Qsf | = | $Qsf_4+Qsf_5$ | | | |
| **Ultimate skin friction Resistance (Rock)** | **Qsf** | **=** | **12129.690** | **kN** | | |
| **Ultimate End Bearing Resistance** | | | | | | |
| **Ultimate End Bearing Resistance** | **Qeb** | = | **CubNcAb** | | | |
| Ultimate shear strength along socket length | Cub | = | 1200.00 | $kN/m^2$ | | |
| Bearing Capacity factor | $N_c$ | = | 9.00 | | | |
| Cross-sectional area of pile | $A_b$ | = | $\frac{\pi D^2}{4}$ | | | |
| | | = | 1.33 | m2 | | |

*(Continued)*

(Continued)

| | | | | |
|---|---|---|---|---|
| **Ultimate End Bearing Resistance** | **Qeb** | = | **14335.09** | |
| Allowable skin friction Resistance (soil) | | = | Qsf/2.5 | |
| **Allowable skin friction Resistance (soil)** | | = | **1243.19** | |
| Allowable skin friction Resistance (Rock) | | = | Qsf/3 | |
| **Allowable skin friction Resistance (Rock)** | | = | **4043.23** | |
| Allowable End Bearing Resistance | | = | Qeb/3 | |
| **Allowable End Bearing Resistance** | | = | **4778.36** | |
| **The ultimate load capacity of pile** | | = | **29572.75** | **kN** |
| **The Allowable load capacity of pile** | | = | **10064.78** | **kN** |

## 3. Pile in Clay Foundation Capacity Calculation

| | | | | | | |
|---|---|---|---|---|---|---|
| Bore Hole No. | BH | **=** | **1** | | | |
| Dredge level/Bed Level | | = | –15.1 | m from CD | | |
| Founding level | | = | –33.3 | m from CD | | |
| Diameter of pile | D | = | 1400 | mm | | |
| Factor of Safety (Soil) | FOS | = | 2.5 | | | |
| Factor of Safety (Rock) | FOS | = | 3 | | | |
| **Level -1** | | | | | | |
| Soil Strata Level | | = | –15.10 | to | 16.50 | m |
| Soil Strata type | | = | **Greyish soft Clay** | | | |
| Length of segment | L | = | 1.4 | m | | |
| Friction capacity | Qf | **=** | **αCAs** | | | |
| N Value | N | = | 30 | | | |
| Average cohesion | C | = | 180 | $kN/m^2$ | | |
| | | = | 180 | $kN/m^2$ | | |
| Reduction factor | α | = | 0.25 | | | |
| Surface area of pile stem | As | = | 3.14x1.4x1.4 | | | |
| | | = | πDL | | | |
| | | = | 6.16 | $m^2$ | | |
| | **$Qsf_1$** | **=** | **277.09** | **kN** | | |
| **Level -2** | | | | | | |
| Soil Strata Level | | = | –16.50 | to | –20.70 | m |

(Continued)

| | | | | | | |
|---|---|---|---|---|---|---|
| Soil Strata type | | = | **Very stiff to hard Clay** | | | |
| Length of segment | L | = | 4.2 | m | | |
| Friction capacity | Qf | **=** | **αCAs** | | | |
| N Value | N | = | 40 | | | |
| Average cohesion | C | = | 240 | $kN/m^2$ | | |
| | | = | 240 | $kN/m^2$ | | |
| Reduction factor | α | = | 0.25 | | | |
| Surface area of pile stem | | | | | | |
| | As | = | 3.14x1.4x4.2 | | | |
| | | = | πDL | | | |
| | | = | 18.47 | $m^2$ | | |
| | **$Qsf_2$** | **=** | **1108.35** | **kN** | | |
| **Level -3** | | | | | | |
| Soil Strata Level | | = | −20.70 | to | −26.50 | m |
| Soil Strata type | | = | **Clay** | | | |
| Length of segment | L | = | 5.8 | m | | |
| Friction capacity | Qf | **=** | **αCAs** | | | |
| N Value | N | = | 43 | | | |
| Average cohesion | C | = | 258 | $kN/m^2$ | | |
| | | = | 258 | $kN/m^2$ | | |
| Reduction factor | α | = | 0.25 | | | |
| Surface area of pile stem | As | = | 3.14x1.4x5.8 | | | |
| | | = | πDL | | | |
| | | = | 25.51 | $m^2$ | | |
| | **$Qsf_3$** | **=** | **1645.38** | **kN** | | |
| **Level -4** | | | | | | |
| Soil Strata Level | | = | −26.50 | to | −31.30 | m |
| Soil Strata type | | = | **Clay** | | | |
| Length of segment | L | = | 4.8 | m | | |
| Friction capacity | Qf | **=** | **αCAs** | | | |
| N Value | N | = | 51 | | | |
| Average cohesion | C | = | 306 | $kN/m^2$ | | |
| | | = | 306 | $kN/m^2$ | | |
| Reduction factor | α | = | 0.25 | | | |
| Surface area of pile stem | As | = | 3.14x1.4x4.8 | | | |
| | | = | πDL | | | |
| | | = | 21.11 | $m^2$ | | |
| | **$Qsf_4$** | **=** | **1615.03** | **kN** | | |
| **Level -5** | | | | | | |
| Soil Strata Level | | = | −31.30 | to | −33.30 | m |
| Soil Strata type | | = | Browny silty sand | | | |
| Length of segment | L | = | 2 | m | | |

*(Continued)*

(Continued)

| | | | | | |
|---|---|---|---|---|---|
| Friction capacity | Qs | = | $KP_{di}\tan\delta A_{si}$ | | |
| Density of soil | γ | = | 20.000 | $kN/m^3$ | |
| Submerged weight of soil | γ' | = | 9.750 | $kN/m^3$ | |
| Coefficient of Earth Pressure | K | = | 1.400 | | |
| | Pdi | = | $\gamma' h_1+\gamma' h2+\gamma' h3+\gamma' h4+\gamma' h4+\gamma' h5/2$ | | |
| Effective Overburden Pressure | | = | 167.70 | $kN/m^2$ | |
| N Value | N | = | 50 | | |
| Angle of internal friction | φ | = | 38.00 | | |
| Angle of wall friction b/w. pile & soil | δ | = | 38.00 | degree | |
| Surface area of pile stem | Asi | = | πDL | | |
| | | = | 3.14x-0.0333x2 | | |
| | | = | 8.79 | $m^2$ | |
| | Qsf5 | = | 1.4x167.7xtan38x8.79 | | m |
| | **Qsf5** | = | **1612.35** | kN | |
| Ultimate skin friction | Qsf | = | $Qsf_1+Qsf_2+Qsf_3$'+ Qsf4'+Qsf5' | | |
| **Ultimate skin friction** | **Qsf** | = | **6258.20** | **kN** | |
| **Allowable skin friction** | **Qsf** | = | **2503.28** | **kN** | |
| **Ultimate End Bearing Resistance** | | | | | |
| **End Bearing Resistance** | **Qeb** | = | **AP (0.5DYNY+PDNq)** | | |
| Diameter of the pile | D | = | $\frac{\pi D^2}{4}$ | m | |
| | | | 1.4 | | |
| Cross-sectional area of pile toe | $A_p$ | = | 1.54 | $m^2$ | |
| N Value | N | = | 50 | | |
| Angle of internal friction | φ | | 38.00 | | |
| | | = | | | |
| Density of soil | γ | = | 20.000 | $kN/m^3$ | |
| Submerged weight of soil | γ' | = | 9.750 | $kN/m^3$ | |
| Bearing Capacity factor depending on angle of internal friction | $N_Y$ | = | 84.860 | | |
| Bearing Capacity factor depending on angle of internal friction | $N_q$ | = | 51.840 | | |
| Effective Overburden Pressure | $P_D$ | = | $\gamma'(h_1+h_2)$ | | |
| | | = | 177.45 | | |
| | | = | 177.45 | $kN/m^2$ | |

(Continued)

| | | | | |
|---|---|---|---|---|
| End Bearing Resistance | Qeb | = | 1.54*((1/2* 1.4*9.75*84.86)+ (177.45*51.84)) | |
| | | = | **15058** | kN |
| Allowable end bearing Resistance | | = | **6023** | kN |
| **Total Ultimate Load Capacity of Pile** | $Q_u$ | = | 21317 | |
| | | = | **21317** | **kN** |
| **Safe Load Capacity of Pile** | | = | **2503.28001919266 + 6023.356** | |
| | | = | **8526.636** | **kN** |

## REVIEW QUESTIONS

1. Describe in brief the types of berthing structure with a neat diagram.
2. Name the factors governing the selection of berthing structure.
3. Write the load combination for an approach jetty as per IS 4651 located in Visakhapatnam, where dead load and live load are acting on the structure.
4. Why must fixity depth be calculated?
5. How is the spring stiffness calculated?
6. What are methods to determine the pile capacity of a pile foundation?
7. Calculate the safe bearing capacity of the pile for the given data: Pile = 1000 mm, Dredge level = −9.5 m from CD, Founding level = −30 m from CD. F.O.S of soil = 2.5, F.O.S of Rock = 3. The borehole diagram is given in Figure 5.21.
8. What are the methods used for design of berthing structures?
9. For a berth of 90 m x 10 m with centre to centre spacing being 5 m in longitudinal direction, design the following structural components:
   a. Pile of diameter 1000 mm
   b. Cast in-situ main beam 1000 x 1500 mm
   c. Precast long beam 1000 x 1500 mm
   d. Cast in-situ pile muff 1 1800 x 1800 x 450 mm
   e. Slab 300 mm (pre-cast 175 mm + in-situ 125 mm)

   Case 1. Add live load of 25 kN/m$^2$ on deck slab

   Case 2. Further add mooring force of 250 kN and 400 kN berthing force for lateral load analysis

   The cross-section of the berth is given in Figure 5.22.
10. Design an anchor wall of width 800 mm and overall depth of 2500 mm with concrete grade of 40 N/mm$^2$ and grade of steel 500 N/mm$^2$. Take maximum moment $M_u$ = 2600 kNm, maximum shear force $V_u$ = 1150 kN, maximum moment in compression face $M_u$= 4060 KNm (limit state of collapse) and maximum moment $M_u$ = 2850 kNm (limit state of serviceability).
11. Write the general procedure for pile load tests.

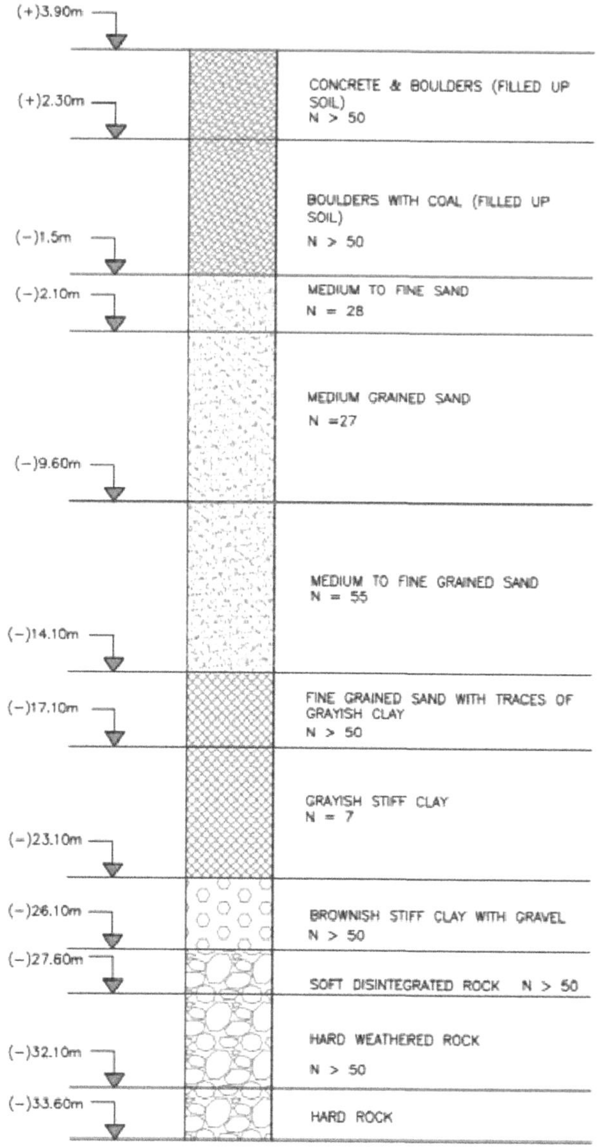


**FIGURE 5.21** Borehole data.

12. Name the types of pile load test done in piles.
13. Describe static pile load test with a neat sketch.
14. What is meant by high strain dynamic test?
15. How is the pull-out test for pile carried out?

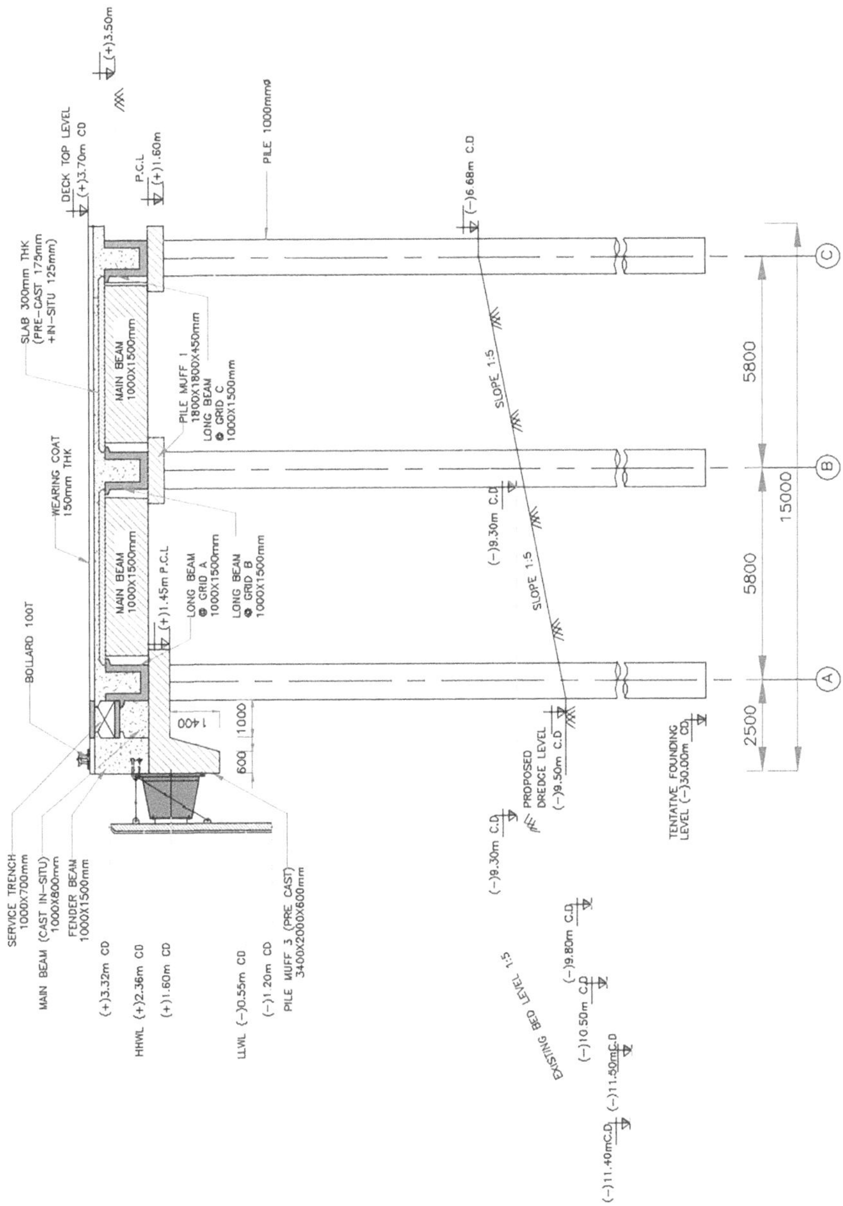


**FIGURE 5.22** Cross-section of berth.

## REFERENCES

**Joseph E. Bowles (1997)** *Foundation Analysis and Design* (5th Edition), McGraw-Hill, New York

**BS 6349–2 (2010)** *Maritime Structures – Part 2: Code of Practice for Design of Quay Walls, Jetties and Dolphins*

**Building Code Requirements for Structural Concrete (ACI 318–95) and Commentary (ACI 318R-95) (1995)** *American Concrete Institute Committee 318*

**John W. Gaythwaite (1990)** Design of Marine Facilities: Engineering for Port and Harbour Structures, Van Nostrand Reinhold, New York

**IS 456 (2000)** *Plain and Reinforced Concrete – Code of Practice* (Fourth Revision)

**IS 800 (2007)** *General Construction in Steel – Code of Practice* (Third Revision)

**IS 2911 (Part 1/Sec 2) (2010)** *Code of Practice for Design and Construction of Pile Foundations Part 1 – Concrete Piles, Section 2 – Bored Cast In-situ Concrete Piles* (First Revision)

**IS 2911 (Part 4) (2013)** *Design and Construction of Pile Foundations – Code of Practice: Part 4 Load Test on Piles* (Second Revision)

**IS 4651 (Part 3) (2020)** *Code of Practice for Planning and Design of Port and Harbours - Loading*

**IS 4651 (Part 4) (2014)** *Code of Practice for Planning and Design of Port and Harbours – General Design Considerations*

**IS 4651 (Part 5) (1980)** *Code of Practice for Planning and Design of Port and Harbours – Layout & Functional Requirements*

**Burt G. Look (2007)** Handbook of Geotechnical Investigation and Design Tables, Taylor & Francis, London

**A.V. Ranga Rao, R. Sundaravadivelu and C. Ganapathy (1999)** A knowledge based expert system for design of berthing structure, *Ocean Engineering*, Volume 26, pp. 653–676.

**SP 16 (1980)** *Design aids for Reinforced Concrete to IS 456:1978*

**M. Srinivas and R. Sundaravadivelu (1987)** *Case Study on Design of Piles for Berthing Structures*, Second National Conference on Dock and Harbour Engineering, IIT, Madras, December, pp. 191–197.

**Carl A. Thoresen (2014)** *Port Designer's Handbook* (3rd Edition), ISBN 978-0-7277-6004-3, http://dx.doi.org/10.1680/pdh.60043.001

**Unnikrishnan Pillai S. and Devadas Menon (2017)** *Reinforced Concrete Design* (3rd Edition), McGraw-Hill Education (India), New Delhi

# 6 Fenders and Bollards

## 6.1 FENDERS AND BOLLARDS

### 6.1.1 Types

The purpose of the marine fendering system is to prevent damage to the vessel and berth, during the berthing process and while the vessel is moored. As the vessel approaches a berth, it possesses kinetic energy by its displacement and motion. As the vessel contacts the berth and is brought to a stop, this kinetic energy must be dissipated. Fendering systems that are being provided are to absorb or dissipate the kinetic energy of the ship. The impact between a berthing vessel and a quay structure could cause damage both to the vessel and to the quay wall unless a fendering system is provided to absorb the impact.

Mooring devices vary progressively in size from those required for small boats to those for large bulk carriers, and include bollards, mooring rings, cleats and quick release hooks. The most important and most commonly used device is the bollard, which should be a low cast-iron post, shaped with lobes such that the ship's ropes may be securely held. Although bollards are classified according to the loads at which they will fail, the design should be such that failure takes place first in the holding down bolts at a predetermined failure load, the structure itself thus being safeguarded.

### 6.1.2 Classification of Fenders

The fenders may be made of rubber, steel, timber, rope, concrete and similar material. Rubber has come into extensive use for fender systems.

- Fenders are fixed or mounted to the berthing structures. The fixed fenders are again subdivided into
  1. Buckling fenders
  2. Non-buckling fenders
- Floating fenders between the ship and the berthing structure

The floating fenders are again subdivided into

1. Pneumatic fenders
2. Foam-filled fenders

The choice of material and the type of fender shall be judiciously made to serve the specific purpose in the case.

DOI: 10.1201/9781003361275-6

### 6.1.3 Types of Fendering Systems

The different types of fendering systems are as follows:

1. Standard pile fenders
2. Rubber fenders
3. Pneumatic fenders
4. Gravity-type fenders
5. Torque fenders

#### 6.1.3.1 Standard Pile Fenders

This system is generally used for low energy absorption. The piles are made of timber, steel, RCC and PSC are driven in front of the berthing structure to absorb the energy from the ship by direct compression and flexure. The energy capacity depends on the size, shape and length of the pile. Wooden piles do not have a long life, while the RCC piles have low energy absorption. Steel piles and PSC piles with rubber buffers are used for larger depths.

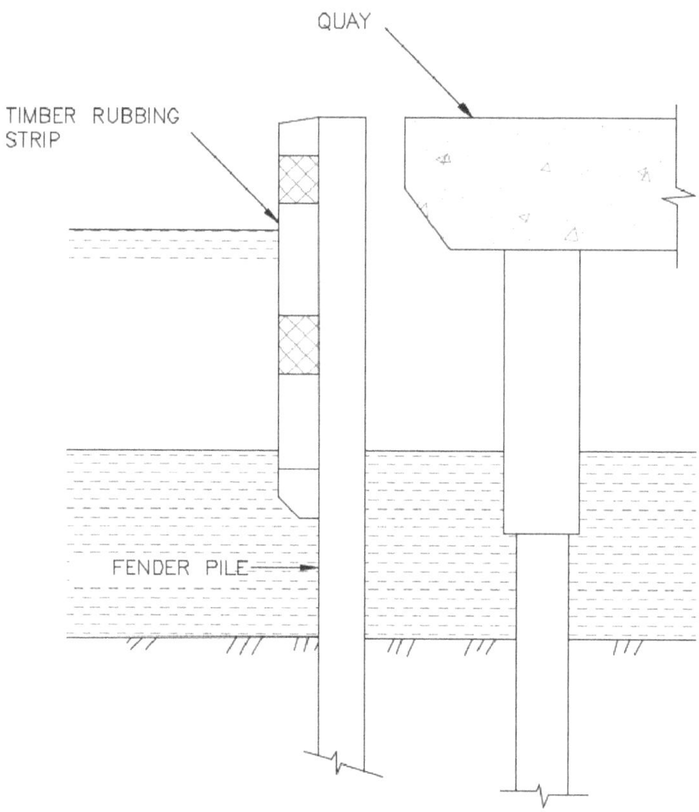


**FIGURE 6.1** Pile fenders.

#### 6.1.3.2 Rubber Fenders

The few types of rubber fenders are shear, compression and buckling fenders. The shear and compression fenders have a linear P – δ relation, i.e. for energies smaller than the rated energy; the fender will be relatively soft. The buckling fenders have a swift increase in loads in the initial stages, but as the fenders are further deflected, the loads are maintained until the rated deflection is reached. A low ratio of load over energy (P/E) is reached for buckling fenders.

#### 6.1.3.3 Pneumatic Fenders

Pneumatic fenders are originally developed for a ship-to-ship transfer, but nowadays are used for jetties and berths also because of their good performance. The

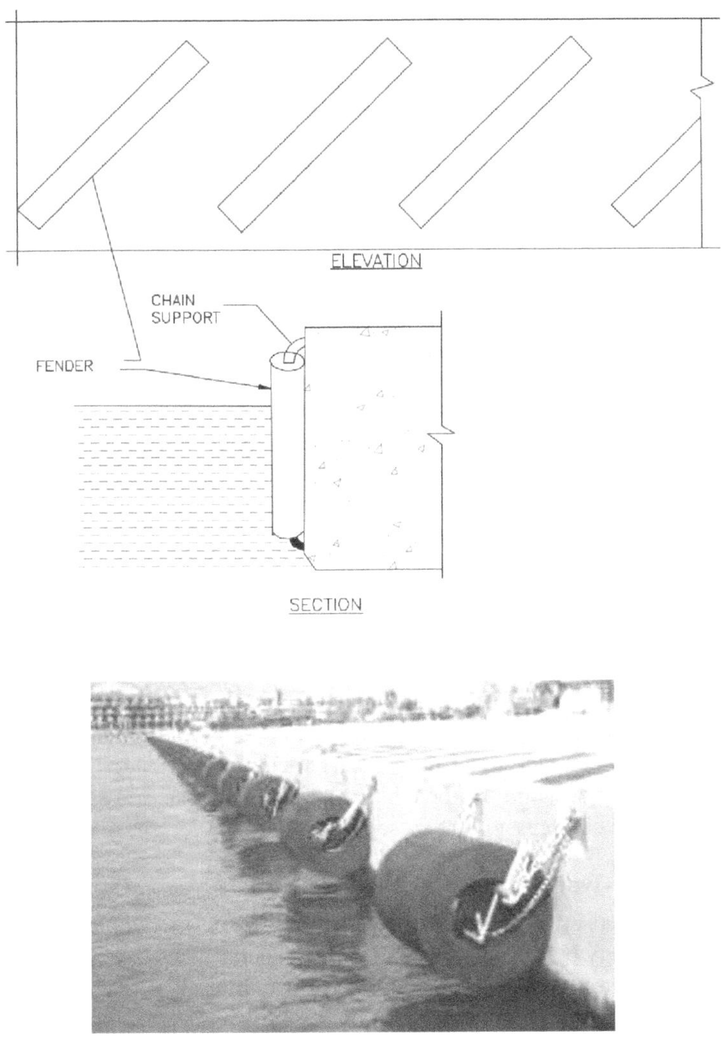


**FIGURE 6.2** Draped hollow cylindrical rubber fender.

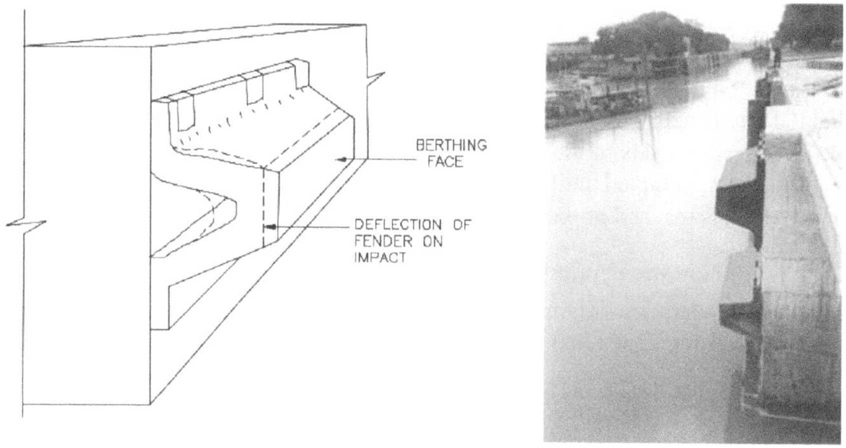


**FIGURE 6.3** Rubber arch type fender.

**FIGURE 6.4** Pneumatic fender.

pneumatic fender systems are pressurized airtight devices designed to absorb impact energy by compression of air inside a rubber envelope. Its dimensions vary from 0.5 m to 4.5 m in diameter and 1 m to 12 m in length. The fender bag is protected by a wire or chain net with tyres or rubber sleeves. The energy absorption does not decline at inclined compression for these fenders.

#### 6.1.3.4 Gravity-Type Fenders

Gravity-type fenders are designed to transform the kinetic energy of the moving ship into potential energy by raising a weight. Three general types have been developed, which act by a system of cables, a pendulum or trunnions. A typical example of the many which have been used is that in which a large concrete block is suspended below the quay deck by two pairs of cables. The front face of the block, suitably faced

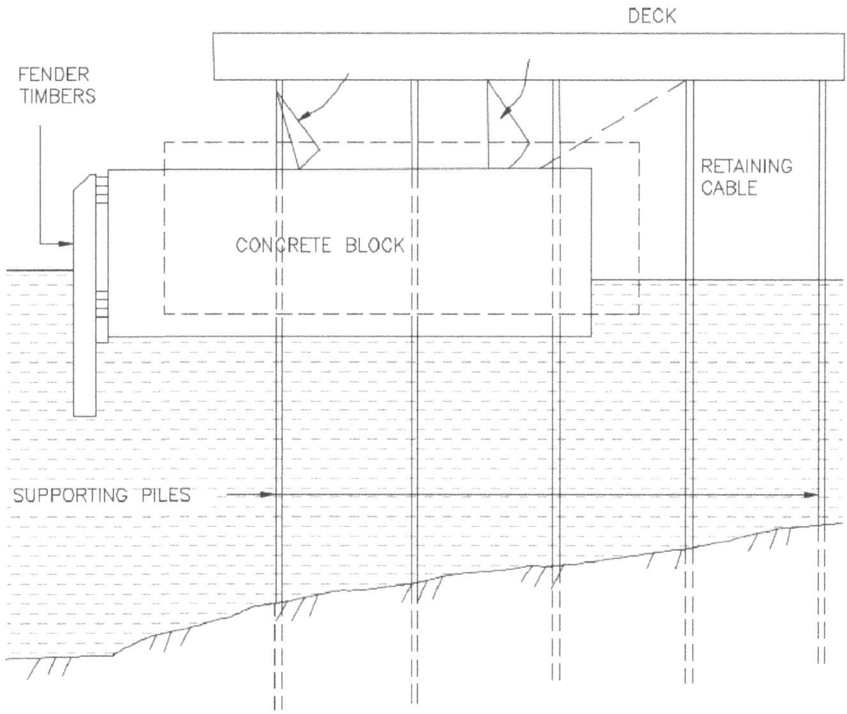


**FIGURE 6.5** Suspended gravity fender.

with a timber rubbing strip, acts as the berthing face. The impact of the ship forces the concrete block to swing backwards and upwards until the vessel has come to rest.

#### 6.1.3.5 Torque Fenders

Torque fender systems are designed to absorb the berthing energy by plastic deformation of metals in torsion. A mild steel torsion bar is used in such a manner that it is twisted by the motion of the ship against the fender.

### 6.1.4 Selection Criteria of Fendering Systems

The selection of an optimum fender for a given service depends on the following factors:

1. The type, size, draft and allowable hull pressure of a vessel
2. Berthing velocity and angle
3. Distance between the berthing point and the vessel's gravity centre measured along the face of the pier
4. Water level, tidal range, wind velocity, the direction of wind and velocity of currents
5. Behaviour and installation pitches of dock fender

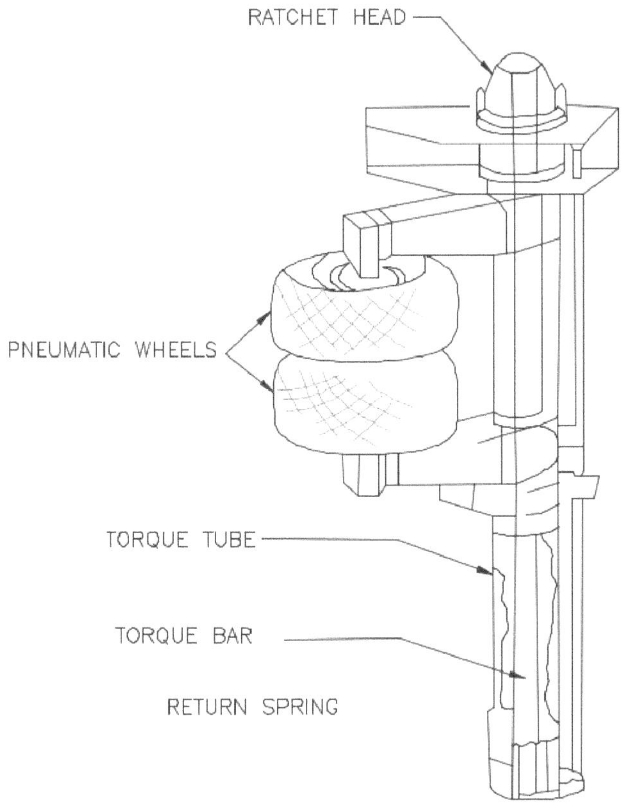


**FIGURE 6.6** Torque fender.

6. Structure and strength of berthing facilities
7. Certain human factors are involved in berthing

### 6.1.5 Fender Design Basis

The fenders should be designed in such a way that

- The berthing of the vessel to the berth facility takes place without damage
- The vessel and the berth (including the fenders) do not get damaged when the vessel is moored
- The periods of operation and safety are extended as much as possible

The design process of a fender system includes the following steps:

- Determine the functional programme of requirements
- Determine operational aspects
- Assess the design criteria

- Calculate the energy to be absorbed by fender (during berthing or when moored)
- Select a suitable fender system and type based on the energy and above criteria
- Determine the reaction force and related friction force
- Check the impact of the forces on the structure and on the vessel and the implications of the selected fender for the face of the structure

The above process may have to be repeated several times to select the most optimal fender for the specific situation. It is the task of the design engineer to choose the fender of which specified characteristics meet the design requirements.

## 6.2 TESTING OF FENDERS

The material which absorbs the berthing energy is termed as solid rubber fenders, air absorbing the berthing energy is termed pneumatic fenders and the foam core that absorbs the berthing energy is termed as foam-filled fenders. PIANC has given specific guidelines for testing fenders. These guidelines for solid-type and pneumatic fenders are discussed briefly in this chapter. The specification of standard testing practices as per international codes is also discussed.

### 6.2.1 PIANC Guidelines for Testing of Solid-Type Fenders

This section briefs the recommended procedures for testing, reporting and verifying the performance of solid-type rubber fenders.

The protocol of testing the fenders consists of three parts:

- Definition of rated performance data (RPD), which defines minimum requirements for manufacturers, published catalogue data
- Description of test apparatus and basic performance-test procedures
- Supporting protocols and other tests include the following
  a. Velocity factor (VF) determination
  b. Temperature factor (TF) determination
  c. Contact angle effect determination
  d. Durability testing
  e. Temperature stabilization
  f. Verification/quality assurance testing
  g. Rubber property testing

#### 6.2.1.1 Rated Performance Data

Fender performance under linearly decreasing or sinusoidal-decreasing deflection velocities to simulate actual vessel-berthing conditions shall be done in all testing.

Rated performance data (RPD), manufacturer's published performance curves and/or tables, shall be based on:

1. Initial deflection (berthing) velocity of 0.15 m/s and decreasing to not more than 0.005 m/s at test end

2. Testing of fully broken-in fenders
3. Testing of fenders stabilized at 23°C ± 5°C
4. Testing of fenders at a zero-degree angle of approach
5. Deflection (berthing) frequency of not less than 1 hour

Catalogues shall also include nominal performance tolerances as well as data and methodology to adjust performance curves and/or tables for application parameters different from RPD conditions. Adjustment factors shall be provided for the following variables:

1. Other initial velocities: 0.001, 0.05, 0.10, 0.20, 0.25 and 0.30 m/s deflection (approaching) velocity, decreasing to not more than 0.005 m/s at test end
2. Other temperatures: +50°C, +40°C, +30°C, +10°C, 0°C, −10°C, −20°C, −30°C
3. Other contact angles: 3, 5, 8, 10, 15 and 20 degrees

In addition, RPD shall contain a cautionary statement that published data do not necessarily apply to constant-load and cyclic-loading conditions. In such cases, designers are to contact the fender manufacturer for design assistance.

Adjustment factors for velocity and temperature shall be provided for every catalogue rubber compound or other energy-absorbing material offered by each manufacturer.

#### 6.2.1.2 Fender Testing

This topic establishes procedures to be able to standardize the performance characteristics of fenders being tested using two methods:

- The traditional and widely used constant velocity (CV) method
- Decreasing velocity (DV) method

Performance testing to establish design data may use either of the two methods.

**METHOD CV:**

Constant-slow-velocity deflection of full-size fenders with performance adjusted by velocity factors developed from scale model tests. This method is the preferred method of most manufacturers.

Test to establish initial performance data by full-size fender under the constant-slow 0.0003–0.0013 m/s (2–8 cm/min) velocity. Establish VF obtaining a scale-model test and calculate the RPD of the full-size fender.

Establish adjustment factors from the scale model test for initial berthing velocities other than 0.15 m/s.

The velocity factor shall be the ratio of performance-test results of scale models under the following conditions:

1. A constant slow strain rate similar to the strain rate of a full-size fender at its test speed

2. Decreasing-speed deflection with initial strain rate similar to that of the full-size fender RPD and other deflection conditions

### Method DV

Linearly or sinusoidally decreasing velocity deflection of full-size fenders. Test parameters shall be as defined for published RPD. RPD tests shall start at 0.15 m/s. Tests to establish adjustment factors for initial berthing velocities other than 0.15 m/s shall begin with those other initial velocities.

### 6.2.1.3 Test Apparatus

The test apparatus shall be equipped with a calibrated load measuring device such as load cell(s) or pressure transducer and linear transducer(s) for measuring displacement capable of providing continuous monitoring of fender performance.

The test apparatus shall be capable of recording and storing load-cell and transducer data at intervals of 0.01 H to 0.05 H, where H is a fender's nominal height, and storing manually entered inputs. The following information shall include, as a minimum:

1. Serial number and description of test item
2. Date, time at start and time at the end of test
3. Location of the test facility and test apparatus ID
4. Stabilization temperature of test specimen
5. Test ambient temperature
6. Graphic plot(s) and tabular printout(s) of:
   - Deflection velocity versus deflection (optional) (If not plotted, deflection velocity and its characteristics shall be separately noted). Applicable to Method DV only
   - Reaction versus deflection
   - Energy versus deflection

For fender tests, all equipment used to measure and record force and deflection shall be calibrated, and certified accurate to within ±1% in accordance with ISO or equivalent JIS or ASTM requirements. Calibration shall be performed within one year of the use of the equipment or less if the normal calibration interval is shorter than one year. Calibration of test apparatus shall be checked annually by a qualified third-party organization, using instrumentation which is traceable to a certified, national standard.

### 6.2.1.4 Test Protocol

The performance test shall deflect specimens according to either of the two methods listed below. Clear and unambiguous calculations must be provided for any adjustments made to the test results.

### Method CV

1. Break-in specimen by deflecting three or more times to its rated deflection or more, as recommended by the manufacturer.

2. Remove load from the specimen and allow it to 'recover' for at least 1 hour.
3. Before conducting the performance test, stabilize the fender temperature. Temperature stabilizing time can include time for preceding steps 1 and 2.
4. Deflect specimen once at constant slow 0.0003–0.0013 m/s (2–8 cm/min) deflection.
5. Stop test when deflection reaches rated deflection or more, as recommended by the manufacturer.
6. Adjust performance to rating temperature (23°C ± 5°C), if required, or to desired application temperature by multiplying both energy and reaction results by the temperature factor (TF).
7. Adjust performance to desired initial berthing velocity by adjusting both energy and reaction results by the velocity factor, VF s.

### Method DV

1. Break-in specimens by deflecting three or more times to its rated deflection or more, as recommended by the manufacturer.
2. Remove load from the specimen and allow it to 'recover' for at least 1 hour.
3. Before conducting the performance test, stabilize the fender temperature. Temperature stabilizing time can include time for preceding steps 1 and 2.
4. Deflect specimen once at a linearly decreasing or sinusoidally decreasing variable deflection velocity as defined in the equations below:

$$V = V_O\left(D\text{-}d\right)/D \text{ or } 0.005 \text{ m/s whichever is greater} \tag{6.1}$$

or

$$V = V_O\cos\left(\pi d/2D\right) \text{or } 0.005\text{m/s whichever is greater} \tag{6.2}$$

where

$V$ = Instantaneous deflection velocity of fender
$V_O$ = Initial deflection velocity (actual berthing condition)
$d$ = Instantaneous deflection of fender
$D$ = Rated deflection of fender

5. Stop test when deflection reaches rated deflection or more, as recommended by the manufacturer.
6. Adjust performance to rated temperature (23°C ± 5°C), if required, or to desired application temperature by multiplying both energy and reaction results by the temperature factor (TF).

## 6.3 SUPPORTING PROTOCOL

### 6.3.1 Temperature Stabilization

1. Test temperature for full-size specimens is defined as the same as the stabilization temperature if an ambient temperature at the test apparatus is

within ±15°C of the stabilization temperature and the testing is completed within 2 hours.
2. To stabilize rubber temperature, store specimen at a constant temperature ±15°C. Record air temperature of the space where the specimen is stored within 3 m of the specimen surface, either continuously or twice a day, no less than 10 hours apart.
3. Stabilization time shall be not less than 20x or 1.5 days or more as recommended by the manufacturer, rounded to the next whole day (x = dimension of greatest rubber thickness, in metres).

### 6.3.2 Velocity Factor (VF)

One of the following protocols shall be followed to determine the VF for every combination of fender configuration, initial velocity other than RPD velocity, fender height and energy-absorbing material.

Specimens for determining VF may be either full-size fenders or models, as noted below, provided they are not smaller than 0.1 m in height.

**Method CV:**

Testing of scale model maintaining strain rates as described in an earlier topic.

The model should be accurately sealed to the model proposed for sale.

1. Test model-sized fenders as per test protocol of method CV at n x 0.15 *m/s* initial velocity at 23°C ± 5°C (n: model height/actual height).
2. Repeat step 1 at other initial velocities (including the constant slow velocity).
3. Derive VFs from the data in steps 1 and 2 per the following method:
   a. Energy velocity factor and reaction velocity factor by method CV

   $VF_{ea}$ and $VF_{ra}$, shall be defined by the following equations:

$$VF_{ea} = E_{v\text{-}M} / E_{RPD\text{-}M} \tag{6.3}$$

$$VF_{ra} = R_{v\text{-}M} / R_{RPD\text{-}M} \tag{6.4}$$

where

$E_{vM}$ = Energy absorption at other initial velocities than that of as method CV

$E_{RPD}$ = Energy absorption at the initial velocity than that of as method CV at the model

$E_{vM}$ = Reaction at other initial velocities per section method CV

$R_{RPD}$ = Reaction at the RPD initial velocity per section method CV at the model

   b. Corrected energy and reaction performance is then calculated by the following formulas:

$$E_a = E_{RPD} \times VF_{ea} \tag{6.5}$$

$$R_a = R_{RPD} \times VF_{ra} \tag{6.6}$$

where

$E_a$ = Energy absorption at alternative initial velocity
$R_a$ = Reaction at alternative initial velocity
$E_{RPD}$ = Energy at RPD initial velocity
$R_{RPD}$ = Reaction at RPD initial velocity

**METHOD DV:**

Testing of full-size fender at decreasing-rate deflection velocity

1. Test full-size fender models as per test protocol of method DV at 0.15 m/s initial velocity 23°C ± 5°C.
2. Repeat step 1 at other initial velocities.
3. Derive the VFIs from the data in steps 1 and 2 per the following method:
   a. Energy velocity factor and reaction velocity factor by method DV

$$VF_{ea} = E_v/E_{RPD} \quad (6.7)$$

$$VF_{ra} = R_v/R_{RPD} \quad (6.8)$$

where

$E_v$ = Energy at other initial velocities per section method DV
$E_{RPD}$ = Energy at the RPD initial velocity per section method DV
$R_v$ = Reaction at other initial velocities per section method DV
$R_{RPD}$ = Reaction at the RPD initial velocity per section method DV

   b. Corrected energy and reaction performance is then calculated by the following formulas:

$$E_a = E_{RPD} \times VF_{ca} \quad (6.9)$$

$$R_a = R_{RPD} \times VF_{ca} \quad (6.10)$$

where

$E_a$ = Energy at an alternative initial velocity
$R_a$ = Reaction at an alternative initial velocity
$E_{RPD}$ = Energy at RPD initial velocity
$R_{RPD}$ = Reaction at RPD initial velocity

### 6.3.3 TEMPERATURE FACTOR (TF)

The following shall be completed for every rubber compound.

1. Tests of the rubber compound and fender type at each of the following temperatures:
   – 30°C, −20°C, −10°C, 0°C, +10°C, +23°C, +30°C, +40°C, +50°C
2. The preliminary temperature factor, TFp, for each rubber at each temperature shall be its shear modulus at that temperature divided by its shear modulus at 23°C.

3. Confirm the TFs by conducting standard performance tests, either method CV or method DV; stabilize specimens at the temperature. Specimens may be either full-size fenders or models not smaller than 0.1 m in height. Test specimens in test apparatus maintained at test temperature for the duration of the test.

Test specimens shall be stabilized at −20°C, 0°C, and +23°C. Divide the results at −20°C and 0°C by the result at + 23°C. If these results corroborate the shear-modulus results within +5%, the TFs shall be the TFs.

If the performance tests do not corroborate the shear-modulus data, further tests shall be conducted on specimens stabilized in temperature-controlled environments at the following temperatures: −30°C, −20°C, −10°C, 0°C, +10°C, +23°C, +30°C, +40°C and +50°C. The TFs for each of these temperatures shall then be calculated by the following formula:

$$TF_1 = R_1/R_{23} \quad (6.11)$$

where

$R_1$ = Reaction at temperature other than 23°C (highest reaction below 35% deflection)

$R_{23}$ = Reaction at 23°C (highest reaction below 35% deflection)

However, in the case of fenders whose reaction does not drop after a peak of around 30% deflection in low temperature, TF shall be based upon the ratio of maximum reactions when energy absorption specified by the fender manufacturer is achieved.

## 6.4 VERIFICATION/QUALITY ASSURANCE TESTING

### 6.4.1 Energy/Reaction Compliance Testing

Verification/quality assurance testing to determine compliance with either RPD or other customer-specified energy and reaction requirements (required performance) shall be performed in a test apparatus, as described earlier. Samples for verification testing shall be actual fender elements fabricated for the project location.

1. Test sample according to method CV or method DV, adjusting performance to required performance.
2. A fender provides required performance (required energy and reaction) within production tolerances if it meets both the following requirements simultaneously at any point during the test:
   a. Velocity and temperature-adjusted energy absorbed is equal to or greater than required multiplied by the nominal energy tolerance (low end) specified in its catalogue data.
   b. Velocity and temperature adjusted reaction are no more than required reaction multiplied by the nominal reaction tolerance (high end) specified in its catalogue data.

3. The fender samples for energy/reaction verification testing shall be selected according to a sample scheme agreed between the customer and the fender manufacturer. If a specific sampling scheme has not been noted, a minimum of 10% of the fender order shall be tested for compliance with energy/reaction requirements.

### 6.4.2 Break-in Deflection

Break-in deflection of actual elements should be at least the manufacturer's rated deflection. At least one cycle should be performed.

Break-in deflection shall be mandatory for all fenders with a catalogue reaction rating of 100 tonnes or more to be installed on monopiles or pile-supported pier structures. Break-in deflection of other fenders shall be as stipulated by the customer.

### 6.4.3 Other Testing

Production fenders may be tested for conformance with specified material properties, alternate performance requirements, and/or durability characteristics.

1. Verification/quality assurance testing of production fenders may be requested by the customer to ensure product conformance with specified contact angle performance, durability, and/or material property characteristics.
2. Other testing requirements, including selection of sampling scheme, shall be as agreed between customer and fender manufacturers.

### 6.4.4 Effect of Contact Angle

Manufacturers shall include graphs or tables defining the effect of deflecting fenders at the contact angles earlier. This data may be generated mathematically or by testing performed on either actual fender elements or on scale models or arrays. It must reflect the effect of angle contact on an entire fender assembly, not just an individual element.

The following is the procedure for defining the effect of each contact angle/configuration combination. The test shall be made on the two major axes of the fender unit:

1. Using a test apparatus as described earlier, execute the steps of the test procedure defined.
2. Determine the base-case energy rating for the zero-degree contact angle of the specimen at the deflection or reaction limit recommended by the manufacturer.
3. Allow the specimen to recover outside the test apparatus for at least 1 hour or more as recommended by the manufacturer.
4. Simulate the desired contact angle and repeat the test cycle only of step 1.
5. Determine the energy rating of the specimen in the contact angle test at the manufacturer's recommended deflection or reaction limit.

6. The contact angle factor is the energy determined in step 5 divided by that determined in step 2.

This factor is applied to energy only. No factor needs to be applied to the reaction, since the maximum reaction is as defined by the zero degree contact angle performance. For combined horizontal and vertical contact angles, multiply the contact angle factor for the horizontal direction by the contact angle factor for the vertical direction if the factors for both directions are different.

### 6.4.5 Durability

1. Each combination of fender type and energy absorbing material shall be tested for durability to ensure its suitability to withstand repeated deflections without enough recovery time to return to original performance characteristics.
2. The test specimen may not be smaller than the smallest fender of the same basic design offered for sale. The specimen L/H ratio may not be less than the lowest ratio of any catalogues model of the same basic design.
3. Before the test begins, stabilize the specimen's temperature to +23°C ± 5°C, temperature stabilization. Do not artificially cool the specimen during the test.
4. The test shall consist of 3 deflections of the specimen to its rated deflection at a maximum period of 150 seconds.
5. The criterion for successful completion of the durability test is no cracks or defects visible to the naked eye after the 3 deflections.

## 6.5 PHYSICAL PROPERTIES OF RUBBER

The properties recommended in this section are those that help assure acceptable resistance to the effects of ageing and environmental attack. The following physical properties of rubber are recommended as standard requirements:

**Resistance to heat ageing:** Test tensile strength, elongation and hardness per JIS, ASTM or ISO standards. Then place the second set of samples into an oven maintained at 70°C ± 1°C for a period of 96 hours. After removing the samples from the oven, repeat the same tensile strength, elongation and hardness tests. Compare the results before and after heat ageing. The following are the minimum requirements for satisfactory ageing resistance:

a. Tensile strength after ageing: not less than 80% of the original value.
b. Elongation after ageing: not less than 80% of the original value.
c. Hardness after aging: not more than 8 points Shore A of increase from original value.

**Resistance to ozone:** Conduct a standard ozone-exposure chamber ozone-exposure test per one of the following standards:

- JIS K 6259 40°C, 20% elongation and 50 pphm ozone level

- ASTM D 1171, ozone exposure, method A, 38°C
- ISO 1431–1, procedure A 40°C, 20% elongation and 50 pphm ozone level

After 72 hours of exposure, there shall be no visible cracking of the test strips.

### 6.5.1 PIANC Guidelines for Testing of Pneumatic Fenders

Pneumatic fenders are designed to absorb the berthing energy primarily by a compressive air contained in the hollow space. This section establishes the recommended procedures for testing, reporting and verifying the performance of pneumatic fenders.

Pneumatic fenders are made by synthetic cord reinforced rubber sheet with compressive air inside. There are two kinds of pneumatic fenders, floating types and fixed types. Depending on the particular design, pneumatic fenders may also include integral components of steel, composites, plastics or other materials. All variations of pneumatic fenders should be performance tested and reported according to this protocol.

The performance of pneumatic fenders whose initial internal pressure shall be set to be specified air pressure in operational shall be specified in terms of guaranteed energy absorption (GEA), reaction force at GEA and internal pressure, which is equal to the hull pressure at GEA deflection. Basically, the performance of pneumatic fenders is stable relating to temperature variations when the initial internal pressure is set to be the specified pressure.

The protocol for testing pneumatic fenders consists of three parts:

- Definition of rated performance data (RPD), which defines minimum requirements for manufacturer's published catalog data
- Definition of test apparatus and basic performance-test procedure
- Supporting protocols:
  a. Temperature stabilization
  b. Velocity factor (VF) determination
  c. Verification testing
  d. Contact angle effect determination
  e. Durability test
  f. Puncture resistance test
  g. Hydraulic pressure test
  h. Air leakage test
  i. Compression recovery test
  j. Physical property of rubber
  k. Dimension

#### 6.5.1.1 Rated Performance Data

All fender performance, except for ship-to-ship operation and/or spacer purpose, shall be defined under linearly decreasing or sinusoidally decreasing deflection velocities to simulate actual vessel-berthing conditions.

Rated performance data (RPD), manufacturer's published performance curves and/or tables, shall be based on the following:

1. Testing of fenders stabilized at the specified initial internal pressure and ambient temperature around 23°C
2. Initial deflection (berthing) velocity of 0.15 m/s and decreasing to no more than 0.005 m/s at test end
3. Testing of fully broken-in fenders
4. Testing of fenders at zero degree angle of approach
5. Deflection (berthing) frequency of not less than 5 minutes

In case of test to evaluate fender performances for the usages involving standoff distance factor, the following tests shall be applied:

1. Testing offenders stabilized at the specified initial internal pressure and ambient temperature around 23°C
2. Constant deflection (berthing) velocity as static during test
3. Testing of fully broken-in fenders
4. Testing of fenders at zero degree angle of approach
5. Deflection (berthing) frequency of not less than 5 minutes

Catalogues shall also include nominal performance tolerances as well as data and methodology to adjust performance curves and/or tables for application parameters different from RPD conditions. Adjustment factors shall be provided for the following variables:

1. Other initial velocities: 0.001, 0.05, 0.10, 0.20, 0.25 and 0.30 m/s deflection (berthing) velocity, decreasing to no more than 0.005 m/s at test end.
2. Other contact angles: 3, 5, 8, 10, 15 and 20 degrees. In addition, RPD shall contain a cautionary statement. If necessary, designers are to contact the fender manufacturer for design assistance.

#### 6.5.1.2 Fender Testing

Performance testing to establish design data, RPD, may use the traditional and widely used method constant velocity (CV) method and decreasing velocity method (DV). Performance testing for the vessel-to-vessel operation and/or spacing purposes may use constant method (C).

***Method CV*** – Constant-slow-velocity deflection of full-size fenders with performance adjusted by velocity factors developed from scale model tests.

This method is the preferred method of a majority of manufacturers.

Test to establish initial performance data by full-size fender under the constant-slow 0.0003–0.0013 m/s (2–8 cm/min) velocity. Establish VF obtaining scale model test and calculate the RPD of the full-size fender. Establish adjustment factors from the scale model test for initial berthing velocities other than 0.15 m/s. Velocity factor shall be the ratio of performance-test results of scale models under the following conditions:

1. Constant slow strain rate similar to the strain rate of a full-size fender at its test speed.

2. Decreasing-speed deflection with an initial strain rate similar to that of the full-size fender RPD and other deflection conditions.

***Method DV*** – Linearly or sinusoidally decreasing velocity deflection of full-size fenders.

Test parameters shall be as defined for published RPD. RPD tests shall start at 0.15 m/s. Tests to establish adjustment factors for initial berthing velocities other than 0.15 m/s shall start at those other initial velocities.

***Method C*** – Constant-velocity deflection without performance adjustment by velocity factor.

#### 6.5.1.3 Test Apparatus

The test apparatus for performance testing shall be equipped with a load cell(s) and linear transducer(s) capable of providing continuous monitoring of fender performance.

The test apparatus shall be capable of recording and storing load-cell and transducer data at intervals of 0.01–0.05 H, where H is a fender's nominal height, and storing manually entered inputs. The following information shall include, as a minimum:

1. Serial number and description of test item
2. Date, time at start and time at the end of test
3. Location of the test facility and test apparatus ID
4. Stabilization of initial internal pressure of test specimen
5. Test ambient temperature
6. Graphic plot(s) and tabular printout(s) of:
   a. Deflection velocity versus deflection (optional) (If not plotted, deflection velocity and its characteristics shall be separately noted). Applicable to method DV only.
   b. Reaction versus deflection.
   c. Energy versus deflection.

For fender tests, all equipment used to measure and record force and deflection shall be calibrated and certified accurate to within ±1%, in accordance with ISO or equivalent IIS or ASTM requirements. Calibration shall be performed within one year of the use of the equipment. Calibration of test apparatus shall be performed by a third-party organization using instrumentation which is traceable to a certified, national standard.

#### 6.5.1.4 Test Protocol

The performance test to establish design data, RPD, shall deflect specimens according to method CV or DY, and performance testing for ship-to-ship operation and/or spacer purpose shall deflect specimens according to method C. Clear and unambiguous calculations must be provided for any adjustments made to the test results.

**Method CV:**

1. Break-in specimen by deflecting one or more times to its rated deflection, or more, as recommended by the manufacturer.

2. Remove load from the specimen and allow it to 'recover' for at least 5 minutes.
3. Before conducting the performance test, stabilize the specified initial internal pressure of the fender.
4. Deflect specimen once at constant slow 0.0003–0.0013 m/s (2–8 cm/min) deflection.
5. Stop test when the deflection reaches rated deflection, or more, as recommended by the manufacturer.
6. Adjust performance to desired initial berthing velocity by adjusting both energy and reaction results by VF.

**Method DV:**

1. Break-in specimen by deflecting one or more times to its rated deflection, or more, as recommended by the manufacturer.
2. Remove load from the specimen and allow it to 'recover' for at least 5 minutes.
3. Before conducting the performance test, stabilize the specified initial internal pressure of the fender.
4. Deflect specimen once at a linearly decreasing or sinusoidally decreasing variable deflection velocity as defined in the equations below:

$$V = V_0(D\text{-}d)/D \text{ or } 0.005 \text{ m/s whichever is greater} \quad (6.12)$$

or

$$V = V_0\cos(\pi d/2D) \text{ or } 0.005 \text{ m/s whichever is greater} \quad (6.13)$$

where

$V$ = Instantaneous deflection velocity of fender
$V_0$ = Initial deflection velocity of actual berthing condition
$d$ = Instantaneous deflection of fender
$D$ = Rated deflection of fender

5. Stop test when deflection reaches rated deflection, or more, as recommended by the manufacturer.

### 6.5.1.5 Method C

1. Break-in specimen by deflecting one or more times to its rated deflection or more, as recommended by the manufacturer.
2. Remove load from the specimen and allow it to 'recover' for at least 5 minutes.
3. Before conducting the performance test, stabilize the specified initial internal pressure of the fender.
4. Deflect specimen once at constant deflection velocity (2–8 cm/min).
5. Stop test when deflection reaches rated deflection or more, as recommended by the manufacturer.

## 6.6 SUPPORTING PROTOCOL

### 6.6.1 Temperature Stabilization

The performance of pneumatic fenders is stable relating to temperature variations when the initial internal pressure is set to the specified pressure as the performance is generated from the compressive elasticity of inside air pressure. The body of a pneumatic fender is covered by the synthetic cord reinforced rubber layer, and these influences of the temperature to the performance are normally negligible.

### 6.6.2 Velocity Factor (VF)

One of the following protocols shall be followed to determine VF for every combination of fender configuration, initial velocity other than RPD velocity, fender height and energy-absorbing material. Specimens for determining VF may be either full-size fenders or models, as noted below, provided they are not smaller than 0.1 m in height.

**Method CV:**

Testing of scale model maintaining strain rates as described earlier.

The model should be accurately scaled with the same rubber compound offered for sale.

1. Test model-size fenders per Section 4.2 at n x 0.15 m/s initial velocity and the specified initial internal pressure of the fender. (n: model height I actual height)
2. Repeat step 1 at other initial velocities. (including the constant slow velocity)
3. Derive the VFs from the data in steps 1 and 2 per the following method:
   a. Energy velocity factor and reaction velocity factor by method CV, $VF_{ca}$ and $VF_{ra}$ shall be defined by the following equations:

$$VF_{ca} = E_{v\text{-}M}/E_{RPD\text{-}M} \tag{6.14}$$

$$VF_{ra} = R_{v\text{-}M}/R_{RPD\text{-}M} \tag{6.15}$$

where

$E_{v\text{-}M}$ = Energy at other initial velocity per method CV at the model
$E_{RPD\text{-}M}$ = Energy at the RPD initial velocity per method CV at the model
$R_{v\text{-}M}$ = Reaction at other initial velocity per method CV at the model
$R_{RPD\text{-}M}$ = Reaction at the RPD initial velocity per method CV at the model

   b. Corrected energy and reaction performance is then calculated by the following formulas:

$$E_a = E_{RPD} \times VF_{ca} \tag{6.16}$$

$$R_a = R_{RPD} \times VF_{ra} \tag{6.17}$$

where

$E_a$ = Energy at alternative initial velocity
$R_a$ = Reaction at alternative initial velocity
$E_{RPD}$ = Energy at RPD initial velocity
$R_{RPD}$ = Reaction at RPD initial velocity

**METHOD DV:**

Testing of full-size fender at decreasing-rate deflection velocity

1. Test full-size fenders per Section at 0.15 m/s initial velocity and the specified initial internal pressure of fender.
2. Repeat step 1 at other initial velocities.
3. Derive the VFs from the data in steps 1 and 2 per the following method:
   a. Energy velocity factor and reaction velocity factor by method A, $VF_{ea}$ and $VF_{ra}$ shall be defined by the following equations:

$$VF_{ea} = E_V/E_{RPD} \quad (6.18)$$

$$VF_{ra} = R_V/R_{RPD} \quad (6.19)$$

where

$E_v$ = Energy at other initial velocity per method DV
$E_{RPD}$ = Energy at the RPD initial velocity per method DV.
$R_v$ = Reaction at other initial velocity per method DV.
$R_{RPD}$ = Reaction at the RPD initial velocity per method CV

   b. Corrected energy and reaction performance are then calculated by the following formulas:

$$E_a = E_{RPD} \times VF_{ea} \quad (6.20)$$

$$R_a = R_{RPD} \times VF_{ra} \quad (6.21)$$

where

$E_a$ = Energy at alternative initial velocity
$R_a$ = Reaction at alternative initial velocity
$E_{RPD}$ = Energy at RPD initial velocity
$R_{RPD}$ = Reaction at RPD initial velocity

## 6.7 VERIFICATION/QUALITY ASSURANCE TESTING

### 6.7.1 ENERGY/REACTION COMPLIANCE TESTING

Verification testing to determine compliance with either RPD or other, customer-specified energy and reaction requirements (required performance) shall be performed in a test apparatus. Samples for verification testing shall be actual fender or the miniature size fender at bigger than 1/5 (one-fifth) scale size of the actual height (or diameter). Test sample according to method CV or method DV, adjusting performance to required performance as specified in earlier sections.

### 6.7.2 Break-in Deflection

Break-in deflection of actual fender, which is mandatory for all solid type fenders, is not required for pneumatic fender, as difference of the first-break-in reaction force and the second-broken-in reaction force is negligible.

### 6.7.3 Other Testing

Production fenders may be tested for conformance with specified material properties, alternate performance requirements, and/or durability characteristics.

1. Verification/quality assurance testing of production fenders may be requested by the customer to insure product conformance with specified contact angle performance, durability, and/or material property characteristics.
2. Other testing requirements, including the selection of sampling scheme, shall be as agreed between customer and fender manufacturers.

## 6.8 OTHER TESTS

### 6.8.1 Effect of Contact Angle

Manufacturers shall include graphs or tables defining the effect of deflecting fenders at the contact angles listed in earlier sections. This data may be generated mathematically or by testing performed on either actual fender elements or on scale models or arrays. It must reflect the effect of angle contact on an entire fender assembly, not just an individual element.

The following is the procedure for defining the effect of each contact angle/configuration combination.

The test shall be made on the two major axes of the fender unit:

1. Using a test apparatus, execute the steps of the test procedure defined in the earlier section.
2. Determine the base-case energy rating for the zero-degree contact angle of the specimen at the deflection or reaction limit recommended by the manufacturer.
3. Allow the specimen to recover outside the test apparatus for at least 5 minutes.
4. Simulate the desired contact angle and repeat the test cycle only of step 1.
5. Determine the energy rating of the specimen in the contact angle test at the manufacturer's recommended deflection or reaction limit.
6. The contact-angle factor is the energy determined divided by that determined in step 2.

This factor is applied to energy only. No factor needs to be applied to the reaction, since the maximum reaction is as defined by the zero-degree contact-angle performance. For combined horizontal and vertical contact angles, multiply the contact-angle factor for the horizontal direction by the contact-angle factor for the vertical direction if the factors for both directions are different.

### 6.8.2 Durability

1. Each combination of fender type and energy-absorbing material shall be tested for durability to ensure its suitability to withstand repeated deflections without enough recovery time to return to original performance characteristics.
2. The test specimen may not be smaller than 0.2 m in height (or diameter) of the same basic design offered for sale. The specimen L/H ratio may not be less than the lowest ratio of any catalogue model of the same basic design.
3. Before the test begins, stabilize the specialized initial internal pressure of the fender and ambient temperature around 23°C. Do not artificially cool the specimen during the test.
4. The test shall consist of 3 deflections of the specimen to its rated deflection at a maximum period of 150 seconds.
5. The criterion for successful completion of the durability test is no cracks or defects visible to the naked eye after the 3 deflections.

### 6.8.3 Puncture Resistance Test

1. A puncture resistance test shall be performed to confirm that the products have puncture resistance strength.
2. The test shall be conducted in accordance with the standard 'ISO 12236 Geotextiles and geotextile related products Static puncture test (CBR test)'.
3. The specimen of puncture test shall be made as same as materials, construction and production method except the number of plies of reinforcement cord layer which shall be the smallest number, i.e. normally two plies for the smallest size fender.
4. The force applied to break through the specimen based on the specification shall be bigger than 15 kN (3.4 kips).

### 6.8.4 Hydraulic Pressure Test

1. Hydraulic pressure test shall be performed to confirm the endurable pressure of pneumatic fenders.
2. The internal pressure of the test shall be set as a pressure at the specified energy absorption in non-compression situation.
3. The test shall be conducted for 10 minutes, and there shall be no leakage of water and no harmful defects during the test.

### 6.8.5 Air Leakage Test

Air leakage tests shall be performed for all pneumatic fenders.

The air leakage test shall be conducted with the initial internal pressure for more than 30 minutes, and the test results shall confirm that there is no air leakage of the initial internal pressure.

### 6.8.6 Compression Recovery Test

1. The fenders are compressed and released repeatedly at the very short period during ship mooring. Therefore, a compression recovery test shall be performed to confirm that the fenders have sufficient compression recoverability.
2. After compression of the fender to guaranteed energy absorption deflection, the fender shall be folded at the deflected condition for 1 minute, then release the load instantaneously. The fender diameter (or height) shall be recovered more than 97% of the original diameter within 5 minutes after the load to the fender released.
3. The test may be performed on a miniature size fender if the fender is too large to be mounted on the testing machine. The reduction scale shall be bigger than 1/30 (one-thirtieth).

### 6.8.7 Physical Property of Rubber

Pneumatic fender shall satisfy requirements of the physical property of rubber, which ensure the durability of the products for the long term.

Resistance to heat ageing: Test tensile strength, elongation and hardness per JIS, ASTM or ISO standards. Then place the second set of samples into an oven maintained at 70°C ± 1°C for a period of 96 hours. After removing the samples from the oven, repeat the same tensile strength, elongation and hardness tests. Compare the results before and after heat ageing. The following are the minimum requirements for satisfactory resistance:

1. Tensile strength after aging: Not less than 80% of original value.
2. Elongation after ageing: Not less than 80% of the original value.
3. Hardness after ageing: Not more than 8 points Shore A of increase from original value.

Resistance to ozone: Conduct a standard, ozone-exposure chamber ozone-exposure test per one of the following standards;

- JIS K 6259 40C, 20% elongation and 50 pphm ozone level
- ASTM D 1171, ozone exposure, method A, 38C
- ISO 1431–1, procedure A 40C, 20% elongation and 50 pphm ozone level

After 72 hours of exposure, there shall be no visible cracking of the test strips.

### 6.8.8 Testing Supplier's Products

1. Before production:
   - The vendor is required to provide a tensile slab measuring 150 mm L × 150 mm W × 2 mm T.
   - The random sample for testing at one or more independent third-party laboratories to determine compound complies with the specification.
   - Testing will include TGA and FTIR testing to ensure compliance with the above rubber compound composition specification.
2. After production:
   - The vendor will be required to submit two samples (approximately 50 g each collected in the presence of the buyer or a third party appointed by the buyer) from the final product. The fender will be selected at randomly for testing.
   - The sample does not necessarily have to be in one piece; it can be thin pieces sliced or scraped by a sharp knife from the fender body without damaging the fender. If the fender sample collection is damaged, it should be repaired before the product is dispatched.
3. Updating quality control practices:

In most cases, only physical properties are tested in marine fenders. This testing is conducted at the manufacturer's laboratory before or after production. Test certificates are produced based on these test results. To guarantee physical property test results, it is essential that a chemical composition test takes place both before and after final production. Each specimen shall be tested for at least for 10 cycles of loading to rated deflection and performance characteristics (reaction deflection and energy deflection curves) at the end of every 5 cycles shall be furnished. The fender shall regain its original shape after 10 cycles of testing and shall not show any sign of damage even after 10 cycles. However, 2% tolerances will be allowed for the height of the fender, i.e., 1800 H certificate to this effect shall be obtained from the inspecting agency.

Table 6.1 details the physical property standards that must be met for the products to be considered of a high enough standard to meet fender performance and long-life cycle.

Grade 0 = no cracks (pass).
Grade 1 = 10 or fewer pinpricks <0.5 mm long (pass).
Grades 2–10 = increasing crack size (fail).

## 6.9 FENDER SPACING AND INSTALLATION REQUIREMENTS

The spacing of the fenders shall depend upon the location of the berths, the type of berthing structure, conditions of berthing, etc. The longitudinal spacing for a continuous quay should not normally exceed 0.15 $L$, where $L$ is the length of the smallest ship. For island berths, the spacing of primary fenders should be in the range of 0.25 $L$ to 0.4 $L$, where $L$ is the length of the smallest ship. If the design vessel range

**TABLE 6.1**
**Specification for Standard Testing Practices**

| Property | Testing Standard | Condition | Requirement |
|---|---|---|---|
| Tensile strength | DIN 53504; ASTM D 412 Die C; AS 1180.2; BS ISO 37; JIS K 6251 | Original<br>Aged for 96 hours at 70°C | 16.0 MPa (min)<br>12.8 MPa (min) |
| Elongation at break | DIN 53504; ASTM D 412 Die C; AS 1180.2; BS ISO 37; JIS K 6251 | Original<br>Aged for 96 hours at 70°C | 350% (min)<br>280% (min) |
| Hardness | DIN 53505; ASTM D 2240; AS1683.15.2; JIS K 6253 | Original<br>Aged for 96 hours at 70°C | 78° Shore A (max)<br>Original +8° Shore A (max) |
| Compression set | ASTM D 395 Method B; AS 1683.13 Method B; BS903 A6; ISO 815; JIS K 6262 | 22 hours at 70°C | 30% (max) |
| Tear resistance | ASTM D 624 Die B; AS 1683.12; BS ISO 34- 1; JIS K 6252 | Original | 70 kN/m (min) |
| Ozone resistance | DIN 53509; ASTM D 1149; AS 1683-24; BS ISO 1431-1; JIS K 6259 | 50 pphm at 20% strain, 40°C, 100 hours | No cracks |
| Sea water resistance | BS ISO 1817; ASTM D 471 | 28 days at 95°C | Hardness: ±10° Shore A (max)<br>Volume: +10/−5% (max) |
| Abrasion | BS903 A9, Method B | 3000 revolutions | 1.5 cc (max) |
| Bond strength | ASTM D429, Method B; BS 903.A21 Section 21.1 | Rubber to steel | 7N/mm (min) |

is large, an outer set of primary fenders will become necessary. The vertical and horizontal disposition of fenders shall be so designed as to prevent the ship's hull and/or berthing structure from being damaged under all tidal conditions. Ship's bow radius and approach, and flair angle and in specific cases, may extend from the cope level to the low water level (LWL).

### 6.9.1 Methodology of Installation Fenders

The fender column and beam shall be designed for the required vessel size. If needed, their size shall be increased by providing additional reinforcements by inserting the new rebar. Then additional required rebar is inserted through the cores, and the gap around the rebar is injected with a chemical compound like Lokfix, Hilti etc., making the rebar intact and the fender is fixed.

## 6.10 FIXING

Fixing of fender is illustrated in Table 6.2.

In the case of installing in existing structures, Existing structures may require retrofit anchors. These are threaded studs which are bonded into drilled holes using high-strength resin grout. Always refer to the drawings to confirm details of hole

**TABLE 6.2**
**Fixing of Cone Fender**

| Figure | Description |
|---|---|
|  | Accurately locate the anchors on the structure according to the design after checking the template against the fender bolt pattern, brackets, and other bolted parts. |
|  | Before installing the main fender system, it is advisable to install auxiliary items, such as brackets. The chain can be connected to the panel or the structure first. Clean the socket and check all threads before offering up the cone fender system. |

*(Continued)*

**TABLE 6.2 (*Continued*)**
**Fixing of Cone Fender**

| Figure | Description |
|---|---|
|  | Prepare a large enough work area to prevent the fender away from any cutting, grinding or shot peening. Place the fender face down to support it on a suitable load to protect the PE pad and paint. |
| 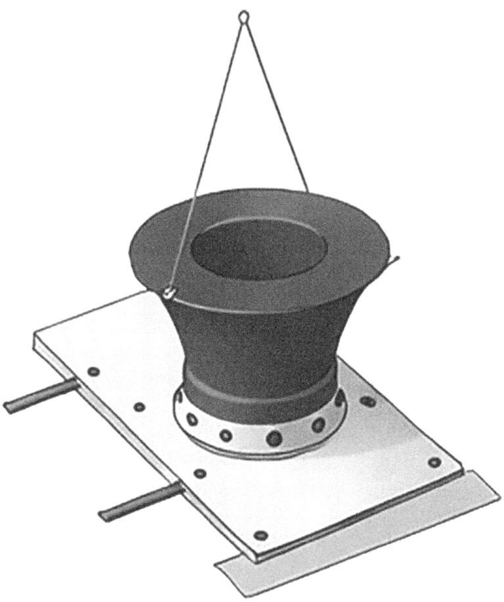 | Raise the conical fender to the proper position with a soft sling or hook, use a special washer to pass all the bolts through the fender flange, and be careful not to damage the cone rubber. |

**TABLE 6.2 (*Continued*)**
**Fixing of Cone Fender**

| Figure | Description |
|---|---|
| 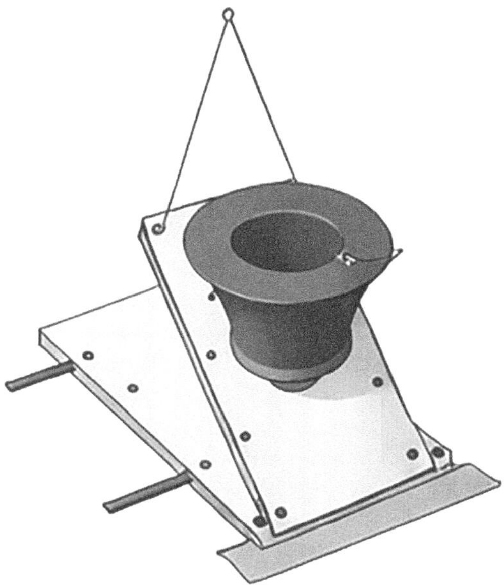 | Attach the appropriate handcuffs to the lifting point on the panel and connect the proper size of the chain sling. The component weight is shown on the drawing or available from www.chinarubberfender.com. |
|  | It is safe to ensure that the promotion area is clear and the promotion is starting; commence the lift and rotate the panel until it is vertical. |

*(Continued)*

**TABLE 6.2 (*Continued*)**
**Fixing of Cone Fender**

| Figure | Description |
|---|---|
| 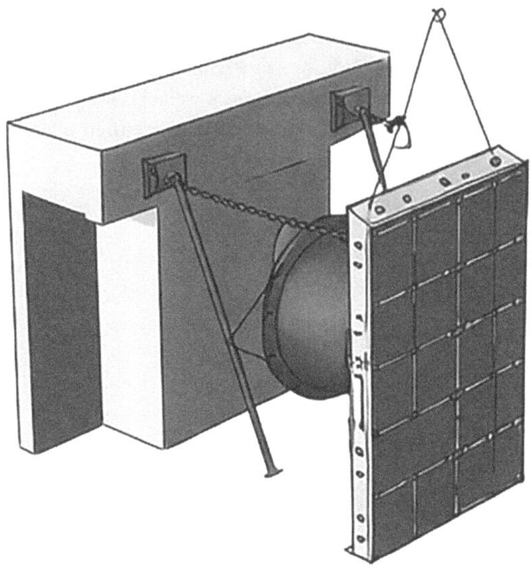 | Make sure that all fixed points are accessible, especially where there are plenty of tides. Use a label line to help guide the fender in place to avoid damage to rubber and paint. |
| 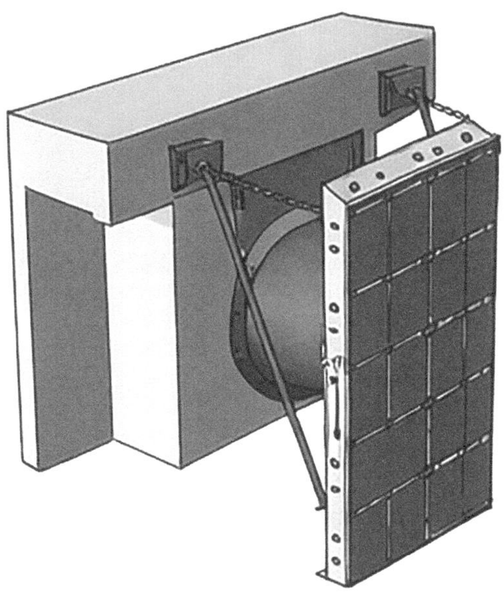 | Align the bolt holes and loosely assemble the bolts (or nuts). Tighten fixings equally, working diametrically until the washer embeds 2–3 mm into the rubber. It is recommended that the crane should support the fender system until the chain is connected and the tension is properly adjusted. |

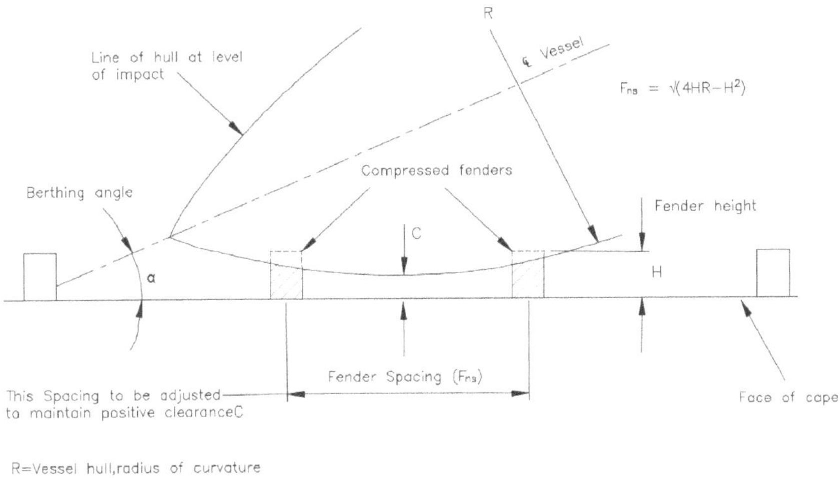


**FIGURE 6.7** Fender spacing.

depth and diameter and for the grout volume required per hole. Manual, pneumatic, electric and battery-powered cartridge guns are available depending on the size of the job and grout volume per hole.

Grouts will not cure if the temperature is too low, or they will cure too fast at high temperatures. The temperature of the concrete is also critical. Not all grouts are suitable for installation in wet or damp locations.

## 6.11 PIANC AND INDIAN CODE COMPARISON

### 6.11.1 Fender Spacing as per PIANC Guidelines

As per PIANC guidelines, the fender spacing should be assessed considering the vessel characteristics and berthing procedure but should not normally be greater than 0.25 $L$, where $L$ is the length of the smallest vessel using the berth. Where a continuous quay is provided, the spacing will usually be closer than this, with similar requirements for checking bow radius, flare and quay overhang as for other vessels.

### 6.11.2 Fender Spacing as per IS 4651 (Part 4) (2014)

The spacing of the fenders shall depend upon the location of the berths, the type of berthing structure, conditions of berthing, etc. The longitudinal spacing for a continuous quay should not normally exceed 0.15 $L$, where $L$ is the length of the smallest ship. For island berths spacing of primary fenders should be in the range of 0.25 $L$ to 0.4 $L$, where $L$ is the length of the smallest ship. If the range of design vessels is large liner, an outer set of primary fenders will become necessary. The vertical as well as the horizontal disposition of fenders shall be so designed as to prevent the ship's hull and/or berthing structure being damaged under all tidal conditions. Ship's bow radius and approach, and flair angle and in specific cases, may extend from the cope level to the low water level.

The design process and the specifications to be carried out is same in both Indian standard code and PIANC guidelines.

## 6.12 CASE STUDY – TESTING OF FENDER

### 6.12.1 Solid Type Fender

The description of cone fender for testing is as follows:

**TABLE 6.3**
**Details of Testing the Cone Fender**

| Item Description | Cone Fendering System |
|---|---|
| Size | 1800H |
| Berthing velocity | 0.15 m/s for 120,000 DWT Vessels<br>0.20 m/s for 20,000 DWT and 80,000 DWT vessels |
| Fender panel size | 5000 × 2500 × 300 mm |
| No. of fenders | 27 nos |
| No. of fenders to be tested | 10% (3 nos) |
| **Load Deflection Test** | |
| Deflection | 72% |
| Reaction force | 221.9 ton (performance tolerance ±10%) |
| Energy absorption | 238.7 ton.mtr (performance tolerance ±10%) |

#### 6.12.1.1 Visual Inspection Results

Visual and dimensional inspection has been done randomly on 10% offer quantity of fixtures, rubber and frontal (see Figures 6.8, 6.9 and 6.10). The dimensions are within limits with tolerance of +4% and −2%. There are no surface defects and damages at the time of inspection.

#### 6.12.1.2 Air Leakage Test Results

Air leakage test at 0.5 kg/cm$^2$ pressure has been done randomly on 10% offer quantity of frontal frames, and no leakages are found. The images of the air leakage test are shown in Figure 6.11.

#### 6.12.1.3 Load Deflection Test Results

Load deflection test of the cone fender has been done as per PIANC guidelines described in 6.2.1. The test results for deflection versus energy absorption and reaction force are shown in Table 6.4 and Figure 6.12. The images of load deflection test are shown Figure 6.13.

The required reaction force and energy absorption as mentioned in the report submitted by the manufacturer versus observed test results are given in Table 6.5.

The tolerance within ±10% is observed as mentioned in the report submitted by the manufacturers. Hence the results are satisfactory.

**FIGURE 6.8** Dimension inspection of cone fender.

**FIGURE 6.9** Dimension inspection of frontal frame – cone fender.

**FIGURE 6.10** Visual inspection of fixtures of the cone fender.

**FIGURE 6.11** Air leakage test of the frontal frame at 0.5 $kg/cm^2$ pressure.

**TABLE 6.4**
**Load Deflection Test Results of Cone Fender**

| Deflection in % | Reaction/Load in Tons (corrected to rated temperature and velocity) (23 degrees and 0.15 m/sec) | Energy Absorption in Ton-Mtr (corrected to rated temperature and velocity) (23 degrees and 0.15 m/sec) |
|---|---|---|
| 0 | 0.00 | 0.00 |
| 5 | 96.47 | 4.48 |
| 10 | 144.27 | 15.67 |
| 15 | 177.22 | 30.60 |
| 20 | 192.39 | 47.78 |
| 25 | 200.14 | 66.01 |
| 30 | 204.21 | 84.80 |
| 35 | 205.64 | 103.84 |
| 40 | 202.22 | 122.79 |
| 45 | 193.18 | 141.16 |
| 50 | 182.82 | 158.63 |
| 55 | 173.44 | 175.18 |
| 60 | 174.77 | 191.36 |
| 65 | 185.00 | 208.12 |
| 70 | 206.74 | 226.35 |
| 72 | 215.45 | 234.20 |

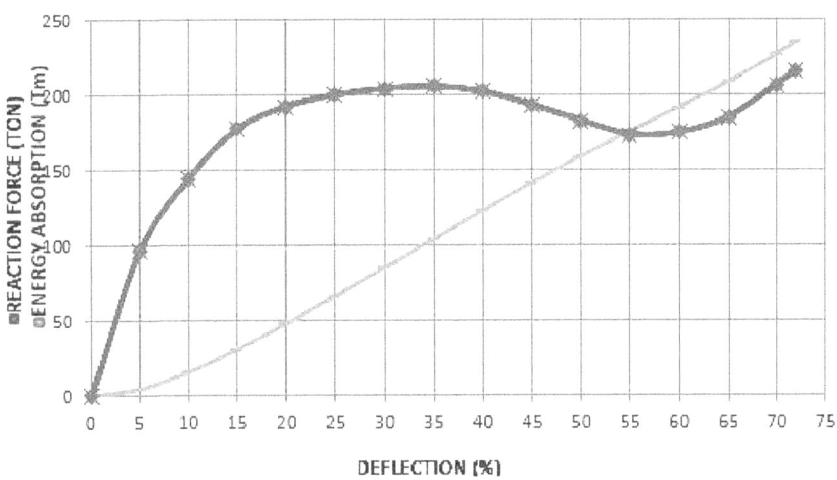


**FIGURE 6.12** Load deflection test results of cone fender (graph).

FIGURE 6.13 Load deflection test images of cone fender.

**TABLE 6.5**
**Comparison of Load Deflection Test Results of Cone Fender**

| Deflection | 72% | 72% |
|---|---|---|
| Reaction force | 221.9 ton | 215.45 ton |
| Energy absorption | 238.7 Tm | 234.20 Tm |

## 6.12.2 Floating Type Fender

The description of pneumatic fender for testing is as follows:

### Air leakage test results

Air leakage test at 0.5 $kg/cm^2$ pressure for 30 minutes pressure holding has been done. No leakage and no pressure drop are found.

### Load deflection test results

Load deflection test of the pneumatic fender has been done as per PIANC guidelines described in section 6.2.2. The test results for deflection versus energy absorption and reaction force are shown in Table 6.7 and Figure 6.14. The images of load deflection test are shown in Figure 6.15.

The required reaction force and energy absorption as mentioned in the report submitted by the manufacturer versus observed test results are given in Table 6.8.

## TABLE 6.6
## Details of Testing the Pneumatic Fender

| Item Description | Pneumatic Fender |
|---|---|
| Size | ø3.3 × 6.5 m long pneumatic fender |
| Berthing velocity | 0.2 m/s for 50,000 DWT vessels |
| No. of fenders | 5 nos |
| No. of fenders to be tested | 5 nos |
| **Load Deflection Test** | |
| Deflection | 60% |
| Reaction force | 190.14 ton (performance tolerance ± 10%) |
| Energy absorption | 310.86 ton.mtr (performance tolerance ± 10%) |

## TABLE 6.7
## Load Deflection Test Results of Pneumatic Fender

| Sl. No. | Fender Diameter (mm) | Initial Internal Pressure (kg/cm²) | Deflection (%) | Load Tester Pressure (kg/cm²) | Reaction Force at Deflection (Tons) | Energy Absorption (Tm) |
|---|---|---|---|---|---|---|
| 1 | 3300.00 | 0.50 | 0% | 0.00 | 0.00 | 0.00 |
| 2 | 2970.00 | 0.60 | 10% | 6.00 | 18.84 | 3.11 |
| 3 | 2640.00 | 0.65 | 20% | 13.00 | 40.82 | 12.95 |
| 4 | 2310.00 | 0.75 | 30% | 21.00 | 65.94 | 30.57 |
| 5 | 1980.00 | 1.00 | 40% | 35.00 | 109.90 | 59.58 |
| 6 | 1650.00 | 1.35 | 50% | 59.00 | 185.26 | 108.28 |
| 7 | 1320.00 | 1.46 | 60% | 99.00 | 310.86 | 190.14 |

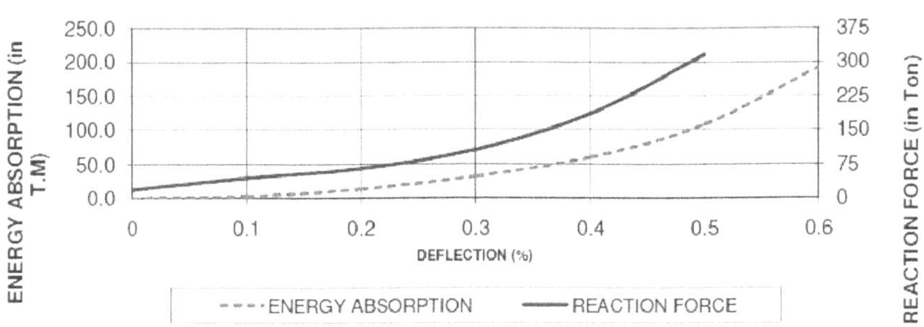


FIGURE 6.14 Load deflection test results of pneumatic fender (graph)

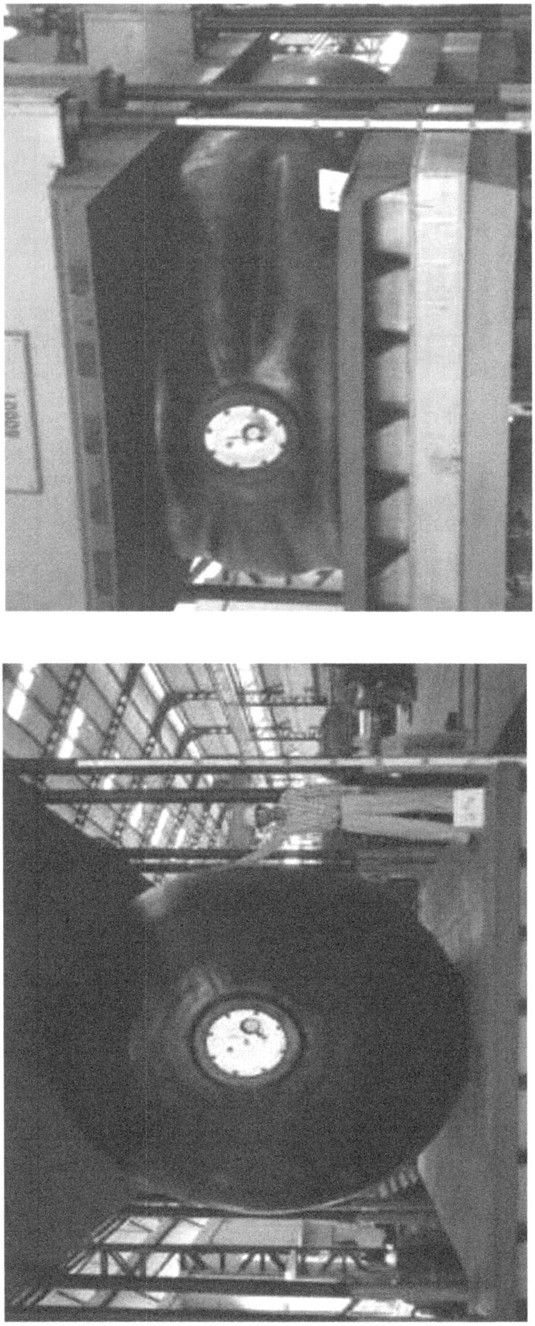

**FIGURE 6.15** Load deflection test images of pneumatic fender.

**TABLE 6.8**
**Comparison of Load Deflection Test Results of Pneumatic Fender**

| Deflection | 60% | 60% |
|---|---|---|
| Reaction force | 184.9 ton | 190.14 ton |
| Energy absorption | 307.44 tm | 310.86 tm |

The tolerance nearly within ± 10% is observed as mentioned in the report submitted by the manufacturers. Hence the results are satisfactory.

## 6.13 BOLLARDS

### 6.13.1 Types of Bollards

The following are types of Bollards

- Tee type
- Twin horn type
- Barrel type

#### Tee type

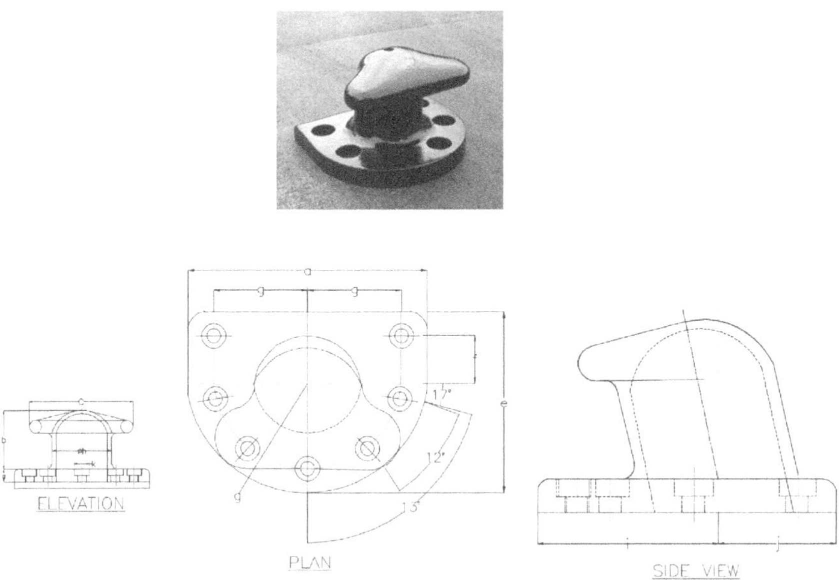


**FIGURE 6.16** Tee type bollard.

## TABLE 6.9
## Size Table for Tee Type Bollards

| | Bollard Capacity (tons) | | | | | | |
|---|---|---|---|---|---|---|---|
| **Dimension** | **15** | **30** | **50** | **80** | **100** | **150** | **200** |
| a | 40 | 40 | 50 | 70 | 80 | 90 | 90 |
| b | 235 | 255 | 350 | 380 | 410 | 435 | 500 |
| c | 340 | 350 | 500 | 550 | 600 | 700 | 800 |
| d | 410 | 450 | 640 | 640 | 790 | 900 | 1000 |
| e | 335 | 375 | 540 | 550 | 640 | 750 | 850 |
| f | 80 | 100 | 150 | 160 | 175 | 200 | 225 |
| g | 155 | 175 | 250 | 250 | 325 | 350 | 375 |
| h | 160 | 200 | 260 | 280 | 350 | 400 | 450 |
| i | 205 | 225 | 320 | 320 | 395 | 450 | 500 |
| j | 130 | 150 | 220 | 230 | 245 | 300 | 350 |
| k | 60 | 60 | 70 | 90 | 100 | 110 | 110 |
| l1° | 30° | 30° | 30° | 15° | 10° | 10° | 0° |
| l2° | – | – | – | 45° | 40° | 40° | 36° |
| l3° | 60° | 60° | 60° | N/A | 80° | 80° | 72° |
| Bolt | M24 | M30 | M36 | M42 | M42 | M48 | M56 |
| Length | 500 | 500 | 500 | 800 | 800 | 1000 | 1000 |
| Qty | 5 | 5 | 5 | 6 | 7 | 7 | 8 |
| CS | As per IS 1030 gr 200–400 N | | | | | | |
| CI | As per IS 210 gr FG 250 | | | | | | |

*[Units:* mm]

### TWIN HORN TYPE

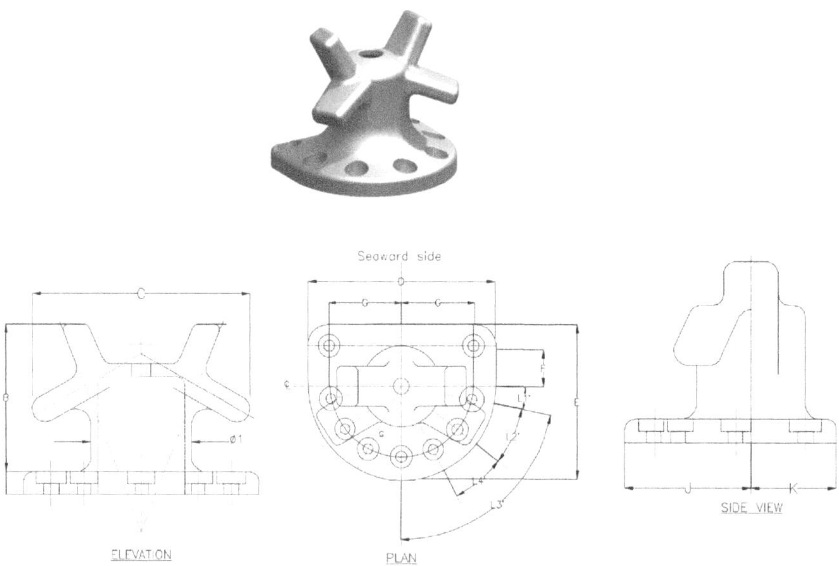


**FIGURE 6.17** Twin horn type bollard.

**TABLE 6.10**
**Size Table for Twin Horn Type Bollards**

| | Bollard Capacity (tonnes) | | | | | | |
|---|---|---|---|---|---|---|---|
| **Dimension** | **15** | **30** | **50** | **80** | **100** | **150** | **200** |
| A | 40 | 40 | 50 | 70 | 80 | 90 | 90 |
| B | 370 | 410 | 500 | 520 | 570 | 585 | 660 |
| C | 400 | 440 | 600 | 660 | 750 | 850 | 930 |
| D | 410 | 480 | 640 | 650 | 800 | 920 | 1000 |
| E | 335 | 405 | 540 | 560 | 650 | 770 | 850 |
| F | 80 | 100 | 150 | 160 | 175 | 200 | 225 |
| G | 155 | 175 | 250 | 250 | 325 | 350 | 375 |
| Ø1 | 160 | 200 | 260 | 300 | 350 | 400 | 450 |
| J | 205 | 240 | 320 | 325 | 400 | 460 | 500 |
| K | 130 | 165 | 220 | 235 | 250 | 310 | 350 |
| L1° | 30 | 30° | 30° | 15° | 10° | 10° | 0° |
| L2° | – | – | – | 45° | 40° | 40° | 36° |
| L3° | 60 | 60° | 60° | N/A | 80° | 80° | – |
| L4° | – | – | – | – | – | – | 36° |
| Bolts | M24 | M30 | M36 | M42 | M42 | M48 | M56 |
| Bolt Length | 500 | 500 | 500 | 800 | 800 | 1000 | 1000 |
| P* | 60 | 60 | 70 | 90 | 100 | 110 | 110 |
| Qty | 5 | 5 | 5 | 6 | 7 | 7 | 8 |

* P= bolt protrusion=recess depth [Units: mm]

## Barrel type Bollard

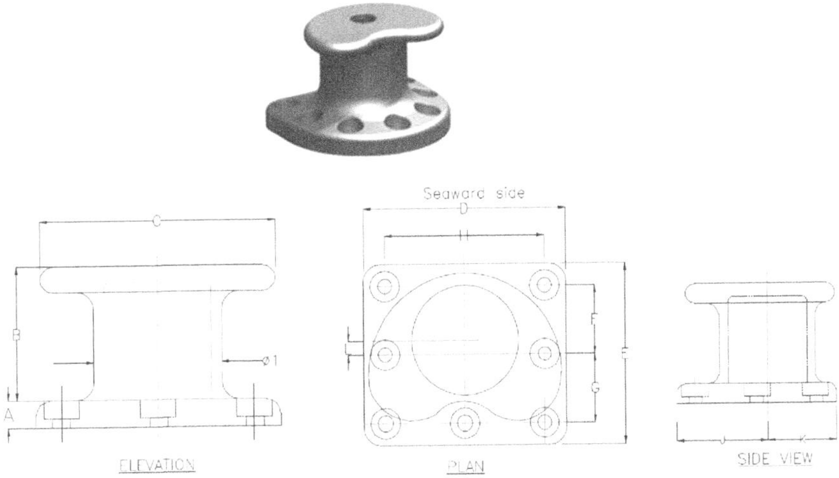


**FIGURE 6.18** Barrel type bollard.

**TABLE 6.11**
**Size Table for Barrel Type Bollards**

| | Bollard Capacity (tonnes) | | | | | | |
|---|---|---|---|---|---|---|---|
| **Dimension** | **15** | **30** | **50** | **80** | **100** | **150** | **200** |
| A | 40 | 40 | 50 | 70 | 70 | 80 | 90 |
| B | 260 | 280 | 320 | 330 | 350 | 405 | 435 |
| C | 340 | 370 | 480 | 530 | 550 | 728 | 800 |
| D | 320 | 360 | 540 | 560 | 590 | 760 | 1000 |
| E | 320 | 360 | 540 | 460 | 490 | 660 | 850 |
| F | – | – | – | – | 175 | 250 | 300 |
| G | – | – | – | – | 175 | 250 | 300 |
| F+G | 220 | 260 | 400 | 320 | 350 | 500 | 600 |
| H | 220 | 260 | 400 | 420 | 450 | 600 | 750 |
| ØI | 160 | 200 | 260 | 280 | 300 | 400 | 450 |
| J | 160 | 180 | 270 | 160 | 295 | 380 | 475 |
| K | 160 | 180 | 270 | 160 | 195 | 280 | 375 |
| L | – | – | – | – | 50 | 50 | 50 |
| Bolts | M24 | M30 | M36 | M42 | M42 | M48 | M56 |
| Bolts length | 500 | 500 | 500 | 800 | 800 | 1000 | 1000 |
| P* | 60 | 60 | 70 | 90 | 90 | 100 | 110 |

* P- bolt protrusion – recess depth [Units: mm]

## 6.13.2 Selection Criteria for Bollards

The selection of bollards for a pier, their spacing and arrangement is governed by published standards and local specifications. To determination proper bollards and locations for a pier one normally has to consider:

- Tidal conditions
- Wind and current forces
- Wave and swell conditions
- Vessel draft changes due to loading or unloading
- Mooring line types and sizes

The appropriate load on the bollard shall then be calculated, which depends upon the layout of harbour, and position of bow line, stern line, spring line and breasting lines; for guidance the bollard pulls independent of the number of laid-on hawsers, may be taken as given in below table (IS 4651 (Part 3) (2020)) since the hawsers are not fully stressed simultaneously.

**TABLE 6.12**
**Bollard Pull**

| Sl. No. | Displacement (tons) | Line Pull (tons) |
|---|---|---|
| 1 | 2000 | 10 |
| 2 | 10,000 | 30 |
| 3 | 20,000 | 60 |
| 4 | 50,000 | 80 |
| 5 | 100,000 | 100 |
| 6 | 200,000 | 150 |
| 7 | Greater than 200,000 | 200 |

NOTE 1 – For ships of displacement tonnage 50,000 and above, the value of line pulls given above should be increased by 25% at quays and berths where there is a strong current. NOTE 2 -Main bollards at the ends of individual large vessel berths at river structures should be designed for a line pull of 250 tons for ships up to 100,000 tons displacement and for double the values given above for larger ships.

### 6.13.3 Testing Plan for Bollards

| A. BOLLARD | Requirement |
|---|---|
| 1. Dimensions and thickness | As per sketch and IS 4897 for cast steel material |
| 2. Other requirements for bollard | As per the clauses of IS 5625 for cast steel material |
| 3. Material to be used in the bollard | Cast steel (carbon steel) as per IS 1030 (1998) |
| a) Grade designation | 230–450 N as per Table 1 of IS 1030 |
| b) Chemical composition | As per Table 1 of IS 1030 related to the specified grade. |
| c) Mechanical properties | As per Table 2 of IS 1030 related to the specified grade. |

**Method of casting:** The steel for the casting would be made from open hearth, electric duplex acid Bessemer, basic oxygen (LD) of a combination of these processes.

The casting would be made under strictly controlled condition to ensure chemical composition, soundness, uniformity, correct grain size to develop shock resistance properties and to avoid any blow-holes.

**B. FIXTURES**

| | |
|---|---|
| 1. Material of bolts, nuts, washers | All fixtures are of stainless steel AISI 316 grade (X 04 Cr 17 Ni 12 Mo 2 conforming to IS 6603 (2001) and IS 6911 (1992). |

*(Continued)*

(Continued)

| | |
|---|---|
| **C. TESTS ON CAST STEEL (Carbon Steel)** | |
| 1. Tensile Test | Tensile test shall be carried out in accordance with IS 1608 (1995). |
| 2. Impact test | Impact Test shall be carried out in accordance with IS 1757 (1988). |
| 3. Bend Test | Bend test shall be carried out in accordance with IS 1599 (1985). |
| 4. Non-destructive tests | Random ultrasonic flow detector test as per IS 3664 (1966) for checking the thickness of the castings and to detect the defect in the casting. |
| 5. Ultra-Sonic Examination | This test shall be carried out in accordance with IS 7666 (1988). |
| 6. Magnetic particle examinations | Magnetic particle inspection shall be carried out in accordance with IS 10724 (1990). |
| 7. Heat Treatment | Magnetic particle inspection shall be carried out in accordance with IS 10724 (1990). |
| 8. Chemical analysis | The manufacturer shall carryout chemical analysis from a sample of each melt and shall supply a test certificate of the chemical analysis of the sample for each melt to satisfy the composition given in Table-1 of IS 1030 related to the specified grade |
| **D. GENERAL** | |
| 1. The bollards shall be designed to with stand sudden jerking effect from ropes connected to vessels and to resist rubbing effect of the ropes on its sides. | |
| 2. The manufacturers/supplier shall ensure that the cast steel bollards are of specified pull capacities and a certificate to that effect shall be submitted by the manufacturers/suppliers. | |

## REVIEW QUESTIONS

1. Describe in brief about fendering systems.
2. Name some of the mooring devices used for mooring of a vessel.
3. How are fenders classified?
4. What are types of fenders used in a fendering mechanism?
5. Explain pneumatic fenders with a neat sketch.
6. What are criteria for selection of optimum fenders?
7. Write the design process for a fender system.
8. What are the protocols for testing of rubber fenders?
9. What is meant by rated performance data?
10. Describe methods CV and DV in brief for test protocol.
11. How is the velocity factor (VF) determined for rubber fenders?
12. Describe in brief the protocols for testing of pneumatic fenders.
13. Compare the spacing of fender provided as per IS 4561 (Part 4) and PIANC guidelines.
14. What are bollards? Brief their testing plan.

## REFERENCES

**IS 4651 (Part 3) (2020)** *Code of Practice for Planning and Design of Port and Harbours – General Design Considerations Loading*

**IS 4651 (Part 4) (2014)** *Code of Practice for Planning and Design of Port and Harbours – General Design Considerations*

**Guidelines for the Design of Fender Systems (2002)** *PIANC, International Navigation Association*

**UNCTAD (1985)** *Port Development: A Handbook for Planners in Developing Countries*, United Nations Conference on Trade and Development, United Nations, New York.

## COURTESY

M/s IRM Offshore Marine Engineers Private Limited
M/s Hitesh Elastomers Limited

# 7 Infrastructure for Shipbuilding and Maintenance

## 7.1 INFRASTRUCTURE FOR SHIPBUILDING AND MAINTENANCE

### 7.1.1 Introduction

The infrastructure for the ship repair facility are slipway, travel lift, ship lift and dry dock. The slipway is for 250–600 T vessels, travel lift for 1000 T vessels, ship lift for 10,000–25,000 T vessels, floating dock for 30,000–60,000 T vessels, wet basin (for outfitting) and dry dock for 30,000–200,000 T vessels. The weight of vessels mentioned are as per current trend, and facilities are built in India and abroad out of this range. The details of infrastructure of ship repair facility are described in upcoming sections.

## 7.2 SLIPWAY

Slipway is a facility provided in ports and harbours for the reconstruction and repair of vessels in dry condition. It is also known as a marine railway. It is a ramp on the shore used to move the vessel from the water. Also, a newly constructed vessel is launched into the sea with the help of a slipway ramp. The slipway is made-up of timber or steel, or concrete material. IS 10020 (Part 4) (1984) provides the details about the planning and design of the slipway.

Following are the components of the slipway:

- Sloping concrete platform or sloping track
- Cradle and roller or transfer bay
- Hauling machinery or dry berth

### 7.2.1 Components of Slipway

#### 7.2.1.1 Sloping Platform System

The sloping platform supports the tracks on which the cradle travels. The track must extend sufficiently far into the water to permit the ship to float over the keel blocks on the cradle and sufficiently far inshore to lift the ship entirely clear at high tide. The track may have two, three or four rails, but two rails are more common. The slope


DOI: 10.1201/9781003361275-7

**FIGURE 7.1** Slipway, Andaman–Port Blair.

of the ways varies from 1 in 12 to 1 in 24. Three types of the cradle are available, namely bottom-transfer cradle, sloping cradle and vessel cradle.

#### 7.2.1.2 Cradle and Roller

It is the framework of timber or iron or keel blocks but usually consists of keel blocks which are much stronger than the other two. The main deck of the cradle provides a working platform for repairing and painting the vessel. The whole structure is mounted on numerous cast iron rollers. The rollers are used to support the cradle on the track.

In general, three types of cradles are in common use:

- Rigid cradle
- Semi-rigid cradle
- Telescopic cradle

#### 7.2.1.3 Hauling Machinery

The modern hauling machinery consists of wire rope drums or chain sprockets or winch mechanism, an endless chain running over a number of sheaves and the operating motor with a gear train. The latest development is the use of electrically driven variable speed gear, comprising of a pump and a hydraulic motor with oil as the medium. Side-haul railways have one chain or rope for each section of the cradle

### 7.2.2 Selection of Site and Orientation of Slipway

The site should be sheltered from winds, strong currents and waves. The wave height should be less than 0.3–0.6 m at the time of slipping. The site should be protected from excessive erosion or siltation. The distance from the high waterline to the approach channel should be adequate to provide safe movement of the vessel. Adequate depth of water should be available to accommodate the proposed vessel, otherwise, dredging should be carried out. The adjacent land area should be available for the provision of pump house, winch house, sewage treatment plant, transfer track and dry repair berths, workshops and office rooms (slipway building). Adequate manoeuvring area, electricity supply, water supply and crane facility should be available.

#### 7.2.2.1 Orientation of Slipway

The slipway should be oriented in such a way that during the manoeuvring or slipping of the vessel, it should not be in beam sea condition for the strong wind or currents.

### 7.2.3 Types of Slipway

Slipways are of two types:

- End haul pattern
- Broadside (side haul) pattern

#### 7.2.3.1 End Haul Pattern

- This is the more common, and moves the ship lengthwise.
- A very elaborate carriage is required for an end on slipway, if the vessel is required to be maintained in a level position.
- The length of the track of an end haul slipway varies from 2.5 to 3.5 times the length of the vessel.

#### 7.2.3.2 Broad Side (Side Haul) Pattern

- In this pattern, the vessel is hauled clear of the water normal in the direction of its length.
- The side haul moves the ship sideways up and down the slipway.
- The side haul railways occupy more waterfront space, but the tracks both above and below water are shorter because the short dimension of the cradle is parallel to the track.

### 7.2.4 Supporting Structures

To repair several vessels simultaneously, the following supporting structures are required:

- Transfer bay
- Dry berth

The transfer bay will run perpendicular to the sloping track, which is used to transfer the vessel from the slipway track to the dry berth. The number of vessels shall be repaired simultaneously with this arrangement. The dry berth shall be oriented perpendicular or parallel to the transfer bay. The dimensions of dry berth and transfer bay shall be arrived based on the required number of work space.

Following are the other mechanical and miscellaneous components required for the operation of slipway:

- Cradle
- Winch house
- Pulley
- Crane
- Workshop building

#### 7.2.4.1 Cradle

The crane rail is selected according to IS 3443 (1980). The cradle is a framework of timber or steel which consist of longitudinal and cross beams to support the keel block and bilge block. The length of the cradle should be one-fourth of the (LBP) length between perpendiculars of the largest vessel proposed to operate. Pawl racks are attached to the beam carrying the rails at an interval of about 6 m to prevent back slip of the cradle.

The cradle is classified as:

- Sloping cradle
- Transfer cradle
- Vessel cradle

#### 7.2.4.2 Winch House

In slipway operations, there will be two winch houses used for hauling down and up operations. The winch capacity is 60–70T, which varies based on the vessel capacity.

#### 7.2.4.3 Pulley

There will be several pulleys to carry forward the hauling operations and other movements between the transfer bay and dry berth.

#### 7.2.4.4 Crane

Crane will be usually provided at the dry berth region to carry out repair works.

#### 7.2.4.5 Workshop Building

It is also called a slipway building, where the administrative and office work will be carried out.

### 7.2.5 Loads

1. Dead load
2. Weight of vessel acting on the cradle
3. Sue load – The concentrated load induced on the cradle and track due to the hauling of the vessel. It normally varies from one-third to one-eighth of the light displacement of the vessel
4. The pull on the hauling chain can be calculated from the following expression:

$$P = W\,(S+C) \quad (7.1)$$

where

P = Pull on the hauling chain in tons

W = Weight of the vessel + the weight of the cradle + the weight of hauling chains and ropes in tons

S = tan θ, where θ is the angle of slope of the slip; and

C = Coefficient of friction which may be taken as 0.035 for wheels and 0.04 for rollers.

## 7.3 TRAVEL LIFT

### 7.3.1 Introduction

A travel lift is a specific type of crane used for lifting vessels out of the water and transporting them to dockyards for the process of anti-fouling, maintenance, inspection and long-term storage of boats. To repair several vessels simultaneously, a transfer bay and dry berth are provided.

### 7.3.2 Components

1. *Outfitting jetty with wet basin*: It is an on-shore structure constructed to float-out/launching of a vessel and lead sea trials.
2. *Approach jetty for travel lift*: Based on site condition, approach jetty shall be proposed. If offshore structures such as dry berth are far away from the outfitting jetty, an approach jetty will be constructed.

3. *Transfer bay*: It is an intermediate structure constructed between the approach jetty and dry berth for transfer purpose. The dimension of the transfer bay depends on the length of the vessel plus 10% of the length of the vessel. If the number of dry berths is increased dimensions of the transfer bay will also be increased.
4. *Dry berth (repair berth)*: It is a platform where the ships are stored for their maintenance and repairing. If several vessels want to be repaired simultaneously, the number of dry berths will be increased. Dimensions of the dry berth shall be as same as the transfer bay.

The two rectangular sides of the travel lift are joined by a single spanning beam at the top of one end. Four groups of steerable wheels, which can turn 360°, is placed one on each corner. Boats are lifted using hydraulic straps (lifting belt) hung between two sides of the travel lift, which can be passed under boats either in the water or on the hard stand. It also has steel wires on the pulley which are used to lift the boat. The components of the travel lift are shown in Figure 7.2.

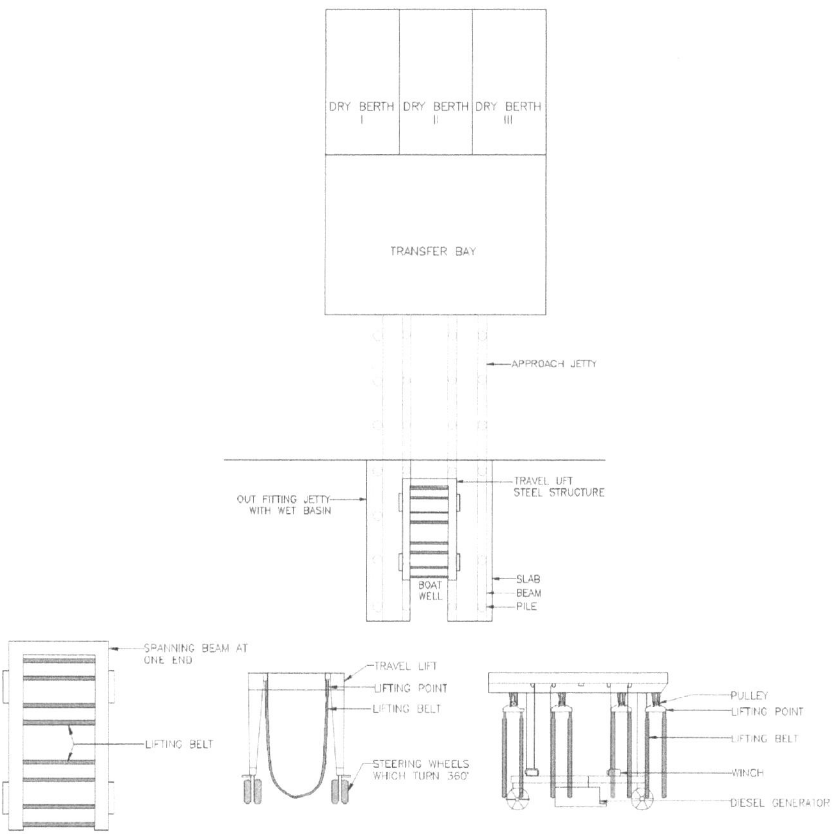


**FIGURE 7.2** Travel lift and its components.

**TABLE 7.1**
**Technical Data for Travel Lift**

| Model | Maximum Lifting Capacity (T) | Clear Width (m) | Clear Height (m) | Shipping Weight (T) | Drive Type | Source |
|---|---|---|---|---|---|---|
| 500 C | 500 | 11.58 | 12.19 | 248.48 | Hydraulic gear | Marine Travel Lift |
| MBH 560 | 560 | 11.3 | 17.8 | 220 | Closed loop hydrostatic type | CIMOLAI Technology |
| MBH 1100 | 1100 | 17 | 23 | 540 | Closed loop hydrostatic type | |

#### 7.3.2.1 Procedure

Removal of the boat from the water is carried out by the following process. Primarily, the boat is placed between a narrow outfitting jetty in the wet basin, which consists of two paths on each side and at the same level. The travel lift is positioned over the boat by driving it on the paths of the jetty structure. The boat can then be lifted out of the water by the travel lift and slowly moved to the place where the boat is to be placed for the maintenance/storage.

#### 7.3.2.2 Technical Data

Travel lifts can have a variety of lifting capacities which can be used for sizeable ships. Technical specifications of travel lift of varying capacity are given by some specialized companies.

### 7.3.3 Travel Lift Operation

**Step I: Controls of the Travel Lift**

Working of the travel lift should be studied before using it.

- Ignition on the right side of the control panel is turned on to start the travel lift.
- Two wheels on the back of the travel lift are used for controlling the turning process.
- Travel lift is moved over the tracks and positioned on the wet basin where the boat is to be lifted.
- Each travel lift has straps on both front and back; it can be raised and lowered by using three levers.
- Once positioned, straps are lowered until they are about 3 feet under the water:
  1. The back strap is controlled by the first and second levers on the left, and the third lever from the left controls the front strap.

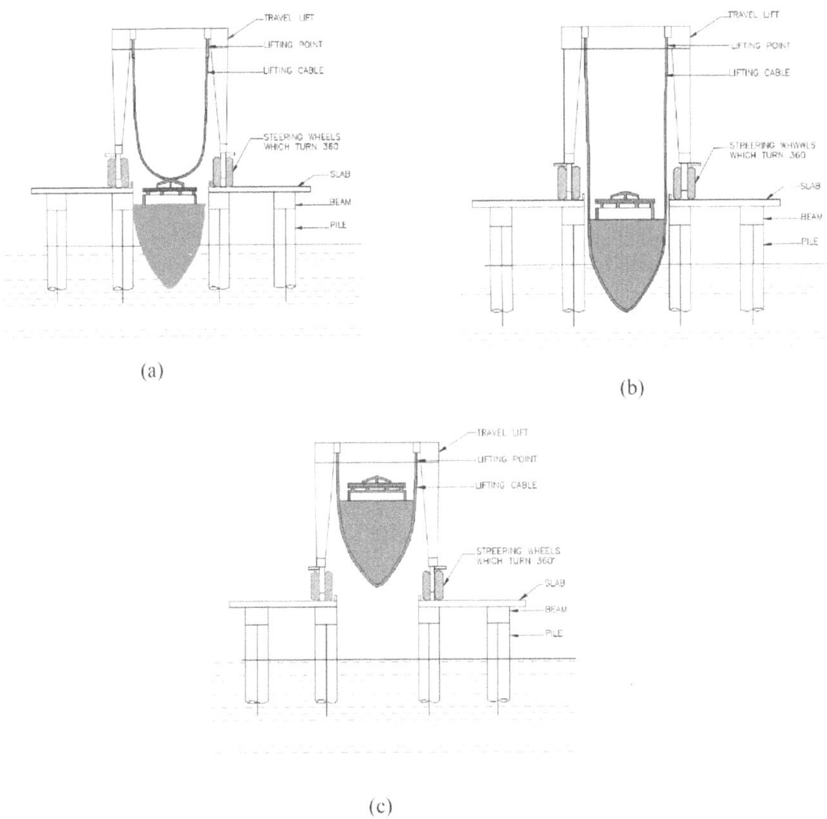


**FIGURE 7.3** Travel lift operation (front view).

2. Back strap is lowered by pushing the two levers upward (they move the straps opposite to the direction of the levers).
3. Front strap is lowered by pushing the third lever downward (it moves the strap the same direction as the lever).

### Step 2: Bring the Vessel into the Wet Basin

- The wet basin is an inlet of water where boats are put in/taken out of the water. The travel lift is located above the well.
- Drive the vessel to the middle of the wet basin and positioned exactly under the travel lift.

### Step 3: Line up the Straps to the Sling Marks

- A name labelled 'sling' should be marked on side of the vessel where the strap should be aligned with when pulling up the boat.

- Sling marks are there to distribute weight of the vessel equally when it is being raised; if not the boat could slide back or forward from the straps.

**Step 4: pull up the vessel**

- Once the straps are lined up in the place of sling marks, you can start to pull up the vessel by moving the levers opposite of how you lowered them.
- The left two levers go down, and the third lever goes up.
- Straps are pulled up at the same level by handling the three levers at the same time.

**Step 5: straighten out the vessel**

- The vessel should be at the same level when pulling it out of the water.
- Since the levers of the front and back straps of the travel lift are independent of each other, you can adjust the straps by moving the levers one or the other to straighten the vessel. (For example, if the front of the boat is leaning down, pull up the front strap until even.)

**Step 6: power wash the vessel**

- Before the boat is put in the stockyard, algae is present in it to be washed off with a power washer.
- While using an industrial power washer, spraying should not be too close to it where the paint of the boat will be washed off.

**Step 7: travel lift is moved to the desired spot**

- Travel lift is moved to the desired spot (i.e., stock yard or dry berth) slowly so that the vessel does not move back and forth too much.

**Step 8: preparing the block platform**

- Block platform is stacked by many keel blocks where the vessel is being placed.
- Make three stacks: one in the middle and two in the corners of the stern and bow.
- After the keel blocks are stacked, add a thin piece of plywood to alleviate pressure between the vessel and the wood.

**Step 9: unpin the straps**

- With the help of the instructor, the travel lift is moved slowly near the block platform of dry berth I.
- Make sure the support stands/blocks on the platforms are flat against the vessel.
- Once the boat is fully supported by the supports on the block platform, the pins holding the straps together can be undone so that the vessel is no longer attached to the travel lift.
- The left and right-side straps of the travel lift can be pinned back together, and the travel lift can be moved back onto the tracks over the well.

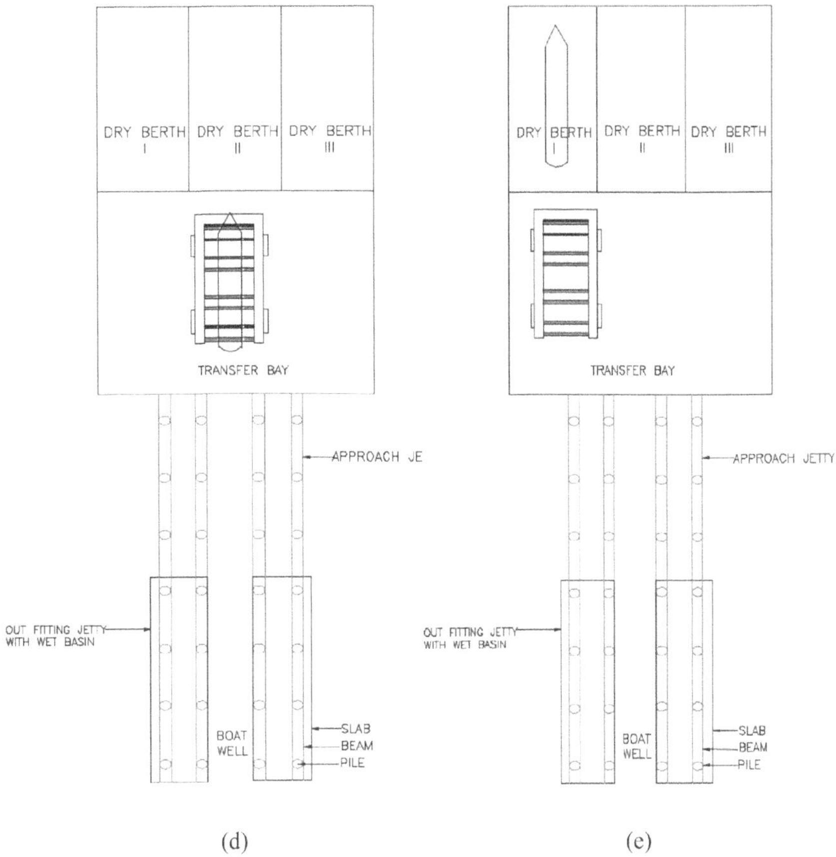


**FIGURE 7.4** Travel lift operation (top view).

### 7.3.4 Advantages of Travel Lift

- Low in maintenance and economical
- Safe, efficient and robust
- Transports the vessel in any direction
- Protects the vessel from heavy wave and bad weather condition

## 7.4 SHIP LIFT

It is an elevated type of dry dock operated by hoists. Hoists are nothing but highly refined machines powered by electric motor with wire rope cables. The number of hoists, capacity of each hoist and their spacing is varied to provide the required dry-docking capacity. The components of hoist are:

- Electric motor
- Reduction gear system

FIGURE 7.5 A view of ship lift.

- Wire rope cable drum
- Double brake arrangement

The railroad-type rails are installed on the platform. The bilge blocks are moved by a system of chains and pulleys. The platform is lowered vertically by optically controlled hoists with wire rope cables and stopped at a predetermined depth below the water. The vessel is positioned over the submerged platform, which is then raised completely above the water level.

### Components of ship lift

- Steel lifting platform
- Piers which consist of hoist with ropes (equal numbers on both sides)

### Supporting components

- Transfer system
- Cradles supported on hydraulic trolleys
- Rails
- Motor with a control mechanism

The ship lift platform area also contains many facilities like an administrative building, warehouse, electrical shop, machine shop and repair shop

The crane rail is selected according to IS 3443 (1980). All rails shall be stored in a clean condition, above ground level, until they are incorporated into the works. In stacking rails, care shall be taken to ensure that the ground is levelled.

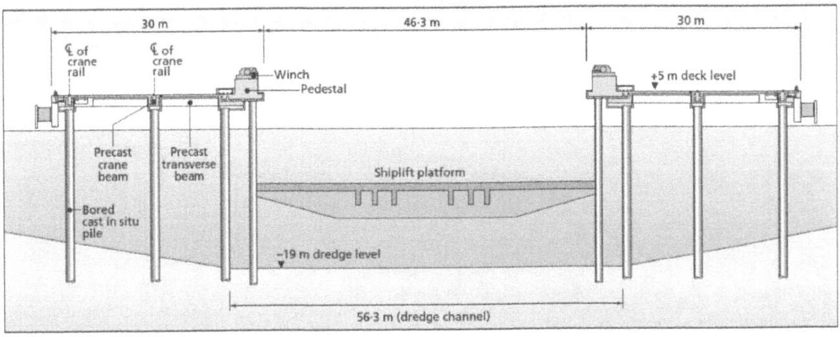


**FIGURE 7.6** Typical cross-sectional view of ship lift.

### Advantages

- Less overall space and maintenance
- The maximum speed of operation
- Requires less manpower
- Platform is raised completely out of water
- Single elevator can be used for handling number of vessels
- Maintenance is simple and requires no specialized technicians
- Damaged vessels can be more easily recovered
- Avoids risk of dock gate and provides easier accessible work areas

## 7.5 FLOATING DOCK

A floating dock is preferable in land restricted site conditions and will be used to carry the vessel which is in damaged condition. The ship which cannot be sailed due to technical issues shall also be transferred to the dockyard by using floating dock. The floating dock also serves as a ship repair and maintenance yard. A floating dock retains itself above water by means of its own buoyancy and is usually power operated. It is a U-shaped cross-section.

The floating dock consists of two side pontoons and one bottom pontoon. The rear side is closed, and the front side can be closed or opened. The side pontoons are ballasted to lower the bottom pontoon. The front can be opened, and the ship can be hauled in. The front gate is then closed, and the side pontoons de-ballasted to raise the ship above the water level along with the bottom pontoon.

### Types of floating dry dock

- Rigid type or non-self-docking type or single-section dock
- Self-docking type or multi-section type

1. Multiple sectional dock
2. End sectional dock
3. Detachable pontoon dock
4. Off-shore dock

FIGURE 7.7 A view of floating dry dock.

# 7.6 DRY DOCK

## 7.6.1 General

A dry dock is a narrow basin used for the construction, maintenance and repair of ships, boats and other watercraft. Routine use of dry docks is for the 'graving' i.e. the cleaning, removal of barnacles and rust, and repainting of ships' hulls. Modern docks are in a box-type shape. It will be an advantage to accommodate the newer, boxier ship designs on porous grounds. The dry dock is usually made of concrete and can be closed by a caisson gate. The dock will be filled with water first, and the vessel may be floated in. The vessel will be supported on keel blocks, and dewatering of the dock will be done.

### 7.6.1.1 Need for Dry Docking

- To expose the underwater portion of the vessels fully and keep it off water during its inspection or repair, or renovation
- To conduct underwater damage surveys as well as periodical surveys for certification of ships sea worthiness
- For bottom cleaning and painting to reduce the drag of water for maintenance of speed
- To attend propeller work and examination or change of tail shaft or other underwater fittings

The dock will be constructed as a concrete structure with the dock floor, walls and gate. The working mechanism of the graving dock is as follows: initially, the dock is filled with water, then the ship is shifted inside the dock and placed on the top of the keel block. After positioning the ship, the gate is closed, and water is removed using the pumping system.

#### 7.6.1.2 Components of Dry Dock

Major components of graving dry dock are as follows:

- Dry dock floors and allied facilities
- Dry dock walls with provision for other amenities
- Dry dock gates
- Dry dock equipment

#### 7.6.1.3 Site Selection of Dry Docks

- A level site of adequate area, either natural or obtained by excavation or reclamation.
- Satisfactory foundations available at a reasonable depth.
- An approach channel as short and as easily maintained as possible. Continuous and expensive dredging must be avoided.
- Easy access by road and preferably, also by rail.
- Provision of lead-in jetties or lead-in dolphins, which can greatly assist ship control during periods of strong winds, may be made if it does not hamper the other navigational requirements of the port near the entrance.

#### 7.6.1.4 Basic Requirements for Functioning of Dry Dock

The following are the basic requirements for the functioning of a dry dock:

- Filling technique of dry dock
- Dewatering pump house
- Collection of floor waste
- Services at dock floor level
- ELL (electrical level luffing) dock crane
- Workshop building
- Admin block
- Capstan
- Bollard
- Drainage facilities
- Firefighting facilities
- Electrical control room

#### 7.6.1.5 Sizing of Dry Dock Components

- The capacity of the dry dock is described in terms of length, width and depth, together with the working space at the sides and ends. The height of keel blocks and tidal conditions is also considered.
- The clear length of the dock should be the minimum horizontal distance between the internal face of the headwall.
- The dock entrance width should not be less than 0.6 m greater than the beam of the largest ship to be docked.

- The working space between the ship and dock should not be less than 1.5 m.
- The height of the dry dock cope line should be higher than the highest high-water level or general ground level surrounding ground to maintain the services.

### 7.6.2 Dock Floor

#### 7.6.2.1 Types of Dock Floor

- Under-drained floors
- Gravity floors
- Tied floors
- Floors bearing directly on the ground
- Floors supported on piling
- Floors constructed under water

#### 7.6.2.2 Loads for Dock Floor

The dry dock floor should be designed by considering the following loads

- Dead load
- Uplift from the ground water
- Concentrated loading from keel block when ship is docked
- Uniform water pressure on top of the floor, when the dock is full
- Horizontal loadings from the earth and water pressure on the walls
- Seismic loadings
- The docking displacement can be calculated from the following expression:

$$D_b = 0.7 \text{ x LBP x B x MDD} \tag{7.2}$$

where

$D_b$ = Docking displacement
LBP = Length between perpendiculars
B = Beam of the ship
MDD = Mean docking draught

Preliminary approximate mean docking draughts should be taken as:

- Container ships: light ship draught plus 3 m
- Passenger vessels: three-quarters of the summer draught
- Tankers and bulk carriers: 5 m

#### 7.6.2.3 Keel Block Load Distribution

1. For ships larger than 30,000 T docking displacement:

$$K_L = \frac{0.6D_b}{LBP} \tag{7.3}$$

where

$K_L$ = Intensity of uniformly distributed load in a large vessel

2. For ships of 30,000 T or less docking displacement

$$K_s = \frac{0.95D_b}{LBP} \tag{7.4}$$

where

$K_s$ = Intensity of uniformly distributed load in a small vessel

#### 7.6.2.4 Bilge Block Load Distribution

1. For ships larger than 30,000 T docking displacement

$$B_L = \frac{0.25D_b}{0.9(LBP)} \tag{7.5}$$

where

$B_L$ = Intensity of uniformly distributed load in a large vessel

2. For ships of 30,000 T or less docking displacement

$$B_s = \frac{0.075D_b}{0.5 \text{ to } 0.7(LBP)} \tag{7.6}$$

where

$B_s$ = Intensity of uniformly distributed load in a small vessel

### 7.6.3 Dock Wall

#### 7.6.3.1 Types of Dock Wall

- Gravity dry dock walls
- Reinforced concrete walls
- Sheet-piled walls
- Diaphragm walls
- Caissons forming dry dock walls
- Soil or rock slope walls

#### 7.6.3.2 Loads for Dock Wall

The dry dock wall should be designed by considering the following loads:

- Earth pressure, including surcharge effects
- Ground water pressure
- Ship hauling gear loading
- Dock arm loading

- Dockside bollard loading
- Crane surge loading
- Access tower loading
- Seismic loading

### 7.6.4 Case Study of Dry Dock

The dry dock is designed to accommodate vessels of up to 330 m in length, with a beam ranging from 50 to 60 m and a ballast draught of approximately 12 m. The key dimensions of the dry dock facilities are provided in Table 7.2.

#### 7.6.4.1 Dry Dock Infrastructure and Associated Systems

The dry dock facility is engineered to accommodate vessels up to 250,000 DWT, with a maximum vessel dimension of 322 m in length, 51 m in breadth, and a ballast draft of 12 m. The infrastructure comprises two dry docks, each with dimensions of 420 m (L) × 100 m (B), facilitating the construction and repair of large-size vessels. The key components of the dry dock complex are as follows:

- Drydock – 420 m × 100 m
- Caisson gate – 90 m long
- Platform – 700 m × 24 m
- Platform – 700 m × 50 m
- Pre-erection shop – 200 m × 100 m

#### 7.6.4.2 Caisson Gate

A floating caisson gate is a type of movable watertight barrier used to seal the entrance of a dry dock. It functions similarly to a lock gate and is designed as a large, hollow, watertight steel or iron structure that operates based on controlled buoyancy.

The caisson is maneuverer into position at the dock entrance and then flooded internally to increase its weight, causing it to sink and form a watertight seal against the dock walls and sill. To reopen the dock, the internal compartments are dewatered using pumps, restoring buoyancy and allowing the caisson to float and be towed away from the entrance.

For this facility, the floating caisson gate is designed with the following dimensions:

- Length: 90 m
- Width: 2.5 m
- Height: 19.2 m

**TABLE 7.2**
**Technical Data for Travel Lift**

| Dock | Length (m) | Width (m) | Depth (m) From CD |
|---|---|---|---|
| Building | 420 | 100 | 15 |
| Pre-erection shop | 200 | 100 | 15 |

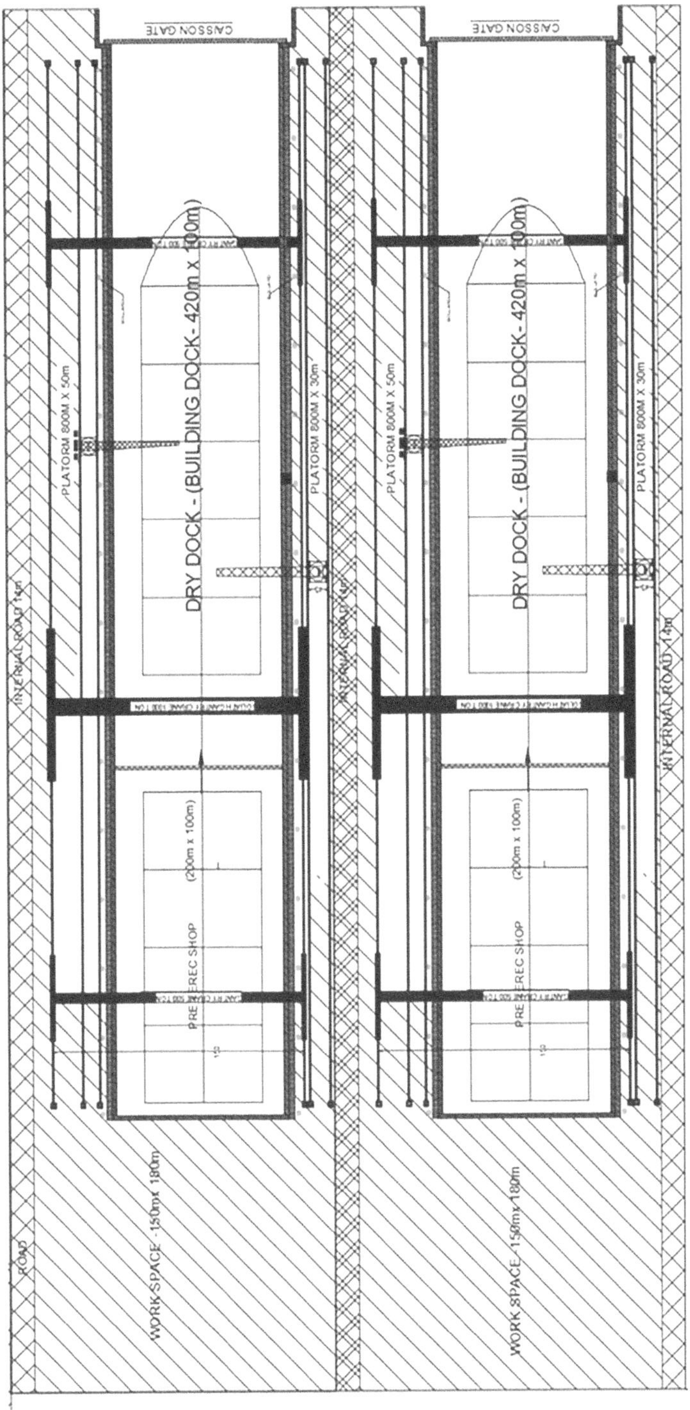


**FIGURE 7.8** Typical plan view of dry dock with cranes.

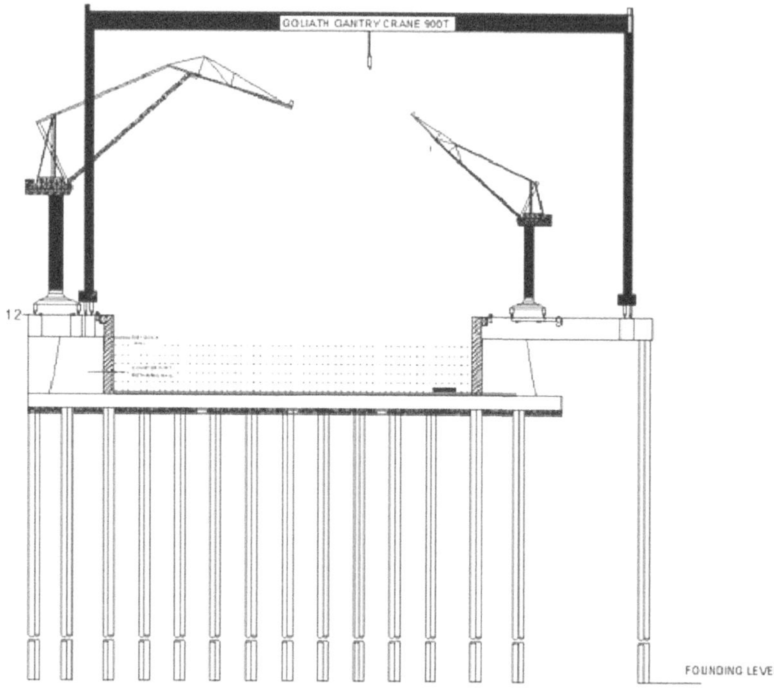


**FIGURE 7.9** Section view of the dry dock with crane.

The structure is engineered to withstand hydrostatic pressure and wave action, while ensuring a reliable seal for maintaining dry conditions within the dock during ship construction or repair. The gate is guided into place with embedded recesses in the dock walls and supported by mechanical and hydraulic systems for positioning and sealing operations.

#### 7.6.4.3 Dry Dock Outfitting Platforms

As part of the dry dock facility layout, four outfitting platforms are strategically planned to support block assembly, pre-outfitting, and final outfitting activities. These platforms are designed to accommodate heavy structural blocks, large equipment modules, and facilitate seamless movement between fabrication shops and the dry docks.

1. Outfitting platforms – 700 m × 24 m (2 Nos.)

Two outfitting platforms, each measuring 700 m in length and 24 m in width, are provided for modular block handling and component assembly. These platforms serve as staging areas for intermediate outfitting, pre-erection and logistics operations. They are equipped with utility connections, foundation slots and crane access for efficient material handling and workflow continuity.

2. Outfitting platforms – 700 m × 50 m (2 Nos.)

In addition, two larger platforms, each measuring 700 m in length and 50 m in width, are designated for heavy outfitting, structural integration and equipment installation. These wider platforms allow simultaneous operations on multiple hull sections or large superstructures and are designed with sufficient load-bearing capacity, embedded rail systems and crane coverage to support large-scale shipbuilding activities.

The combined outfitting platform layout ensures streamlined block movement from the pre-erection shops to the dry docks and outfitting quay, enhancing productivity and reducing turnaround time in ship construction cycles.

#### 7.6.4.4 Pre-erection Shop

The pre-erection shop, with a footprint of 200 m in length and 100 m in breadth, is a critical facility in the modular shipbuilding process. It serves as the intermediate staging area where sub-assemblies and fabricated blocks are aligned, welded and outfitted prior to their transfer to the dry dock or assembly platform.

The shop is equipped with:

- Heavy-duty overhead gantry cranes for material handling and block movement
- Welding and fabrication bays with appropriate fixtures for block alignment
- Dedicated zones for outfitting, painting touch-ups, and quality inspections
- Logistics access for efficient movement of prefabricated sections to the final erection site

The pre-erection shop enhances productivity by enabling parallel construction activities, minimizing dry dock occupancy, and ensuring seamless integration of blocks during final assembly.

#### 7.6.4.5 Crane Requirements

To support the efficient handling of heavy shipbuilding blocks and modules, each dry dock shall be equipped with a dedicated set of high-capacity cranes as per the following specifications:

- Goliath crane
- Each dry dock will be served by one ultra-high-capacity Goliath crane with a minimum lifting capacity of 1,000 tonnes. This crane is critical for lifting and positioning large hull blocks and superstructure modules during the assembly process.
- High-capacity gantry cranes
- In addition to the Goliath crane, each dry dock will be equipped with two high-capacity gantry cranes, each having a lifting capacity in the range of

400–500 tonnes. These cranes will support block erection, hull assembly and major equipment installation.

- Jib cranes (dry docks)
- Each dry dock will also be fitted with two jib cranes with a lifting capacity between 80 and 100 tonnes, providing localized lifting support for outfitting and modular alignment operations.
- Jib cranes (outfitting jetties)
- The outfitting jetties shall be provisioned with a minimum of two jib cranes per jetty, each with a lifting capacity of 50 tonnes, to handle equipment loading, outfitting components and logistics support during final vessel completion stages.

This crane configuration ensures optimal coverage, redundancy and load-handling efficiency across the dry dock and outfitting areas, enabling safe and timely execution of shipbuilding and repair operations.

#### 7.6.4.6 Ballast Draft Versus Light Draft – Definitions and Considerations for Port Design

The draft of a vessel is defined as the vertical distance between the keel (bottom of the hull) and the waterline. This parameter is critical for determining the required water depth in navigational and operational areas of a port facility.

*Light Draft*

The light draft refers to the draft of the vessel when it is empty or in ballast condition without cargo, carrying only the minimum fuel, stores and crew necessary for operation. According to Thoresen's reference, the light draft is generally estimated as approximately 0.55 times the fully loaded draft. This condition is typically encountered when the vessel is in transit to a loading port or under dry-docking and maintenance operations.

*Ballast Draft*

The ballast draft corresponds to the draft of the vessel when it carries ballast water in dedicated tanks to ensure stability, manoeuvrability, and proper immersion of the propeller and rudder when operating without cargo. For single-screw vessels, a commonly adopted thumb rule is that the ballast draft is approximately two-thirds of the fully loaded draft or roughly equal to the diameter of the propeller, to ensure efficient propulsion and steerage.

*Design Depth and Under-Keel Clearance*

The required water depth for port basin, approach channel and dock operations must be determined based on the expected vessel drafts and regulatory guidelines. As per IS 4651 (Part 5) (1980) Cl.4.6.1, the following minimum under-keel clearances (UKC) are recommended:

- 10% of vessel draft within the navigational channel
- 15% of vessel draft at the turning circle

- 20% of vessel draft at the channel entrance, particularly in unsheltered or exposed areas

These clearances are vital to ensure safe navigation, especially under varying tidal conditions, wave actions, and sedimentation tendencies.

#### 7.6.4.7 Work Shop Facilities for Dry Docks

The landside infrastructure of a shipyard is a critical enabler in the ship construction, repair and maintenance value chain. These facilities are strategically located and functionally diversified to support modular fabrication, outfitting, logistics and overall operational efficiency of the dry dock complex.

The following are the key functional components of the landside facilities designed to meet the operational requirements of large-scale shipbuilding:

1. *Panel block shop*
   Dedicated to the fabrication of large flat panel structures using automated welding and panel line systems.
2. *Steel stock pre-treatment and cutting shop*
   Includes shot blasting, priming and CNC cutting of raw steel plates, ensuring material is prepped for fabrication.
3. *Pre-outfit shop and block stock area*
   Supports the outfitting of blocks with pipes, cables, foundations, etc., before they are transferred for final assembly.
4. *Curved block shop*
   Specialized facility for fabricating complex curved hull sections using advanced forming techniques.
5. *Outfitting jetty*
   A waterfront facility for outfitting operations post-launch, including installation of heavy equipment and commissioning.
6. *Compressor station*
   Provides centralized compressed air supply to various shops and operational zones.
7. *Painting and blasting shop*
   Controlled environment for surface preparation (blasting) and coating (painting) of structural blocks and components.
8. *Deck house shop*
   Manufactures deckhouses and superstructure elements using steel and outfitted modules.
9. *Tool centre*
   Centralized tool storage, calibration and distribution centre to ensure readiness of equipment for fabrication tasks.
10. *Field shop office*
    Provides administrative and supervisory support directly within fabrication zones for efficient workflow management.

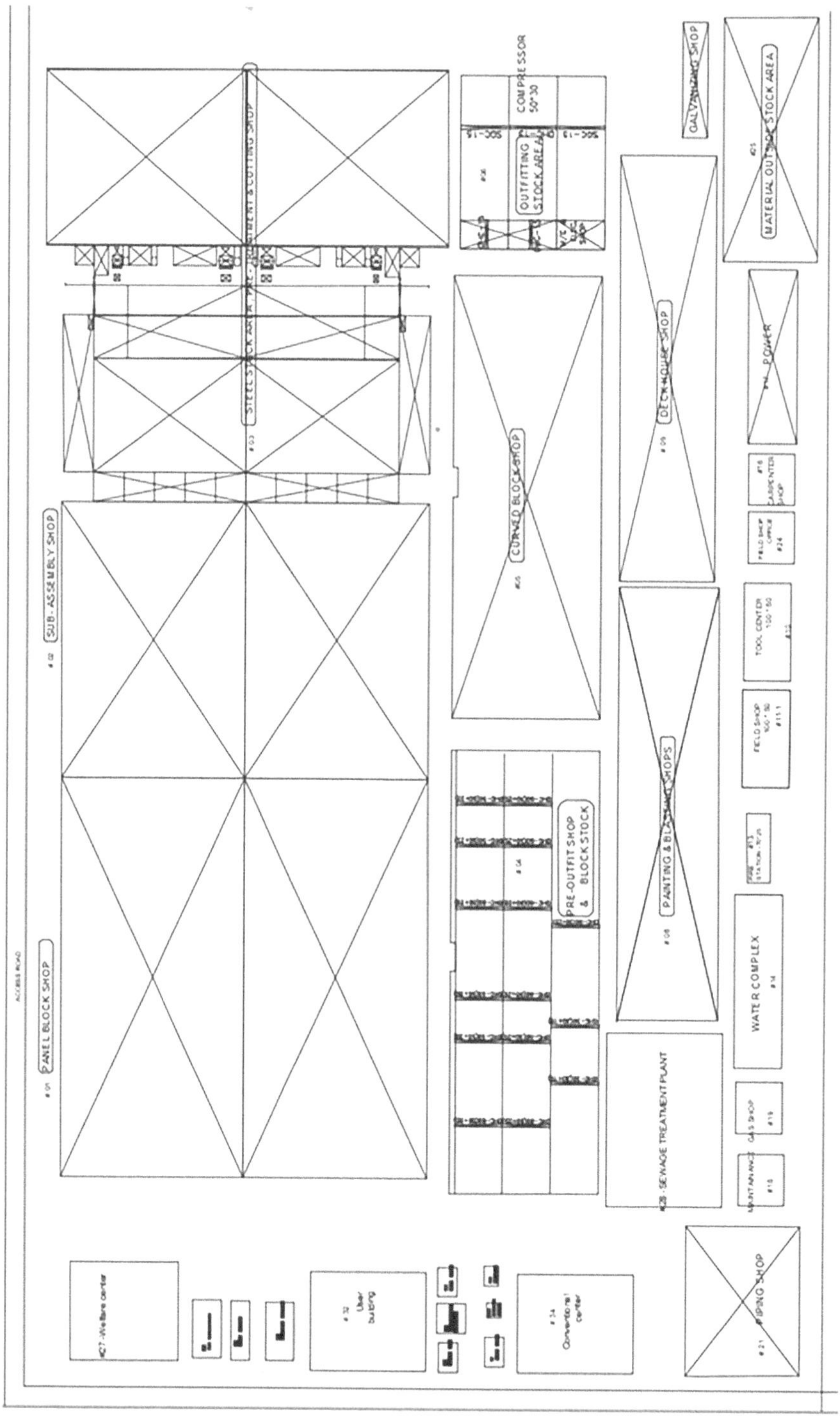


**FIGURE 7.10** Landside facilities.

11. *Erection equipment stock area*
    Storage and maintenance yard for cranes, lifting gear, jigs and assembly tools used during block erection.
12. *Fire station*
    Equipped with firefighting vehicles and response units to ensure industrial fire safety compliance.
13. *Water complex*
    Supplies treated water for industrial and domestic use across the facility.
14. *Pre-erection shop*
    Supports the final stage of block preparation before they are transferred to the dry dock for hull erection.
15. *Carpenter shop*
    Fabrication of wooden fixtures, formworks and temporary structures used during ship construction.
16. *Galvanizing shop*
    Provides anti-corrosion coating of metal components via hot-dip or electro-galvanizing methods.
17. *Piping shop*
    Dedicated for the prefabrication and testing of various piping systems used in marine installations.
18. *Outfitting Shop*
    Performs outfitting works including electrical systems, HVAC, instrumentation and accommodation modules.
19. *Warehouse*
    Central storage of raw materials, spare parts, consumables and tools with inventory control systems.
20. *Field shop house*
    Temporary and field-level shelter for workers and supervisors engaged in on-site fabrication.
21. *Material outside stock area*
    Open yard for bulk material storage such as steel plates, pipes and structural sections.
22. *Block stock area*
    Dedicated area for temporary storage of fabricated hull blocks prior to erection.
23. *Welfare centre*
    Includes facilities for workers' welfare – canteen, rest areas, locker rooms and sanitation.
24. *Guest house*
    Accommodation facility for visiting technical personnel, consultants, and shipowners.
25. *Security office*
    Ensures surveillance, access control and overall security of the yard premises.

26. *Job training centre*
    Equips workers with technical skills, safety training and process knowledge required for shipbuilding.
27. *User building*
    Facility designated for clients, classification society representatives and ship inspectors.
28. *Hospital*
    Medical centre with emergency care and occupational health services for employees.
29. *Convention centre*
    Multipurpose space for meetings, reviews, stakeholder presentations and events.
30. *Petrol bunk*
    Fuel dispensing station for internal logistics vehicles and construction machinery.
31. *Administrative block*
    Houses senior management, planning, procurement, HR and engineering offices.
32. *Sewage treatment plant*
    Provides environmental compliance through collection, treatment and discharge of domestic and industrial wastewater.

This landside area requirement for constructing a 250,000 DWT vessel is appropriately 140 hectares. The major landside facilities are block fabrication and outfitting operations. Its design ensures seamless integration of storage, manufacturing, administrative functions and safe logistics.

#### 7.6.4.8 Other Ancillary Facilities in Dry Dock

To ensure the seamless functioning of shipbuilding and repair operations, a range of ancillary facilities are integrated within the dry dock complex. These facilities are critical for supporting operational safety, workforce welfare, administrative control and skill development. The key ancillary components include:

- Medical and emergency response infrastructure
- A fully equipped on-site hospital and a dedicated fire station are established to provide immediate medical attention, emergency care and firefighting capabilities. These facilities are designed to comply with industrial safety standards and marine regulatory requirements, ensuring a rapid response in case of occupational injuries, health emergencies or fire incidents.
- Welfare centre
- A centralized welfare centre is provided to support employee well-being, incorporating rest areas, dining halls, sanitation facilities and recreational zones. This setup promotes occupational health, morale and productivity, especially critical during large-scale shipbuilding activities involving a significant workforce.

**TABLE 7.3**
**Area for Each Facility**

| Sl. No. | Description | Nos. | Length (m) | Width (m) | Area ($m^2$) |
|---|---|---|---|---|---|
| 1 | Panel block shop | 2 | 405 | 386 | 156,330 |
| 2 | Sub-assembly shop | 2 | 280 | 386 | 1,08,080 |
| 3 | Steel stock pre-treatment and cutting shop | 2 | 440 | 420 | 184,800 |
| 4 | Pre-outfit shop and block stock area | 2 | 450 | 155 | 69,750 |
| 5 | Curved block shop | 2 | 450 | 155 | 69,750 |
| 6 | Outfitting jetty | 2 | 177 | 150 | 26,550 |
| 7 | Compressor station | 2 | 118 | 26 | 3,068 |
| 8 | Painting and blasting shop | 2 | 440 | 105 | 46,200 |
| 9 | Deck house shop | 2 | 435 | 100 | 43,500 |
| 10 | Tool centre | 4 | 100 | 50 | 20,000 |
| 11 | Field shop office | 4 | 100 | 50 | 20,000 |
| 12 | Erection equipment stock area | 2 | 180 | 120 | 21,600 |
| 13 | Fire station | 2 | 70 | 25 | 1,750 |
| 14 | Water complex | 2 | 175 | 50 | 8,750 |
| 15 | Pre-erection shop | 4 | 200 | 100 | 80,000 |
| 16 | Carpenter shop | 2 | 50 | 50 | 2,500 |
| 17 | Power | 2 | 180 | 50 | 9,000 |
| 18 | Maintenance | 2 | 50 | 50 | 2,500 |
| 19 | Gas shop | 2 | 50 | 50 | 2,500 |
| 20 | Galvanizing shop | 2 | 120 | 25 | 3,000 |
| 21 | Piping shop | 2 | 150 | 120 | 18,000 |
| 22 | Outfitting shop | 2 | 170 | 50 | 8,500 |
| 23 | Warehouse | 2 | 250 | 100 | 25,000 |
| 24 | Field shop house | 2 | 50 | 50 | 2,500 |
| 25 | Material outside stock area | 2 | 250 | 100 | 25,000 |
| 26 | Block stock area | 2 | 454 | 90 | 81,720 |
| 27 | Welfare centre | 1 | 100 | 115 | 11,500 |
| 28 | Guest house | 1 | 30 | 20 | 600 |
| 29 | Security office | 1 | 15 | 15 | 225 |
| 30 | Job training centre | 2 | 100 | 120 | 12,000 |
| 31 | User building | 1 | 100 | 120 | 12,000 |
| 32 | Hospital | 1 | 20 | 15 | 300 |
| 33 | Petrol bunk | 1 | 250 | 100 | 25,000 |
| 34 | Dewatering pumphouse | 1 | 200 | 100 | 20,000 |
| 35 | Administrative block | 4 | 454 | 90 | 40,860 |
| 36 | Conventional centre | 2 | 180 | 120 | 21,600 |
| 37 | Outfitting structures (office & work shop) | 2 | 200 | 100 | 20,000 |
| 38 | Main sub station | 1 | 60 | 30 | 1,800 |
| 39 | Staff canteen | 1 | 60 | 20 | 1,200 |
| 40 | Worker's canteen | 1 | 60 | 30 | 1,800 |
| 41 | Sewage treatment plant | 2 | 180 | 120 | 21,600 |

- Administrative block
- The administrative facility houses management offices, project coordination units and operational command centres. It is strategically located for effective monitoring, communication and coordination across all yard functions, ensuring efficient decision-making and resource deployment.
- Training and skill development centre
- A dedicated training facility is incorporated within the complex to support continuous skill development, technical upskilling and safety training of personnel. The centre is equipped with simulation tools, classrooms and workshops to ensure workforce competence in modern shipbuilding technologies and procedures.

These ancillary facilities are planned and implemented in accordance with international best practices and occupational safety norms, thereby contributing to a safe, efficient and sustainable shipyard ecosystem.

## 7.7 WET BASIN

The wet basin is like a dry dock without a dewatering facility and is used for outfitting works and underwater inspection. A wet basin is generally built for vessels less than 250 T.

## 7.8 SUMMARY

The parameters governing the choice of ship repairing facility are:

- DWT of vessel
- Tidal levels
- Tranquility
- Soil characteristics
- Availability of land for shore facilities
- Maintenance cost
- Operational cost and manpower requirement
- Capital cost
- Construction time

The construction of a dry dock is challenging in certain situations, and floating docks are a feasible technique in such conditions. The travel lift capital, maintenance and operation cost is less than the slipway and is the preferred option. An amphibian-type travel lift is also now available. The advantage of a floating dock is that the facility can be used for damaged ships. The most sophisticated, easily constructed and the operatable facility is ship lift and is preferred by navy and shipyards for bigger vessels.

# Annexure – Case Study

## MAJOR SHIPYARDS IN INDIA

India hosts a number of significant shipbuilding and ship repair facilities that contribute to both defence and commercial maritime sectors. The prominent shipyards in the country include:

1. *Garden Reach Shipbuilders & Engineers Ltd. (GRSE), Kolkata*
   A leading defence public sector undertaking (DPSU), GRSE specializes in the construction of warships and auxiliary vessels for the Indian Navy and Coast Guard. It is renowned for its indigenous ship design and engineering capabilities.
2. *Hindustan Shipyard Ltd. (HSL), Visakhapatnam*
   Located on the east coast, HSL is one of the oldest shipyards in the country. It plays a vital role in defence shipbuilding and submarine refits, operating under the Ministry of Defence.
3. *Larsen & Toubro Shipyard, Kattupalli*
   Operated by Larsen & Toubro Ltd., the Kattupalli shipyard near Chennai is a state-of-the-art facility with capabilities in both shipbuilding and offshore platform fabrication, catering to defence and commercial projects.
4. *Cochin Shipyard Ltd. (CSL), Kochi*
   As one of the largest shipbuilding and maintenance facilities in India, CSL is a key player in constructing complex vessels including aircraft carriers, and is also involved in ship repair and offshore structure fabrication.
5. *Goa Shipyard Ltd. (GSL), Goa*
   A DPSU under the Ministry of Defence, GSL is known for building medium-sized ships primarily for defence applications, including patrol vessels, missile crafts and survey ships.
6. *Mazagon Dock Shipbuilders Ltd. (MDL), Mumbai*
   MDL is India's premier warship and submarine builder, with a strong track record of delivering frontline naval platforms, including stealth frigates and conventional submarines.
7. *SWAN (Reliance Naval and Engineering Limited), Pipavav*
   Located in Gujarat, this private sector shipyard was once among the largest in the country. It has seen reduced activity due to financial difficulties but remains a significant asset in terms of infrastructure and location.

In addition to these operational shipyards, there are a few facilities currently classified as insolvent or non-operational:

- Sembmarine Kakinada
- A joint venture that has faced operational and financial setbacks, limiting its activity in the Indian shipbuilding market.

- Bharati Shipyard Ltd., Ratnagiri
- Once a major private shipbuilder, Bharati Shipyard experienced insolvency due to financial mismanagement and project delays.
- Hooghly Dock & Port Engineers Ltd. (HDPEL), West Bengal
- Formerly a key riverine and coastal shipbuilder, HDPEL is largely inactive today, undergoing restructuring due to prolonged financial challenges.

India presently ranks 20th in the list with about 0.05% of the overall shipbuilding output of the world. Over the past 10 years, the Indian shipyards have demonstrated a high of about ~97,000 GT per annum in 2017. This peak may be considered to be the total annual capacity of the Indian ecosystem at present. In term, there is little of building large cargo vessels, infrastructure capacity as well as capabilities are very limited. Therefore, even if demand is spurred through targeting increased domestic ownership of cargo vessels

The global commercial shipbuilding market, with an annual output of 64.7 million GT in FY 2023–2024, is estimated to be approximately INR 5.8 lakh crores (USD 70 billion), and is primarily dominated by China, South Korea and Japan which together account for about 85% of global commercial shipbuilding. Indian shipbuilding output for 2023–2024 was 0.04 million GT (0.06% of global shipbuilding) and India ranks 22nd in terms of annual output capacity.

## REVIEW QUESTIONS

1. What are the infrastructures available for a ship repair facility?
2. What is a slipway? Explain the components of a slipway.
3. How is the selection and orientation of the site for the slipway carried out?
4. Describe in brief the types of slipway with a sketch.
5. Name the loads considered in the design of the slipway.
6. Explain the travel lift and its components with a neat sketch. Write its advantages.
7. Describe the travel lift operation with the schematic diagram.
8. What are the components of a ship lift? Illustrate with diagrams.
9. What is meant by floating dock? Name the types of floating docks.
10. What is the purpose of a dry dock? Write its components.
11. What are the basic requirements needed for the functioning of a dry dock?
12. Write the types of dock floors available.
13. What are loads to be considered for the design of the dock floor?
14. Write the types of dock wall and the loads acting on it.
15. What is a wet basin? State its purpose.

## REFERENCES

**IS 2911 (Part I/Sec 2) (2010)** *Code of Practice for Design and Construction of Pile Foundations – Bored Cast in Situ Piles*

**IS 3443 (1980)** *Specifications for Crane Rail Sections* (First Revision)

**IS 4651 (Part 1) (2020)** *Code of Practice for Planning and Design of Ports and Harbours – Site Investigation* (First Revision)

**IS 4651 (Part 2) (2020)** *Code of Practice for Planning and Design of Ports and Harbours – Earth Pressure*

**IS 4651 (Part 3) (2020)** *Code of Practice for Planning and Design of Port and Harbours – Loading*

**IS 4651 (Part 4) (2014)** *Code of Practice for Planning and Design of Port and Harbours – General Design Considerations*

**IS 4651 (Part 5) (1980)** *Code of Practice for Planning and Design of Port and Harbours – Layout and Functional Requirements*

**IS 10020 (Part 4) (1981)** *Recommendations for Design and Construction of Port and Harbour Components, Part 4: Slipways*

**R. Sundaravadivelu, S.R. Gandhi, R. Natrajan, and C. Thilakavathy (2000)** Design of Slipway facility for repair and maintenance of port crafts, *Proceedings of 4th International Conference on Coasts, Ports & Marine Structures*, November, Shahid Port Complex, Bandar Abbass

# 8 Dredging and Disposal

## 8.1 DREDGING AND DISPOSAL

### 8.1.1 General

National Ocean Service defines dredging as the act of removing silt and other material from the bottom of water bodies. In general, dredging is an activity of removal of sediments/debris at the bed level of lakes, rivers, harbours, and other water bodies. Dredging activity is a routine necessity adopted in waterborne infrastructure worldwide due to the overall natural process of siltation and sedimentation.

Dredging often is focused on maintaining or increasing the depth of navigation channels, anchorages, or berthing areas to ensure the safe passage of boats and ships. Vessels require a certain amount of water depth to float and not hit the sea bed. This water depth continues to increase over time as larger ships are deployed. Since massive ships carry the bulk of the goods imported into the country, dredging plays a vital role in the nation's economy.

Dredging is also performed to reduce the exposure of fish, wildlife, and people to contaminants and to prevent the spread of contaminants to other areas of the water body. This environmental dredging is often necessary because sediments in and around cities and industrial areas are frequently contaminated with various pollutants. These pollutants are introduced to waterways from point sources such as sewer overflows, municipal and industrial discharges, and spills; or may be introduced from nonpoint sources such as surface runoff and atmospheric deposition.

### 8.1.2 Site Investigation

The site investigations particularly required for dredging work should provide data on tides and bathymetry; on wind, waves and currents; and on the nature of the materials to be dredged, their in-situ strengths, particle size distribution, degree of compaction and, in the case of silts, settling characteristics once disturbed. In all cases, standard classifications of material to be dredged should be used.

### 8.1.3 Dredge Operation

The selection of the most suitable dredger depends upon the material to be dredged, the depth of dredging, the quantity and disposition of the material, the location of the dumping ground, the rate of production required and also on whether the dredger may have complete or partial possession of the waterway.

The programme and sequence of the dredging operations require careful consideration. If possible, dredging should commence in a location which would be least liable to siltation from later dredging. Timing of the dredging activity relative to the construction of other parts of the project, some of which, such as breakwaters


DOI: 10.1201/9781003361275-8

or groynes, may give shelter or affect siltation, must be considered in the overall planning of the project. Timing may also depend on local seasonal variations of tidal currents and winds, which can render dredging activities much easier in one season than in another, especially in estuaries. A further factor which must be borne in mind is the date of the start of operations at a new port, which may be in the advanced stage of completion of the whole project so that the construction requirements of, or the need for access to, particular berths will influence the sequence of dredging.

The removal of solid rock constitutes a specialized dredging activity in that some means must be found to fragment the rock before it can be lifted from the sea bed by one of the conventional dredging methods described above. Usually, a bucket or grab dredger is used to remove rock debris, but suction dredgers have also been employed to remove the sufficiently fragmented rock.

The most commonly employed method of rock-breaking underwater is drilling and blasting. However, jointed rock with thin bedding layers may also be fragmented by a heavy chisel or a pneumatic hammer on the rock surface. Drilling and blasting underwater is a specialized, slow and expensive operation, and many trials may be needed to obtain the right results for the dredgers to be used. The dredging of coral or cemented sand causes frequent problems. These can sometimes be easily fragmented and dredged by a powerful cutter suction dredger. However, only careful investigation will show whether this is likely, and massive formations may need to be treated as a rock before dredging.

### 8.1.4 Economic Factors

In assessing dredging costs, it is important to remember that in many places, large equipment must be brought from far afield to do the work, with a consequent heavy mobilization cost. The quantity of dredging required for any particular scheme has a pronounced effect on the overall unit rate. For a modest amount of work, it may be found economical to employ simple equipment such as excavators on barges. Even though their working costs and productivity are less favorable, the savings in mobilizing the less sophisticated equipment can more than offset the less efficient performance in the field. Such judgments are difficult to make and are often best resolved when tenders for the work are invited, but during the project study, the various factors must be considered so as to permit a realistic estimate of dredging costs to be prepared.

## 8.2 TYPES

The following types of dredger are available for dredging.

### 8.2.1 Bucket Dredger

The modern bucket dredger comprises a continuous chain of buckets mounted on a ladder adjustable for depth. Each bucket discharges its load at the top of the ladder into chutes which direct the material into a hopper barge. Bucket dredgers are best confined to work in sheltered locations and are useful for fairly accurate trimming of

FIGURE 8.1 Dredger in basin.

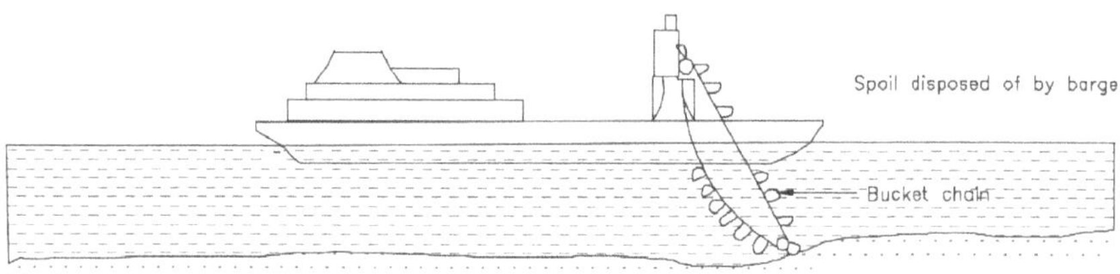


FIGURE 8.2 Self-propelling bucket dredger.

the bed. They can deal with some hard materials, but large pieces in the bucket can cause severe delays.

### 8.2.2 Grab Dredger

The grab dredger is usually a self-propelled vessel with a hopper and a grab crane. A simpler version which requires attendant barges is simply a crane on a pontoon.

### 8.2.3 Dipper Dredger

Excavation is achieved by means of a forward- or rearward-facing shovel mounted on the arm suspended in front of the pontoon body, which operates by digging into the underwater material.

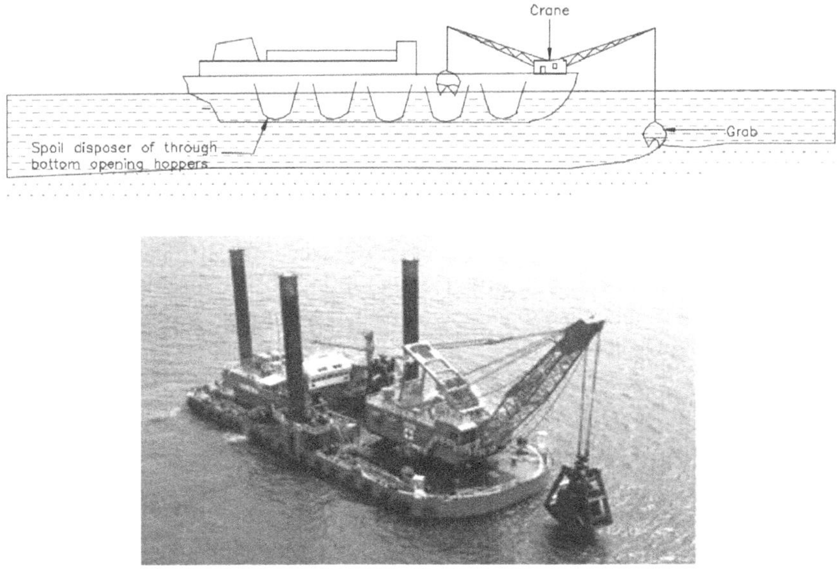


**FIGURE 8.3** Grab hopper dredger.

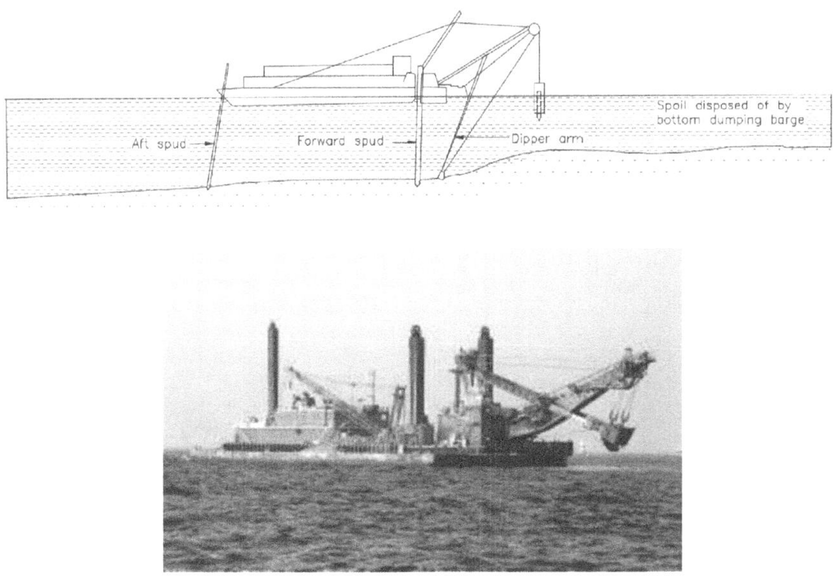


**FIGURE 8.4** Dipper dredger.

### 8.2.4 Suction Dredger

The basic components of this type of dredger are a pontoon hull which houses the pumps and engines, a suction pipe suspended from the vessel to the sea bed, and a delivery pipe from the pumps to a dumping area or, in some instances, to barges.

Only small granular material can be dredged by suction, and it is common for this type of dredger to be equipped with a rotating cutter on the end of the suction pipe. Cutters can be designed to suit the particular material to be dredged, and cutter suction dredgers can remove sand and gravels, medium-stiff clays and, with the addition of special cutting teeth, stiff clays and soft or broken rock. Output, however, varies considerably according to material type and conditions. Nevertheless, this cutter suction dredger is now the most usual dredging equipment for capital work and is particularly suited for reclaiming purposes.

### 8.2.5 Cutter Suction Dredger

The cutter suction dredgers are used to dredge the areas which has a low water depth. The cutter suction dredger (CSD) can dredge close to the platform. The working methodology is based on two anchor spuds anchored at the rear end of the dredger. A cutter anchored for the steadiness of dredged materials is transported through the floating pipeline for a lead up to 1–2 km. For disposal of more than 2 km, a booster

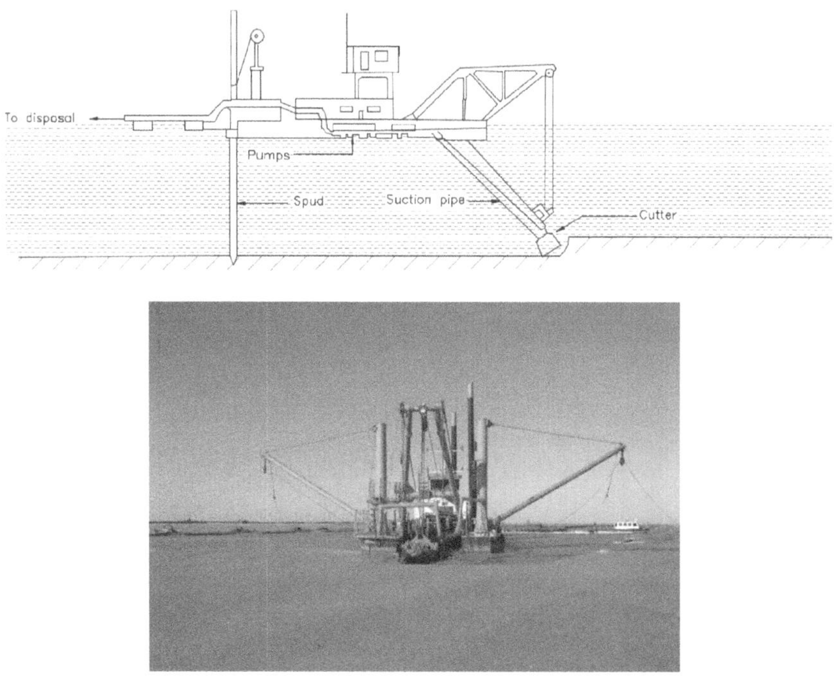


**FIGURE 8.5** Cutter suction dredger.

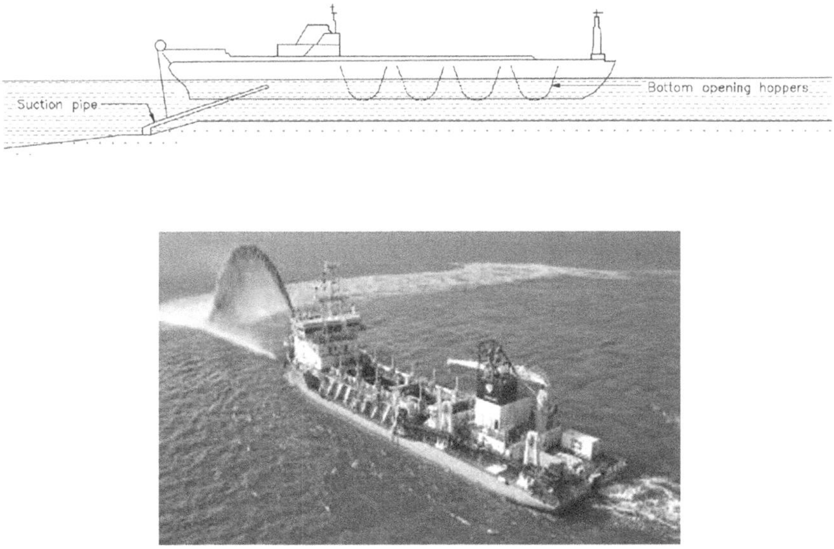


FIGURE 8.6 Trailer suction hopper dredger.

station is deployed, which can dispose of the material up to 4 km. The other way of disposal for cutter suction dredgers is to load into self-propelled barges. These barges will dump the dredged material into a deeper contour.

Heavy-duty CSD can work in open sea conditions with a barge loading system with the combination of four to six hopper barges of 2000–3000 $m^3$ capacity.

### 8.2.6 Trailer Suction Hopper Dredger

The trailer suction hopper dredger (TSHD) is a self-propelled vessel with suction pipes suspended from one or from both sides. The dredged material is delivered through the suction pipes to the hopper. When the hopper is full, the vessel proceeds to the dumping ground. This type of dredger is widely used in the maintenance of channels, where its ability to manoeuvre as a ship is a distinct advantage. A further advantage of this type of vessel, when compared with the other types discussed, is its ability to remain effective in rough water and offshore locations. It is, however, suitable only for relatively loose materials as would be found in capital dredging and also in large quantities of maintenance dredging.

The TSHD with hopper capacity varies from 8,000 $m^3$ to 50,000 $m^3$ with a dredging draft of about 8–12 m. The only disadvantage is that TSHD cannot dredge in lesser water depth like the cutter suction dredger.

### 8.2.7 Backhoe Dredger

The backhoe dredger is used with the hopper barge of 1000 $m^3$ capacity, and it shall be placed about the self-propelled barge having the bucket capacity of 1.50–2.00 $m^3$.

**FIGURE 8.7** Backhoe in front of sharf.

**FIGURE 8.8** Backhoe discharging in barge.

## 8.3 CAPITAL AND MAINTENANCE DREDGING

- Capital dredging is to provide a suitable depth of water both in the approach channel and the area within the breakwaters. It is envisaged that most of the channel dredging will be carried out by large trailer suction hopper dredgers.
- Cutter suction dredgers or shallow draught trailer dredgers will initially be required to dredge and deepen the shallower areas to permit access for the large trailer dredgers.
- Dredge depth for the outer approach channel and harbour basin is considered for technical evaluation and cost estimates.
- Any unsuitable material (silts and clays) will need to be either disposed of offshore or onshore in landscaped areas or areas where there will be no subsequent loadings (e.g. low-lying areas).
- Surveys should be carried out to indicate that there is no rock to be dredged and the material likely to be encountered is sand and clay.
- For the purposes of this study dredged side slope of 1 in 5 has been assumed, although this may be refined when more data is available.
- The construction of breakwaters at the port would cause erosion on the northern side of the port and accretion (deposition) on the southern side. The major quantum of maintenance dredging at the port will be in the form of dredging the approach channel.
- For transport of the dredged material to the northern side, it is proposed to use a combination of groin field and beach nourishment using the rain bowing technique.

**FIGURE 8.9** Split barge.

FIGURE 8.10 Split barge in reclamation area.

## 8.4 DREDGE DISPOSAL

- The best dredge disposal option available for the project is to dispose the dredge material at a deeper contour.
- The dredge disposal can be with hopper barges which will load the materials from the dredging and dump them into the deeper contour position.
- The dredged material from the harbour basin and approach channel will be disposed on to the northern/southern part of the shoreline, where erosion may likely take place for beach nourishment.
- Dredged material will be disposed in the sea at a 20 m depth contour, and it is very likely that suitable disposal sites will be readily available to authorities.

## 8.5 RECLAMATION

It is of considerable advantage if the dredged material from a port project can be used for reclamation projects, but only certain granular soils are suitable for this purpose. Silts and clays are generally much more difficult to use. A thorough geological and geotechnical survey of the extension area is required to ensure that the existing subsoils can support future buildings.

The settlement behaviour of the fill material should be carefully analysed and monitored before construction is permitted. Pre-loading through the construction of embankments may be needed to achieve settlement within a reasonable period before

**FIGURE 8.11** Reclamation progress.

buildings are erected. The geotechnical analysis will generally provide the correct solutions.

An important aspect in project planning is the balance of dredging and reclamation, and this must be assessed in each case with a view to the economy.

## REVIEW QUESTIONS

1. What is dredging? Explain the need for dredging.
2. What are the study and investigations to be done before dredging?
3. Describe in brief the dredging operation.
4. What are the types of dredgers used in dredging operation?
5. What is the difference between a bucket dredger and a grab dredger?
6. Describe in brief the types of suction dredger with a neat sketch.
7. What is a backhoe dredger? Write its advantages and disadvantages.
8. What are the best methods available for the disposal of dredged materials?
9. What is meant by reclamation of land and how the dredged materials are used?
10. Why is capital and maintenance dredging done?

## REFERENCES

**IS 4651 (Part I) (1974)** *Code of Practice for Planning and Design of Ports and Harbours – Site Investigation* (First Revision)

**IS 4651 (Part 2) (1989)** *Code of Practice for Planning and Design of Port and Harbours – Earth Pressures* (First Revision)

**IS 4651 (Part 5) (1980)** *Code of Practice for Planning and Design of Port and Harbours – Layout & Functional Requirements*

**National Ocean Service**, *Dredging*. National Oceanic and Atmospheric Administration U.S. Department of Commerce. Available at https://oceanservice.noaa.gov/facts/dredging.html

**UNCTAD (1985)** *Port Development: A Handbook for Planners in Developing Countries* (2nd Edition Revised), United Nations Conference on Trade and Development, New York

# 9 Environmental Impact Assessment and Environmental Management Plan

## 9.1 ENVIRONMENTAL IMPACT ASSESSMENT (EIA) AND ENVIRONMENTAL MANAGEMENT PLAN (EMP)

The environment plays a vital role in the overall development of the country. Recognizing the importance of environmental protection and sustainable development, the Ministry of Environment and Forest, Government of India has formulated policies and procedures governing the industrial and other developmental activities to prevent indiscriminate exploitation of natural resources and to promote the integration of environmental concern in developmental projects.

Ports and harbours are the gateways of maritime trade. The need to enhance the standard of living calls for continual growth in the economy and for rapid industrialization, as well as commensurate development of all the associated sectors of the country. Inasmuch as maritime transport is economical and the only means to transport larger volumes of cargo across oceans, ports and harbours are therefore called upon to handle larger volumes of cargo throughputs of both raw material as well as products. Thus, demand for handling bigger vessels and deploying state of the art cargo handling systems many times require augmentation/expansion of facilities at existing ports and development of new ports and harbours. However, port development and operational activities may create a wide range of impacts on the environment through activities like dredging, reclamation, construction work, development of utilities and services, discharges from ships and waterfront industries, cargo operations and other port related activities. The potential adverse effects of port development encompass water pollution, contamination of bottom sediments, damage to marine ecology and fisheries, beach erosion/accretion, current pattern changes, waste disposal, oil leakage and spillage, emission of hazardous gases, air pollution, noise pollution, floodlight effect and other socio-cultural impacts. The preparation of an EIA (environmental impact assessment) report and implementation of an EMP (environmental management plan) are essential to effectively address these adverse effects.

DOI: 10.1201/9781003361275-9

### 9.1.1 Major Sources of Environmental Impacts in Respect of Port and Harbour Projects

Major sources of the adverse effects on environment because of development of port and harbour projects can be categorized into the following types:

- Location of port
- Construction
- Port operation, including ship traffic and discharges
- Cargo handling, storage and land transport

## 9.2 COASTAL REGULATION ZONE (CRZ)

### 9.2.1 General

The purpose of coastal regulations in India is to preserve the coastal environment by regulating the use of land near the Indian coastline. Developmental activities along the Indian coast are governed by the provisions of the Coastal Regulation Zone (CRZ) Notification 1991, under the Environment Protection Act 1986. This notification provides guidelines for protection and use of the land within 500 m of the coast and 100 m along tidal-influenced water bodies.

The Central Government declared the coastal stretches of seas, bays, estuaries, creeks, rivers and backwaters which are influenced by tidal action (in the landward side) up to 500 m from the high tide line (HTL) and the land between the low tide line (LTL) and the HTL as Coastal Regulation Zones (CRZ) on 19 February 1991. On 7 January 2011, the Indian Ministry of Environment and Forests (MoEF) released Coastal Regulation Zone (CRZ) Notification 2011 to replace the CRZ Notification of 1991. Also for the first time, an Island Protection Zone Notification (IPZ) was released to cover Andaman & Nicobar Islands, Lakshadweep.

#### 9.2.1.1 Objectives of Coastal Regulation Zone

The main objectives of the CRZ Notification 2011 are:

- To ensure livelihood security to the fishing communities and other local communities living in the coastal areas.
- To conserve and protect coastal stretches.
- To promote development in a sustainable manner based on scientific principles considering the dangers of natural hazards in the coastal areas and sea level rise due to global warming.

### 9.2.2 CRZ Mapping and Classification of CRZ

The CRZ maps indicating the HTL and LTL, demarcated by one of the authorized agencies and the project layout superimposed on the map should be submitted on 1:5000 scale map. The state/union territory Coastal Zone Management Authority (CZMA) shall recommend this map.

The CRZ Map delineates the LTL, HTL, HTL+200 m, HTL+500 m and LTL to 12 nautical miles.

**Necessary details to be provided in this map:**

- A map specifying locations of the state, district and project location.
- A map of project area and 10 km area from boundary of the proposed/existing project area, delineating protected areas notified under the Wild Life (Protection) Act, 1972/critically polluted areas as notified by the Central Pollution Control Board (CPCB) from time to time/notified eco-sensitive areas/interstate boundaries and international boundaries.
- A map covering aerial distance of 15 km on the landward side from the proposed project boundary delineating environmental sensitive areas as specified in Form I of EIA notification dated 14 September 2006.
- Site layout plan of the proposed development should be submitted on a scale of 1:5000, clearly marking the layout of breakwaters, navigation channels, harbour basin, Single Buoy Moorings (SBMs), moorings, berths number and dimensions, water depths, dry docks, workshops, container freight station, cargo handling systems, conveyors, pipe lines, covered and uncovered storage yards with their capacities, warehouses, transit sheds, roads, railway tracks, water supply, effluent disposal point, drainage, power, administrative and operational buildings, utilities, welfare and community centres, townships, greenbelt, dredged material disposal, etc. Boundaries of the proposed port shall be shown therein with latitude and longitude.
- Area drainage contour map of the project area and 2–5 km from the proposed project area shall be clearly indicated. In case of any proposed diversion of nallah/canal/river, the same should be shown in the map.
- Hydrographic charts of the offshore area giving the general morphology of the coastal stretch to a scale of 1:50,000 shall be submitted covering water depth up to 10 m beyond the maximum proposed dredging depths of the project and covering 5 km along the coast from the project limits on both sides.

Coastal Regulation Zones are classified as follows (Natami-Centre for Policy Research, Environmental Justice Program):

- Zone I CRZ I (ecologically sensitive)
- Zone II CRZ II (built-up area)
- Zone III CRZ III (rural area)
- Zone IV CRZ IV (water area)

The Coastal Regulation Zones are described in the following sections.

#### 9.2.2.1 CRZ I (Zone I)

CRZ IB is from LTL to HTL, and port and harbour development is a permissible activity.

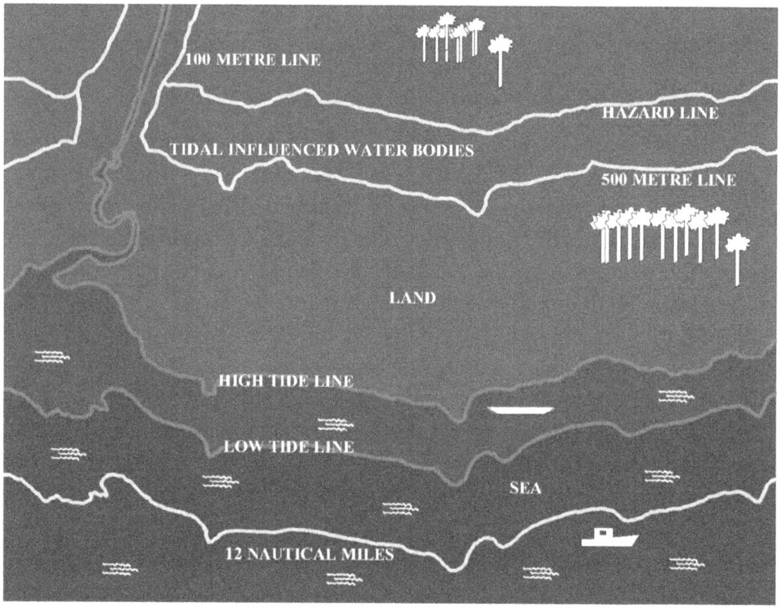


**FIGURE 9.1** Area Covered by CRZ I, II, III and IV.

**Source: Namati-Centre for Policy Research.**

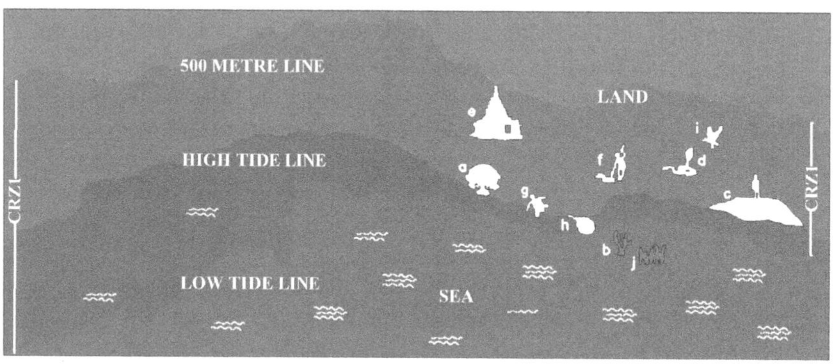


**FIGURE 9.2** CRZ I.

**Source: Namati-Centre for Policy Research.**

The areas (CRZ IA is from LTL to HTL+500 m) that are ecologically sensitive and the geomorphologic features which play role in the maintaining the integrity of the coast:

- If mangrove areas are more than 1,000 m$^2$, a buffer area of 50 m along the mangroves shall be provided.
- Corals and coral reefs and associated biodiversity.

- Sand dunes.
- Mudflats which are biologically active.
- National parks, marine parks, sanctuaries reserve forests, wildlife habitats and other protected areas under the provisions of Wild Life (Protection) Act, 1972 (53 of 1972), the Forest (Conservation) Act, 1980 (69 of 1980) or Environment (Protection) Act, 1986 (29 of 1986); including Biosphere Reserves.
- Salt marshes.
- Turtle nesting grounds.
- Horseshoe crab habitats.
- Sea grass beds.
- Nesting grounds of birds.
- Areas or structures of archaeological importance and heritage sites.

No construction including port and harbour structure is permitted in CRZ IA. Special permission is required for any construction activities.

1. The area between LTL and HTL

### Norms for regulation of activities permissible in CRZ I

1. No new construction shall be permitted in CRZ I except:
    - Projects relating to the Department of Atomic Energy
    - Pipeline conveying systems including transmission lines
    - Facilities that are essential for activities permissible under CRZ I
    - Installation of weather radar for monitoring of cyclone movement and prediction by the Indian Meteorological Department
    - Construction of transharbour sea link and without affecting tidal flow of water between LTL and HTL
    - Development of greenfield airport already approved at only Navi Mumbai
2. Areas between LTL and HTL which are not ecologically sensitive and necessary safety measures will be incorporated while permitting the following:
    - Exploration and extraction of natural gas
    - Construction of dispensaries, schools, public shelter, community, toilets, bridges, roads, jetties, water supply, drainage and sewerage which are required for traditional inhabitants living within the biosphere reserves after obtaining approval from concerned CZMA
    - Necessary safety measures shall be incorporated while permitting such developmental activities in the area falling in the hazard zone
    - Salt harvesting by solar evaporation of sea water
    - Desalination plants
    - Storage of non-hazardous cargo such as edible oil, fertilizers and food grain within notified ports
    - Construction of transharbour sea links, roads on stilts or pillars without affecting the tidal flow of water

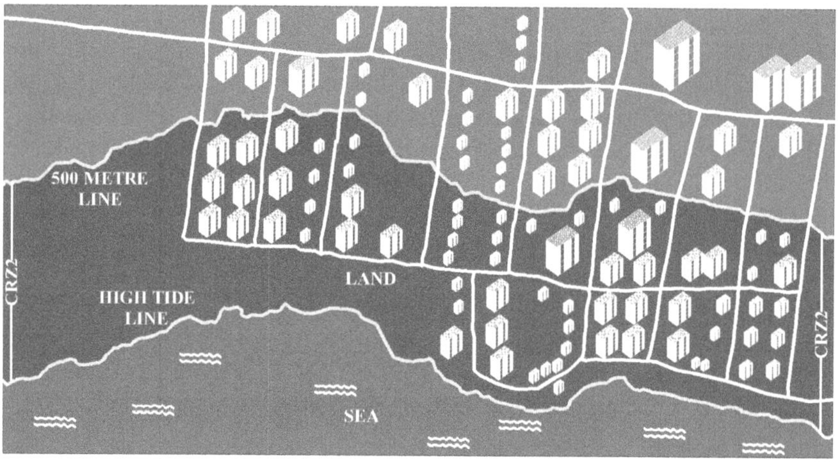


**FIGURE 9.3** CRZ II.

**Source: Namati-Centre for Policy Research.**

### 9.2.2.2 CRZ II (Zone II)

The area that have already been developed up to or close to the shoreline. For this purpose, developed area is referred to as that area within the municipal limits or in other legally designated urban areas which is already substantially built up and which has been provided with drainage and approach roads and other infrastructural facilities such as water supply sewerage mains.

**Norms for regulation of activities permissible in CRZ II:**

- Buildings shall be permitted only on the landward side of the existing road, or on the landward side of the existing authorized structures.
- Buildings permitted on the landward side of the existing and proposed roads or existing authorized structures shall be subject to the existing local town and country planning regulations including the 'existing' norms of floor space index or floor area ratio, provided that no permission for construction of buildings shall be given on the landward side of any new roads which are constructed on the seaward side of an existing road.
- Reconstruction of authorized building to be permitted subject with the existing floor space index or floor area ratio norms and without change in present use.
- Facilities for receipt and storage of petroleum products and liquefied natural gas appended to this notification and facilities for regasification of liquefied natural gas.
- Desalination plants and associated facilities.
- Storage of non-hazardous cargo, such as edible oil fertilizers and food grain in notified ports.
- Facilities for generating power by non-conventional power sources and associated and associated facilities.

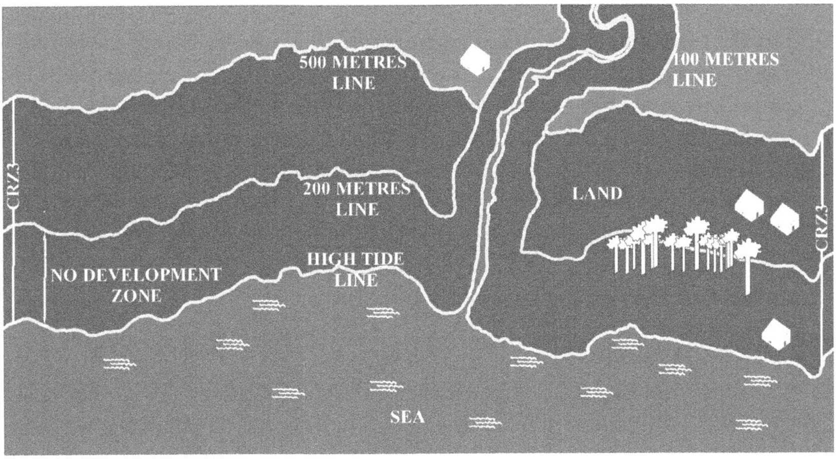


**FIGURE 9.4** CRZ III.

Source: Namati-Centre for Policy Research.

### 9.2.2.3 CRZ III (Zone III)

Areas that are relatively undisturbed and those which do not belong to either CRZ I or CRZ II. These will include coastal zone in the rural areas (developed and undeveloped) and also areas within municipal limits or in other legally designated urban areas which area not substantially built up.

**Norms for regulation of activities permissible in CRZ III:**

- Building permission for such construction/reconstruction will be subject to conditions that the total number of dwelling units shall not be more than twice the number of existing units, total covered up area on all floors shall not exceed 33% of the plot size.
- The overall height of the construction shall not exceed 9 m and construction shall not be more than two floors (ground floor plus one floor).

**Area between 200 meters to 500 meters:**

The following activities shall be permissible in the above areas:

- Development of vacant plot in designated areas for construction of hotels or beach resorts for tourists or visitors subject to the conditions as specified in the guidelines at Annexure-III of CRZ Notification 2011.
- Facilities for receipt and storage of petroleum products and liquefied natural gas as specified in Annexure-II of CRZ Notification 2011.
- Storage of non-hazardous cargo such as edible oil, fertilizers and food grain in notified ports.
- Foreshore facilities for desalination plants and associated facilities.
- Facilities for generating power by non-conventional energy sources.

- Construction or reconstruction of dwelling units so long it is within the ambit of traditional rights and customary uses such as existing fishing villages. Building permission for such construction or reconstruction will be subject to local town and country planning rules with overall height of construction not exceeding 9 m with two floors (ground floor + one floor).
- Construction of public rain shelters, community toilets, water supply drainage, sewerage, roads and bridges by CZMA who may also permit construction of schools and dispensaries for local inhabitants of the area for those panchayats, the major part of which falls within the CRZ if no other area is available for construction of such facilities.

### 9.2.2.4 CRZ IV (Zone IV)

The water area from the LTL to 12 nautical miles on the seaward side. It shall include the water areas of the tidal influenced water body from the mouth of the water body at the sea up to the influence of tide which is measured as 5 ppt (parts per thousand) during the driest season of the year.

**Norms for regulation of activities permissible in CRZ IV:**

- In CRZ IV areas, there is no restriction on the traditional fishing and allied activities undertaken by local communities.
- However, no untreated sewage, effluents or solid waste shall be let off or dumped in these areas.
- A comprehensive plan for treatment of sewage generating from city must be formulated within a period of one year from the date of issue of this notification and be implemented within two years thereafter.

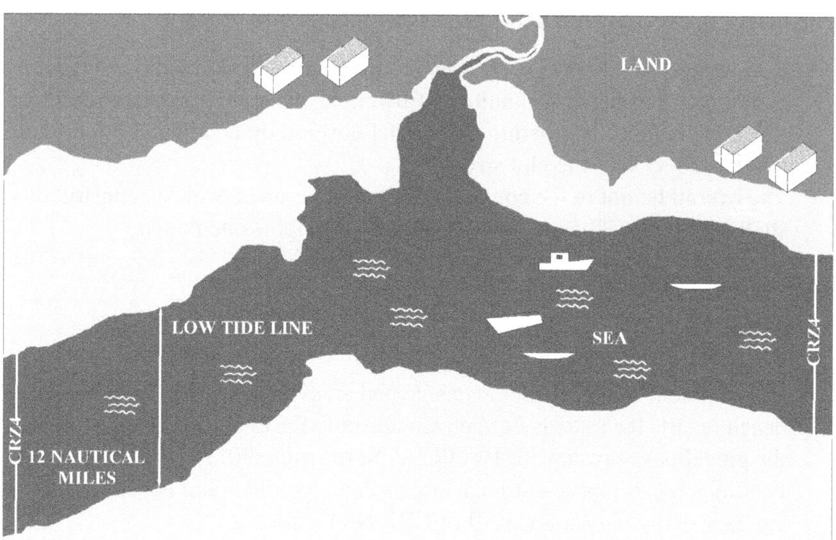


FIGURE 9.5 CRZ IV.

Source: Namati-Centre for Policy Research.

## Areas Requiring Special Considerations

Areas requiring special consideration for protecting the critical coastal environment and difficulties faced by local communities:

1. The CRZ areas falling within municipal limits of Greater Mumbai.
2. The CRZ areas of Kerala including the backwaters and backwater islands.
3. The CRZ areas of Goa.
4. Critically vulnerable coastal areas (CVCAs) such as the Sunderbans region of West Bengal and other ecologically sensitive areas identified as under the Environment (Protection) Act, 1986 and managed with the involvement of coastal communities including fisher folk.

For instance, special dispensation is given to Sunderbans and other ecologically sensitive areas:

- Sunderbans is the largest mangrove area in the country. Nearly 500,000 people live within the biosphere itself.
- Sunderbans is being greatly affected by the rising sea levels and changing weather patterns caused by climate change.
- Further, the local communities face immense hardship due to lack of infrastructure facilities.
- An integrated management plan is proposed to be drawn up, under the notification, for Sunderbans and such ecologically important areas such as the Gulf of Khambhat and Gulf of Kutch in Gujarat, Ratnagiri in Maharashtra, Karrwar and Coondapur in Karnataka, Vembanad in Kerala, Bhaitarkanika in Orissa, Coringa, and East Godavari and Krishna in Andhra Pradesh.
- These areas would be declared as CVCAs, and an integrated management plan would be prepared for each of these areas in consultation with the local communities.
- Until the integrated management plans are drawn up and initiated, all necessary infrastructure developments for the local inhabitants in these areas shall be permitted on a case to case basis by the CZMA.

## List of Petroleum and Chemical Products Permitted for Storage in CRZ except CRZ IA:

1. Liquefied petroleum gas
2. Crude oil
3. Motor spirit
4. Kerosene
5. Aviation fuel
6. High-speed diesel
7. Lubricating oil
8. Butane
9. Propane
10. Compressed natural gas
11. Naphtha

12. Furnace oil
13. Low sulphur heavy stock
14. Liquefied natural gas
15. Fertilizers and raw materials for manufacture of fertilizers

## 9.3 ENVIRONMENT CLEARANCE

The steps for construction of major project on marine work include the following:

- Preliminary project report/technical feasibility report.
- Form I duly filled in along with terms of reference (TOR).
- Clearance from the state government:
  a. Clearance from the State Pollution Board to be obtained for giving consent for establishment (COE) of port facility.
  b. Clearance from the state Forest Department for establishment of port facility (Chief Wild Life Warden of the state government).
  c. CRZ approval based on public hearing and state CZMA presentation.
- The expert committee of MoEF will after scrutiny of the Form I will send final terms of reference (TOR) to be submitted to the committee covering any additional TOR they may require apart from those already mentioned in Form I by the developer.
- Detailed project report.

### Documents required for the proposed project in chronological order:

- Form I
- Rapid EIA report
- Comprehensive EIA report
- Disaster management report
- Risk assessment report
- Management plan
- No objection certificate
- Four separate CRZ maps

### After submission of documents:

- Recommendations from the CZMA
- Clearance from the MoEF, EIA, CRZ within 60 days
- Processing fees: charges towards CZMA

Project construction commences within five years of clearance being issued. Half yearly compliance reports submitted to CZMA (1 June and 31 December). A compliance report will be displayed on the website of regulatory authority post clearance monitoring.

The stages of environmental clearance process for new projects:

1. Screening – The clearance depending upon the nature and location of project.
2. Scoping – All relevant environmental concerns for the preparation of an EIA report in respect of the project or activity for which prior environmental clearance is sought.

3. Public consultation – The process by which the concerns of local affected persons and others.
4. Appraisal – Detailed inspection by the Expert Appraisal Committee (EAC) and State Expert Appraisal Committee (SEAC) of the EIA report.

The validity of environmental clearance allowed for ports and harbours sector is valid for a period of five years. The regulatory authority concerned may extend this validity period by a maximum period of five years.

## 9.4 TERMS OF REFERENCE

Development of port facilities can make a significant contribution to the economic development and the growth of maritime transport. At the same time, it may also create adverse impacts on the surrounding environment. Port development may create a wide range of impacts on the environment through activities like construction work, dredging, reclamation, landfills, discharges from ships and cargo operations, and other port-related activities. Port development and operation should, therefore, be planned with careful consideration of their environmental impacts. The preparation of EIA report and implementation of EMP is essential for effectively managing these adverse effects.

EIA-EMP report should be based on maximum rated capacity of the project in terms of cargo handling, technology, equipment, manpower, resource use, etc. The report should be based on generic structure given in upcoming sections to the EIA notification 2006 for the project or its expansion based on proposed peak rated capacity. The report should incorporate the page numbers of various chapters, sections and sub-sections, tables, appendices, drawings, figures, etc., with titles shall be clearly indicated under the heading contents.

Ports and harbours with ≥5 million TPA of cargo handling capacity (excluding fishing harbours) are classified as category A projects and with <5 million TPA cargo-handling capacity and/or ≥10,000 TPA of fish-handling capacity are classified as category B projects, subject to the applicability of General Conditions as stipulated in the EIA Notification, 2006.

The upcoming sections shall be given in EIA report.

### 9.4.1 Introduction

This section should cover the following.

- Purpose of the project, project proponent, brief description of the project (name, nature, size, location of the project, its importance to the country and the region).
- Land description: plot/survey nos./village, tehsil, district, state and extent of the land. Profile of the project proponent, name and contact address with e-mail, implementing organization, organizational chart, project consultants etc. and whether the project attracts the provisions of General Conditions of EIA Notification 2006. If so, applicability should be discussed.
- The proponent should confirm that the project meets the central/state/local environmental regulations and standards applicable for the project.

**TABLE 9.1**
**Ministry of Environment and Forests (MoEF) Notification – 2006**

| Project or Activity | Category With Threshold Limit A | B |
|---|---|---|
| Ports & Harbours* | ≥5 million tonne per annum (TPA) of cargo-handling capacity (excluding fishing harbours) | <5 million TPA of cargo-handling capacity and/or ports/harbours ≥10,000 TPA of fish-handling capacity |
| Authority for approval of TOR &issue/reject of EC | MoEF, GOI on the recommendations of Expert Appraisal Committee (EAC) | State/Union territory Environmental Impact Assessment Authority (SEIAA) on the recommendations of State or Union territory level Expert Appraisal Committee (SEAC) |

General conditions shall apply*

General Conditions (GC): Any project or activity specified in Category 'B' will be treated as Category A, if located in whole or in part within 10 km from the boundary of: (i) protected areas notified under the Wild Life (Protection) Act, 1972; (ii) critically polluted areas as notified by the Central Pollution Control Board from time to time (Annexure-2); (iii) notified eco-sensitive areas; or (iv) inter-state boundaries and international boundaries

- Any litigation pending against the proposed project and/or any direction/ order passed by any court of law against the project, and if so, details thereof.
- In case of expansion/modernization of the project, the environmental compliance status for the existing project should be explained.

### 9.4.2 Project Description

In this section the following details are to be included.

- The broader details of the basic activities, location, layout and implementation schedule of the project.
- Type of the project: new, expansion, modernization, container cargo handling facility, fishing, minor/major protection.
- Relevance of the project in the light of the existing development plans of the region.
- Project coverage, master plan, phasing and scope.
- Description of a project site, geology, topography, transport and connectivity, demographic aspects, socio-cultural and economic aspects, villages, settlements.
- Capacity of the port, types of cargo proposed for handling, cargo handling equipment, ancillary operations, housing, truck parking details, etc.
- Technologies involved for design, construction, equipment and operation.
- Use of existing public infrastructure: road, railway and inland waterway networks, water supply, electrical power, etc.

- Estimated water budget for the proposed project: during construction/operation stages.
- Estimated cost of development of the project, environmental cost, funding agencies (whether governmental or based on BOT etc.).
- Details of land acquisition, rehabilitation of community's/village's present status of such activities.
- Resources, manpower and time frame etc., required for project implementation.

### 9.4.3 Analysis of Alternatives (Technology and Sites)

In case, the scoping exercise results in need for alternatives this chapter shall include:

- Description of various alternatives like locations or layouts or technologies studied
- Description of each alternative
- Summary of adverse impacts of each alternative
- Selection of alternative

### 9.4.4 Description of the Environment

*Study area*: As a primary requirement of EIA process, the proponent should collect primary baseline data in the project area as well as in the area falling 5 km from the proposed project boundary, and secondary data should be collected within 15 kms aerial distance from the project boundary, as specifically mentioned at column 9(iii) of Form I of EIA Notification 2006. The study areas mentioned in this document shall be considered for guidance purposes, but the exact study area for different environmental attributes (water, air, noise, soil, etc.) is to be submitted considering the proposed activities and location, along with proper reasoning for review and approval by the expert appraisal committee.

*Land environment*: Availability of land for earmarking for the port without causing any due hardship to local habitat and their socio-cultural and economic aspects is very important. Data on the land availability is to be ascertained from local authorities, revenue records etc. Justification for the proposed quantum of the area is to be given.

*Topography*: Baseline data to be given on description of existing situation of the land at the proposed project area including description of terrain: hill slopes, coastal and inland topography, coastal features (lowland, beaches, littoral areas, shoal areas). Study of land use pattern, habitation, cropping pattern, forest cover, HTL, environmentally sensitive places, etc. by employing remote sensing techniques (if available) and through secondary data sources.

*Geology*: Baseline data to be provided on rock types, regional tectonic setting (reported fractures/faulting, folding, warping), and history of any volcanic activity, seismicity and associated hazards, mainly in the coastal area.

Information on quarry yields, strengths of rock, distance of quarries from habitat, restrictions for quarrying, environmental controls, statutory permissions, etc., should be provided.

*Soil*: Soil data including type, classification, characteristics, soil properties, etc., are important from engineering considerations for design of structures, loading capacities of cargo stockpiles, green belt development, etc. Changes in parameters of soil also may affect plantation and vegetative growth, which in turn may endanger the health of the local habitat. Baseline data of the soil, results of investigations carried out to be provided for the project area.

*Meteorological data*: Meteorological data covering the following should be incorporated in the EIA report. The data for at least a 10-year period should be presented from the nearest meteorological station, except for the history of cyclones and tidal surges for which 100-year data is required.

- Wind speed and direction
- Rainfall
- Relative humidity
- Temperature
- Barometric pressures
- History of cyclones

#### 9.4.4.1 Water Environment

*Ground water*: Baseline data of ground water including data of pH, dissolved solids, suspended solids, Biochemical Oxygen Demand (BOD), Dissolved Oxygen (DO), coliform bacteria, oil and heavy metals (depending upon the type of cargo) is to be collected at least for one season. Usage purpose of the ground water, if any, is to be indicated.

*Surface water*: Baseline data on location of surface water like lagoons, lakes, tidal inlets, streams, rivers, their details, present quality and their utility, if any, is to be provided. Details of water bodies in the project area shall be described specifically. Water quality is to be monitored for one season.

#### 9.4.4.2 Marine Environment

*Coastal hydrology/geomorphology*: Coastal hydrology requires collection of oceanographic data during the study period, covering the following parameters:

- Tides
- Waves (wind waves and swells)
- Storm surges
- Currents
- Salinity
- Sea water temperature
- Suspended load
- Sea bed bathymetry

Baseline oceanographic data should extend at least to depths more than 10 m of proposed deepening of the harbour approach and basin as per master plan proposed. A study on likely changes in the sediment transport and littoral drift due to the construction of the port, particularly the breakwater, should to be taken up.

Details of mangroves, marshes and other coastal vegetation, sand dunes, coastal stability, seismic characteristics, history of any endangered species, coastal erosion and shoreline changes should be furnished.

#### 9.4.4.3 Bed Sediment Contamination

Baseline data on bottom sediments and the associated bottom biota and other physical habitat at the proposed project area and the neighbourhood areas must be collected and analysed.

#### 9.4.4.4 Sea/Harbour Water Quality

Baseline data shall be collected on chemical parameters in the open sea and in the proposed port area for understanding hydrochemical characteristics in the marine environment (sea water temperature, BOD, DO, pH, Total Suspended Solids (TSS), salinity, heavy metals depending upon the cargo, etc.).

#### 9.4.4.5 Biological Environment

*Marine/Coastal Ecology*

Baseline data of aquatic flora and fauna at the project area, including the coastal area, is to be ascertained by proper surveys including mangroves, marshes, other coastal vegetation and sand dunes. Data on coastal stability, seismic characteristics, history of any endangered species, coastal erosion and shoreline changes, if any, is also necessary.

*Flora and Fauna in the Neighbourhood*

Details on secondary data on the existing flora and fauna in the study area as well as 15 km from its boundary, carried out by an university/institution under the relevant discipline (Botanical Survey of India (BSI), Zoological Survey of India (ZSI), Wildlife Institute of India (WII) etc.) shall be included in the list of flora and fauna along with classification as per Schedule given in the Wild Life (Protection) Act, 1972 (for fauna) and in the Red Book Data (flora) and a statement clearly specifying whether the study area forms a part of an ecologically sensitive area or migratory corridor of any endangered fauna.

*Air Environment*

Baseline data of ambient air parameters, namely Respirable Suspended Particulate Matter (RSPM), nitrogen dioxide, sulphur dioxide, carbon monoxide, heavy metals and other harmful air pollutants depending upon the type of the cargo, should be monitored.

This data should be collected in an area extending at least 5 km from the project boundary by observation at a number of locations. Specific importance should be attached to areas near the project, say up to 1 km. One season data should be monitored other than monsoon as per the CPCB norms. One station should be in the upwind/non-impact/non-polluting area as a control station.

*Noise*

Baseline data on noise pollution at the project area and the neighbourhood up to 1 km or nearest residential areas is to be monitored as per the CPCB norms.

*Existing Solid Waste Disposal Facilities*

Details of authorized municipal solid waste facilities, biomedical treatment facilities and hazardous waste disposal facilities in the area should be inventoried, in case it is proposed to utilize the same.

*Socio-economic and Occupational Health Environments*

Baseline data at the project area shall include the demography, particularly on human settlements, health status of the communities, existing infrastructure facilities in the proposed area and area of impact due to the proposed activity. Present employment and livelihood of these populations and awareness of the population about the proposed activity shall also be included.

*Public Utilities*

Baseline data of existing public utility infrastructure shall be ascertained and reported to assess the impacts of the project on these public utilities to incorporate desired methods in the EMP and monitor the same during the construction and operational phases of the port.

### 9.4.5 Anticipated Environmental Impacts and Mitigation Measures

This section describes the likely impact of the project on each of the environmental parameters, methods adopted for assessing the impact such as model studies, empirical methods, reference to existing similar situations, reference to previous studies, details of mitigation methods proposed to reduce adverse effects of the project, best environmental practices and conservation of natural resources. The identification of specific impacts followed with mitigation measures should be done for different stages, such as location of the port, construction including dredging, ship traffic including discharges from vessels and cargo operations.

#### 9.4.5.1 Land Environment

*Anticipated Impacts*

Impact of project construction/operation on the land requirement/land use pattern should be assessed. Effect of future growth of the port facility and/or of the ancillaries should be carefully assessed by preparing master plans for the port and the ancillaries. Impact on the public utilities arising out of the utilities for the project activities and impact on the natural drainage system are equally important. Prediction of impacts should include impacts on the existing infrastructures like road network, housing, ground water/surface water, and loss of productive soil and impact on natural drainage pattern.

*Mitigation Measures*

Mitigation measures to reduce adverse effects like adopting soil improvement techniques and adopting suitable design methods to reduce land requirement. Where land acquisition and consequential Rehabilitation and Resettlement (R&R) methods are called for, it should be implemented duly adhering to the norms and complying with pertinent statutory requirements for such land acquisition. Strengthening of road and rail network infrastructure to handle the increase in traffic and truck parking arrangements, as well as integration of port development with the local land use plan, should be planned.

### 9.4.5.2 Topography, Geology and Soil

*Anticipated Impacts*

Impact of port construction/operation on the topography due to activities like depletion of hills due to large-scale quarrying, filling of low-lying area with dredged spoil and borrowed material, damage to existing vegetation/green belt and plantation, changes in land use patterns, disturbance to existing protected areas like mangroves, forests and environmentally sensitive areas/zones should be assessed. Flooding due to filling up of low-lying areas should be assessed. Impacts on the surrounding land use pattern, on infrastructure like housing, ground water, etc., should be assessed.

Impact of the project construction on the geology and vice versa should be studied in detail. Impact of project construction/operation on the soil parameters, probability of settlement, subsidence, slides, surface drainage, etc. are to be estimated.

*Mitigation Measures*

Mitigation measures to reduce adverse effects include study of alternative sites, improving green belt, obtaining construction materials from other sources, usage of alternative construction materials like fly ash, where possible, storm water management etc. Adopting soil improvement techniques and adopting suitable design methods, ground covering etc.

### 9.4.5.3 Water Environment

*Ground Water*

**Anticipated impacts:**

Discharge of trade effluent and sewage and its impact is studied. Impact of project construction/operation on the ground water because of leachate, run off from material, cargo storages and toxic or harmful substances, percolation, sea water intrusion etc.

**Mitigation measures:**

Mitigation measures to reduce adverse effects like impervious paving the cargo areas, impervious roads, lined drains, routing surface drainage to settlement tanks/pits, etc. Treatment of effluent, recycle/reuse and disposal should be planned. Ground water study on leachate should be carried out periodically and should be correlated with baseline data. Remedial measures should be taken in case of any deviation. Based on the total water budget of the project, the use of ground water should be reviewed and alternatives should be presented.

#### 9.4.5.4 Surface Water

*Anticipated Impacts*

Impact of port operations on surface water sources, contamination due to cargo operations, impact on utility of surface water resources by the neighbouring colonies, impact on surface water flow for instance flooding due to anticipated obstructions, etc.

*Mitigation Measures*

Protection measures to surface water resources to prevent reduction in their quality due to construction and operational activities and choice of alternative resources. Proposals to treat effluents confirming to standards notified under the EP Act 1996 should be submitted.

#### 9.4.5.5 Marine Environment

*Coastal Hydrology*

**Anticipated impacts:**

Impact of the project construction/operation on the coastal hydrology on account of port construction should be assessed by suitable model studies.

**Mitigation measures:**

Careful site selection and port design should be planned to minimize impacts due to changes in current patterns and other coastal hydrology. Model experiments or computer simulations of these changes are useful in developing an appropriate design. Shore protection works like construction of sea walls, groynes, sand bye passing, or beach nourishment should be studied.

#### 9.4.5.6 Bed Sediment Contamination

**Anticipated impacts:**

Impact of the project construction/operation on the bed sediment contamination because of port construction/operations is to be assessed by suitable empirical/model studies.

**Mitigation measures:**

A survey of contamination of bottom sediments should be undertaken before dredging.

#### 9.4.5.7 Sea/Harbour Water Quality

**Anticipated impacts:**

Impact of the project construction/operation on the sea/harbour water quality because of port construction is to be assessed by suitable empirical studies.

**Mitigation measures:**

Proper collection and disposal of liquid and solid waste from shore establishment and ships should be planned. Regular monitoring of water quality should be carried out within the port and in adjacent waters during operation to identify adverse environmental changes.

### 9.4.5.8 Biological Environment

*Anticipated Impacts*

Impacts of the project construction/operation on the marine/coastal ecology on account of port construction should be assessed by suitable empirical/model studies. Impacts due to floodlights on the nesting of sea turtles and other species should be studied.

*Mitigation Measures*

Mitigation measures to reduce adverse effects should be provided. Plan for minimizing impacts on local flora and fauna, and screen for the presence of rare, threatened or endangered species that are indigenous to the project location.

### 9.4.5.9 Air Environment

*Anticipated Impacts*

Impact of project construction/operation on the ambient air quality on account of emissions of dust during construction and cargo handling as well as emission of gases from equipment deployed for construction and cargo handling should be assessed. Prediction due to emissions during cargo handling/emissions from the ships in the port area/emissions due to increased traffic, emission inventory for critical pollutants with and without mitigation measures, prediction of the impact due to the existing activity on the proposed project, prediction of impacts due to sanctioned/ongoing projects in the surrounding area on the proposed project and the ambient environment shall be carried out.

*Mitigation Measures*

Mitigation measures proposed during the construction stage should include dust suppression measures by suitable techniques. Mitigation measures proposed during the operation stage should include alternative solutions such as closed conveyor system, closed silos, closed vehicles to transport dusty cargo, etc., mitigation measures to lower the emissions from the ships and green belt development.

### 9.4.5.10 Noise Pollution

*Anticipated Impacts*

Impact of project construction/operation on the noise and vibration on account of construction equipment, cargo handling equipment and road traffic should be studied.

*Mitigation Measures*

Mitigation measures includes proper installation of machines such as providing rubber paddings etc. Mufflers should be provided wherever feasible to muffle the sounds from engines, motors etc. Head phones and earplugs can be provided to workers working in noisy environments.

### 9.4.5.11 Solid Waste Management

*Anticipated Impacts*

Impact due to non-hazardous and hazardous solid waste generated during the construction and operational stages should be assessed.

*Mitigation Measures*

Mitigation measures to comply the norms should be planned. Options for minimization of solid waste and environmentally compactable disposal/recycling of waste to conserve natural resources should be planned. Management and disposal of temporary structures made during construction phase should be planned.

#### 9.4.5.12 Socio-economic and Occupational Health Environment

*Anticipated Impacts*

Predicted impact on the communities of the proposed activity. Details of public and private land in the proposed and immediate surroundings, socio-economic status of the affected owners of the private land shall be properly complied. Present status of health, housing, public utilities, commercial structures and transportation should be collected. Impact of the project on socio-cultural aspects should be assessed. Socio-economic impacts due to displacement of fishing settlements and population influx due to increased activities should be assessed.

*Mitigation Measures*

Mitigation measures to reduce adverse effects including satisfactory R&R methods should be planned.

### 9.4.6 Environmental Monitoring Program

This section shall include details of the EMP. It should include the technical aspects of monitoring the effectiveness of mitigation measures including measurement methodologies, data analysis, reporting schedules, emergency procedures, detailed budget and procurement schedules.

### 9.4.7 Additional Studies

### 9.4.8 Project Benefits

This section shall include benefits accruing to the locality, neighbourhood, region and nation.

- Improvements in the physical infrastructure by way addition of project infrastructure, ancillary industries that may come up because of the project.
- Improvements in the social infrastructure like roads, railways, townships, housing, water supply, electrical power, drainage, educational institutions, hospitals, effluent treatment plants improved waste disposal systems, improved environmental conditions, etc.
- Employment potential – skilled, semi-skilled and unskilled labour both during construction and operational phases of the project with specific attention to employment potential of local population as well as necessity for imparting any specialized skills to them to be eligible for such employment in the project on a long-term basis, i.e., during operational and maintenance stages of the project and other tangible benefits like improved standards of living, health, education etc.

**TABLE 9.2**
**Required Studies for Specific Conditions**

| Specific Condition | Study Required |
|---|---|
| Studies identified by the proponent and the Regulating Authority | Studies directed by the Expert Appraisal Committee while deciding the TOR for the project |
| Studies identified by the public and other stakeholders | Public hearing with the issues raised by the public and the response of the project proponent in tabular form shall be discussed |
| Risk Analysis and Disaster Management Plan (DMP) | • Risk analysis<br>• Safety measures for handling bulk liquid substances<br>• Personal protection equipment<br>• DMP<br>• Oil spill contingency plan<br>• Emergency response procedures |
| Natural resource conservation and optimization | • Plan of action for conservation of natural resources such as utilization of fly ash and other suitable waste materials availability for the construction of the project. Dredged material utilization and disposal plan should be furnished.<br>• Water conservation measures should be addressed. Energy efficiency measures in the activity are to be drawn up. |
| R & R action plans | • Detailed R&R plan with data on the existing socio-economic status of the population in the study area and broad plan for resettlement of the displaced population, site for the resettlement colony, alternative livelihood concerns/employment and rehabilitation of the displaced people, civil and housing amenities being offered. etc., and the schedule of the implementation of the project specific R&R plan.<br>• Details of provisions (capital and recurring) for the project specific R&R plan. |

Specific studies requirement depending on the site and activity proposed shall be discussed in addition to the above studies.

### 9.4.9 Environmental Cost Benefit Analysis

If recommended by the Expert Appraisal Committee at the scoping stage, this section shall include the environmental cost benefit analysis of the project.

## 9.5 DETAILED PROJECT REPORT

The detailed project report includes the following:

- Meteorological and oceanographic
- Geotechnical data
- Topographical information
- Transport linkages
- Traffic study

- Mathematical model study
  a. Wave tranquility studies
  b. Sedimentation studies
  c. Navigation simulation studies
  d. Disposal study for dredging material
- Ship size design
- Expected cargo details
- Port facility requirement general
  a. Berth requirement
  b. Cargo handling rates
  c. Cargo handling equipment
  d. Operation time
  e. Berth occupancy
  f. Berth requirement for master plan
- Storage requirement of different cargo
- Buildings
- Receipt and evacuation of cargo
  a. Port access road
  b. Railway system
  c. Water and power requirement
- Master plan of port
- Onshore port layout
- Engineering civil work
  a. Breakwater
  b. Berths
  c. Dredging
  d. Site grading
  e. Open and covered storage
  f. Port buildings, railway and roads
- Engineering of mechanical handling system
  a. Shop loading and unloading system, conveyors, stackers, reclaimers
  b. Port handling system
  c. Break bulk cargo handling system
- Engineering electrical works
  a. Electrical power source, requirement substations
  b. Illumination and cables
- Utilities
  a. Buildings, roads, railways and railway exchange yard
- Water supply, stormwater drainage, and sewerage
  a. Bunkering and security system port crafts
- Port crafts
  a. Tugs, pilot and mooring launches
- Navigation: fairway buoy, channel marker buoys, beacons, transit lights, Vessel Traffic Management System (VTMS)
- Fire and pollution control
- Port administration and management (terminal operations and organization structure)

- Cost estimates
  a. Capital cost estimates
  b. Operation and maintenance costs
- Project implementation and phasing of investment
- Financial and economic analysis
  a. Clearance from state coastal zone management authority (CRZ) to be provided
  b. Details of construction activities to be taken in CRZ area

## 9.6 ONE SEASON AND COMPREHENSIVE STUDY

The one season and comprehensive study of a location for construction of new port includes the following:

- CRZ classification of LTL/HTL/CRZ
- Risk analysis and disaster management
- Terrestrial EIA study
- Rapid marine environment impact assessment study
- Bathymetry survey
- Study on migration of olive turtle
- Study on effect of mangrove growth
- Disposal study of dredged material and its impact on marine environment
- Shoreline evaluation study
- Littoral drift study
- Geotechnical study
- Side scan study of sea bed
- Shallow seismic study
- Public hearing
- Environmental impact on dredging activities
- Study on marine environment on construction of berths
- Details of cargoes to be handled and their impacts on air, water and noise pollution due to such handling
- Details of drainage system in berth and stock yard of effluent treatment to treat the discharge/runoff from stock yard
- Details of monitoring programme
- Details of mangrove plantation year wise with action plan for implementation with expenditure
- Details of waste generation along with treatment facilities including sewage treatment plant
- Details of EMP report with financial outlays proposed
- Details of water resources for construction and operation of project and approval of competent authority for drawl of ground water
- Details of transportation of stones for breakwaters and groynes and their impact
- Details of stone quarries proposed for construction materials
- Details of reclamation and its impact on terrestrial ecology to be studied
- Details of monitoring on impact of marine ecology and shoreline changes during operation of project

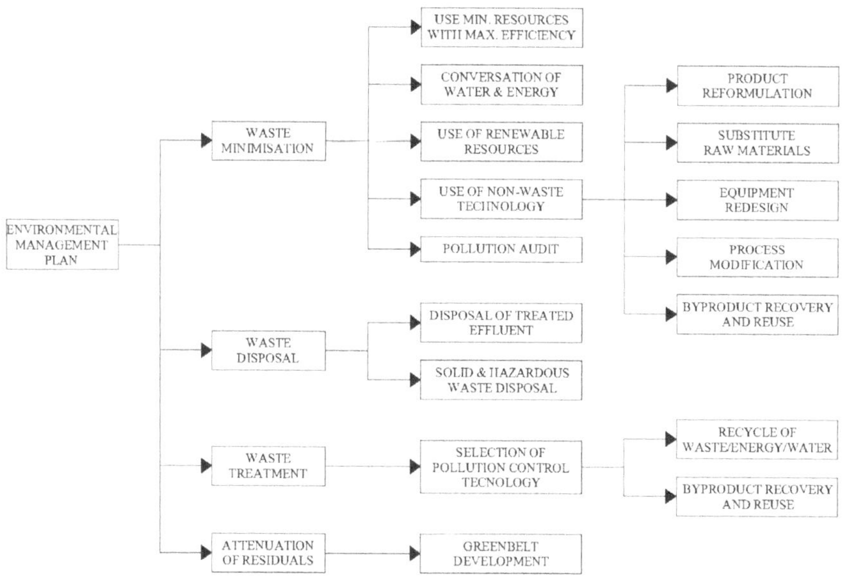


**FIGURE 9.6** Components and strategies for environmental management plan (EMP).

## 9.7 EMP AND MONITORING

EMP and monitoring includes the following:

- Summary of potential impacts and recommended mitigation measures
- Allocation of resources and responsibilities for plan implementation
- Administrative and technical setup for management of environment
- Institutional arrangements proposed with other organizations/government authorities for effective implementation of environmental measures proposed in the EIA
- Safeguards/mechanism to continue the assumptions/field conditions made in the EIA environmental specifications for contractors should cover the required safeguards during the design and construction stage

## REVIEW QUESTIONS

1. What are the adverse effects of port development on environment?
2. What is meant by CRZ? What are the objectives of CRZ?
3. What are the necessary details to be provided in a CRZ map?
4. How are CRZ zones classified as per the Namati-Centre for Policy Research, Environmental Justice Program?
5. What are the areas which are defined under CRZ I?
6. Write the norms for regulation of activities permissible in CRZ II.
7. How are the areas defined in CRZ IV?

8. Write the steps involved in the construction of major project on marine work.
9. What are the documents required for environmental clearance?
10. What is TOR? State its purpose.
11. Explain in brief the contents of TOR.
12. What are the components to be included in DPR?
13. Why is a one season comprehensive study carried out?
14. Describe in brief EMP and its components.

## REFERENCES

**Environmental Impact Assessment Guidance Manual Ports & Harbours (2009)** *Ministry of Environment & Forest*, Government of India, New Delhi

**Rasha E.I. Gohary and Tamer E.I. Gohary (2017)** Environmental Impact Assessment (EIA) Guidelines for development of ports, harbours and marines along Egyptian Mediterranean coast-A case study, *International Journal of Current Engineering and Technology*, Volume 7, Issue 3, pp. 1079–1091, E-ISSN: 2277-4106, P-ISSN: 2347-5161

**Rasha E.I. Gohary and Naggy A.A. Hassan (2012)** Environmental Impact Assessment (EIA) of marine structures – A case study, *KSCE Journal of Civil Engineering*, Volume 16, Issue 5, pp. 689–698, https://doi.org/10.1007/s12205-012-1347-5

**A.K. Gupta, S.K. Gupta and Rashmi S. Patil (2005)** Environmental management plan for ports and harbours, *Clean Technologies and Environmental Policy*, Volume 7, pp. 133–141, https://doi.org/10.1007/s10098-004-0266-7

**MoEF (2011)** *The Environment (Protection) Act Notification-Coastal Zone Regulations. S.O.19 (E)*, Ministry of Environment and Forest, Government of India, New Delhi, India

**Hiren B. Soni, V. Vijay Kumar and Pankaj N. Joshi (2019)** Environmental impact assessment and environmental management plan: A case study of Kachchh, Gujarat, India, *Current World Environment Journal*, Volume 14, Issue 1, pp. 99–127, ISSN: 0973-4929

# 10 Rehabilitation and Retrofitting

## 10.1 REHABILITATION AND RETROFITTING

As per the Comité Euro-International du Béton – Fédération Internationale de la Précontrainte (CEB – FIP) guidelines, the definition of rehabilitation is bringing the structure to its original or higher level of function, including durability and strength, while retrofitting is the action to modify the functionality or form of a structure or its components and to improve future performance. It relates particularly to the strengthening of structure against seismic loading as a means of minimizing the damage during specified earthquake or to increase load carrying capacity. Visual inspection, non-destructive tests (NDTs) and extraction of cores are carried out to identify the cause and extent of damage. Ultrasonic pulse velocity (UPV) tests, core tests, half-cell potential tests, carbonation tests and evaluation of chlorides and pH are the commonly carried out NDT tests to evaluate the condition of the concrete in the structures.

## 10.2 ULTRASONIC PULSE VELOCITY TEST

- Ultrasonic pulse velocity (UPV) test is an in-situ NDT to assess the quality of concrete. This test is done to assess the quality of concrete by ultrasonic pulse velocity method as per IS 516 (Part 5/Sec 1) (2018) (First Revision).
- In this test, the quality of concrete is assessed by measuring the velocity of an ultrasonic pulse passing through a concrete structure.
- This test is conducted by passing a pulse of the ultrasonic wave through concrete to be tested and measuring the time taken by the pulse to get through the structure.
- Higher velocities indicate good quality and continuity of the material, while lower velocities may indicate concrete with many cracks or voids.
- This test cannot be used to determine the compressive strength of concrete. Only a core test can give the compressive strength of concrete.

### 10.2.1 Principle of Ultrasonic Pulse Velocity Test

The underlying principle of this test is as follows:

- UPV is a wave propagation test.
- The method measures the time of travel of an ultrasonic pulse passing through the concrete being tested. Comparatively higher velocity is obtained when concrete quality is good in terms of density, uniformity, homogeneity, etc.


DOI: 10.1201/9781003361275-10

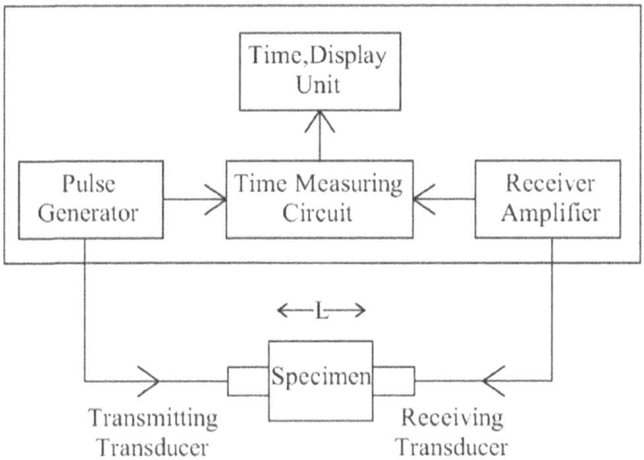


**FIGURE 10.1** Line sketch of principle of working of ultrasonic pulse.

**FIGURE 10.2** Ultrasonic pulse velocity in progress in a structure.

- It consists of transmitting ultrasonic pulses of 50–60 kHz frequency through a concrete medium and measuring the travel time of ultrasonic pulses for known or measured length.
- Velocity is indirectly correlated to concrete quality.

### 10.2.2 Testing Procedure

- Ultrasonic testing equipment includes a pulse generation circuit consisting of the electronic circuit for generating pulses and a transducer for transforming the electronic pulse into a mechanical pulse with an oscillation frequency of 40–50 kHz and a pulse reception circuit that receives the signal.
- The transducer, clock, oscillation circuit and power source are assembled for use.

- After calibration to a standard sample of the material with known properties, the transducers are placed on opposite sides of the material.
- Pulse velocity is measured by a simple formula:

$$\text{Pulse velocity} = \frac{\text{Path Length}(\text{L})}{\text{Transit Time}(\text{T})} \quad \text{(From IS 516 (Part 5/Sec 1) (2018))}.$$

### 10.2.3 Ways of Measuring UPV

Three ways of measuring pulse velocity include:

- Direct transmission
- Semi-direct transmission
- Indirect or surface transmission

#### 10.2.3.1 Direct Transmission

- The direct method is considered the most reliable for transit time measurement because the maximum pulse energy travels perpendicularly (at right angles) from the transmitter to the receiver.
- This method is applicable when opposite faces of the concrete element are accessible, allowing the transducers to be placed directly facing each other.
- It is generally preferred as it ensures maximum energy transfer to the receiving transducer, thereby yielding high sensitivity and accuracy in the measurement.

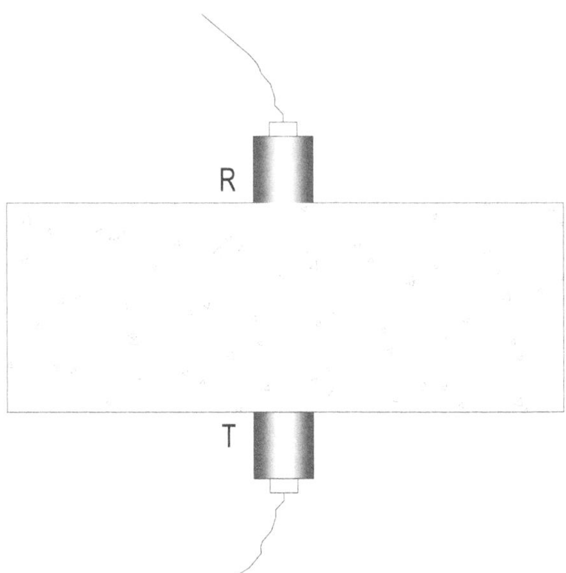


**FIGURE 10.3** Direct transmission.

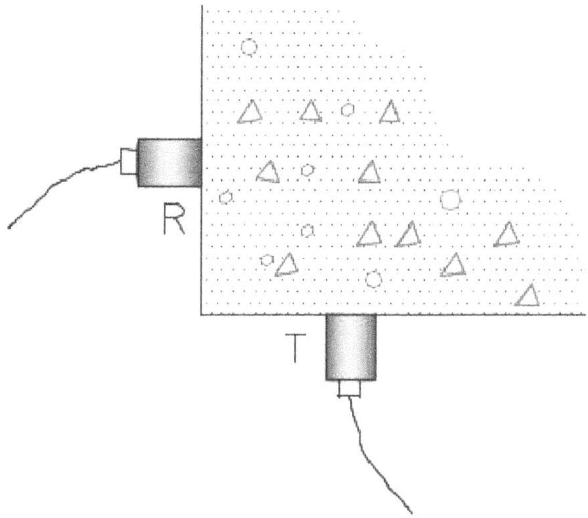


FIGURE 10.4 Semi-direct transmission.

- The pulse path is well-defined and straight, making the distance easy to measure accurately, which enhances the reliability of the calculated pulse velocity.
- Wherever feasible, the direct method should be used for assessing concrete quality, as it provides the most dependable results compared to semi-direct or indirect methods.

#### 10.2.3.2 Semi-direct Transmission

- When two adjacent sides of a member are accessible, this semi-direct method of measuring is used.
- The semi-direct method can sometimes be used satisfactorily if the angle between the transducers is not too great and if the path length is not too large.
- The sensitivity will be smaller, and if these requirements are not met, it is possible that no clear signal will be received because of the attenuation of the transmitted pulse.
- The path length is also less clearly defined due to the finite transducer size, but it is generally regarded as adequate to take this from centre to centre of transducer faces.

#### 10.2.3.3 Indirect or Surface Transmission

- When only one side of the member is accessible, this indirect method of measuring is used.
- The indirect method is the least satisfactory, since the received signal amplitude may be less than 3% of that for a comparable direct transmission.

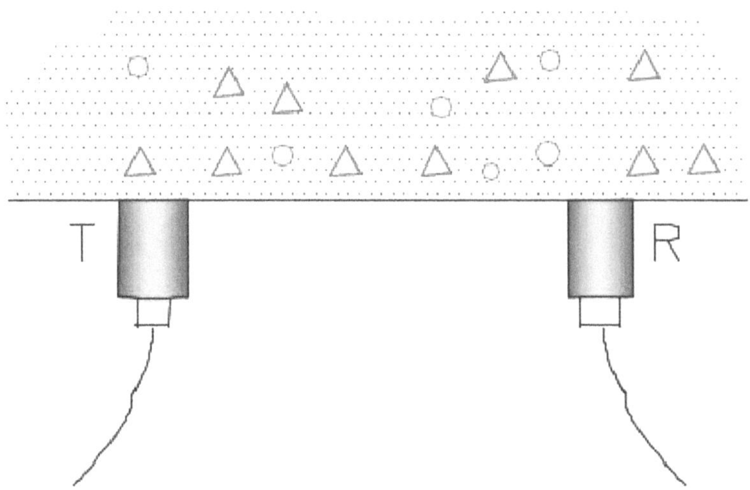


**FIGURE 10.5** Indirect or surface transmission.

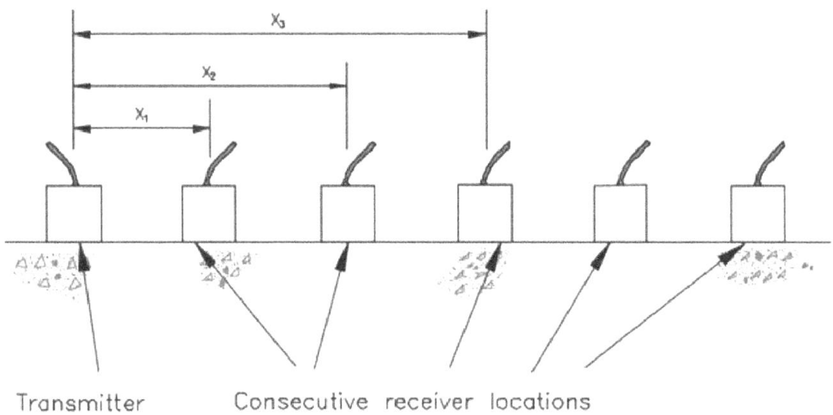


**FIGURE 10.6** Indirect reading transducer arrangement.

- The received signal is dependent upon the scattering of the pulse by discontinuities and is thus highly subject to errors.
- The pulse velocity will be predominantly influenced by the surface zone concrete, which may not be representative of the body and the mean pulse velocity is given by the slope of the best straight line.

If there is a discontinuity in this plot, it is likely that either surface cracking or an inferior surface layer is present. Unless measurements are being taken to detect such features, this method should be avoided if possible and only used where just one surface is available.

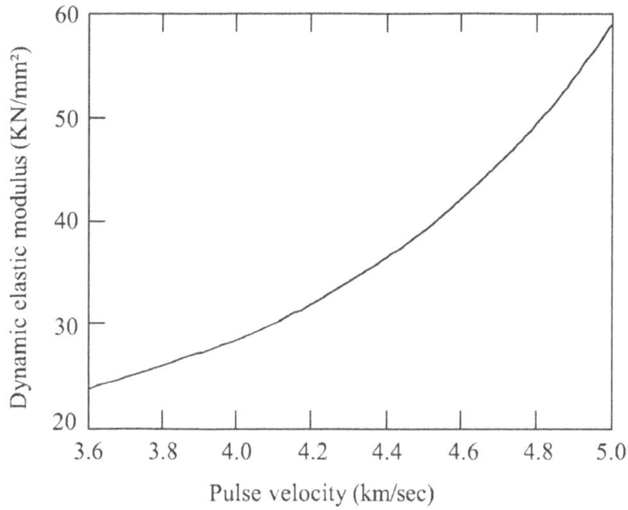


**FIGURE 10.7** Pulse velocity versus dynamic elastic modulus.

The general guidelines for quality assessment of concrete on UPV values 'V' (velocity) km/sec are given below.

As per IS 516 (Part 5/Sec 1) (2018) (First Revision)
For a concrete grade of less or equal to M25:
'V' greater than 4.50 km/sec = Excellent quality
'V' in the range of 3.50 to 4.50 km/sec = Good quality
'V' below 3.50 km/sec = Doubtful quality

For a concrete grade of above M25:
'V' greater than 4.50 km/sec = Excellent quality
'V' in the range of 3.75 to 4.50 km/sec = Good quality
'V' below 3.750 km/sec = Doubtful quality

- The relationship between pulse velocity and dynamic elastic modulus of the composite material measured by resonance tests on prisms is reliable, as shown in Figure 10.7.
- Although this relationship is influenced by the value of dynamic Poisson's ratio for most practical concretes made with natural aggregates, the estimate of modulus of elasticity should be accurate within 10%.

### 10.2.4 Interpretation of Results

- Quality of concrete in terms of homogeneity, trapped air, internal flaws, cracks, segregation, honeycombing, compaction, workmanship and durability can be concluded from this test.

**FIGURE 10.8** Measurement of UPV at site.

- The test can also be used to evaluate the effectiveness of the repair.
- Ultrasonic testing is indicative, and other tests such as destructive testing must be conducted to evaluate the compressive strength.

A valuable application of the UPV technique, which does not require a detailed correlation of pulse velocity with any other property of the material, is in the detection of honeycombing and cracking.

Since the pulse cannot travel through the air, the presence of a crack or void on the path will increase the path length (as it goes around the flaw) and increase attenuation so that a longer transit time will be recorded. The apparent pulse velocity thus obtained will be lower than for the sound material.

In crack detection and measurement, even micro-cracking of concrete will be enough to disrupt the path taken by the pulses.

### 10.2.5 Reliability and Limitations

- UPV measurement has been found to be a valuable and reliable method of examining the interior of a body of concrete in a truly non-destructive manner.
- Modern equipment is robust, reasonably cheap and easy to operate, and reliable even under site conditions; however, it cannot be overemphasized that operators must be well trained and aware of the factors affecting the readings.
- It is similarly essential that results are properly evaluated and interpreted by experienced engineers who are familiar with the technique.
- The method has few limitations, other than when two opposite faces of a member are not available.

- The method provides the only readily available form of determining the extent of cracking within concrete; however, the use for detection of flaws within the concrete is not reliable when the concrete is wet.
- The least reliable application is for the strength estimation of concrete.

## 10.3 HALF-CELL POTENTIAL TESTS

### 10.3.1 General

- Corrosion of reinforcing steel is an electrochemical process, and the behaviour of the steel can be characterized by measuring its half-cell potential.
- The greater the potential, the higher the risk that corrosion is taking place.
- An electrode forms one-half of the cell and the reinforcing steel in the concrete the other.
- The preferred reference electrode for site use is silver/silver chloride in potassium chloride solution, although the copper/copper sulphate electrode is still widely used.
- The survey procedure is initially to locate the steel and determine the bar spacing using a cover meter.
- The cover concrete is removed locally over a suitable bar, and an electrical connection is made to the steel.
- It is necessary to check that the steel is electrically continuous by measuring the resistance between two widely separated points.
- The reinforcing bar is connected to the half-cell via a digital voltmeter.
- Readings of half-cell potential are taken over a regular grid of points (say 0.5 m apart) to give a potential map of the area.
- Contour lines may be plotted between points of equal potential to indicate those areas that have the greatest risk of corrosion.
- Locally exposing and inspecting the reinforcement in areas where both high and low risks of corrosion are indicated can be used to approximately to calibrate the potential readings for the structure with respect to its present corrosion and the need for further investigation or repair.

### 10.3.2 The Basics of Concrete Corrosion

- In reinforced concrete structures, there is a natural protective oxide film that forms on the surface and prevents the bar from corroding.
- With time, chlorides (from de-icing salts or marine exposure) and/or $CO_2$ penetrate the concrete and break down the protective layer.
- Chlorides destabilize the passive film leading to its localized breakdown, while $CO_2$ lowers the pH of the concrete below the level of stability of the passive film.
- In the presence of oxygen and water, an electrochemical reaction initiates the process of corrosion.
- Corrosion is illustrated as shown in Figure 10.9, where the metal (rebar) reacts in the solution (available in the concrete pores) and gives away

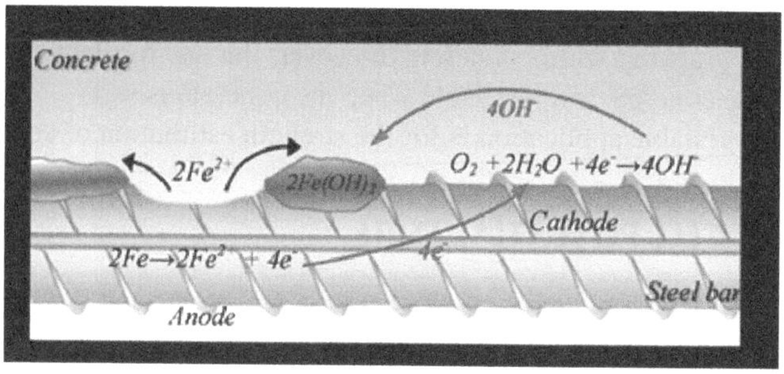


**FIGURE 10.9** Reinforced concrete corrosion reaction.

electrons from the anode (where oxidization occurs) to the cathode (where reduction occurs). The positive ions formed at the surface of the anode will react and create corrosion by-products.

- This electrochemical reaction creates a potential difference, and consequently a corrosion current, between the anodic and cathodic areas at the surface of the steel reinforcement. This current, or the potential distribution on the reinforcement surface, is what is of interest when measuring half-cell potential.

### 10.3.3 The Half-Cell Potential Technique

- The schematic in Figure 10.9 represents a cell where each side is referred to as a half-cell.
- Each half-cell is represented by an electrode in a solution (electrolyte) and both half-cells are connected. Since one of the electrodes has a higher tendency to corrode compared to the other, that electrode (anode) will oxidize and in turn will donate electrons.
- To keep the system in equilibrium and balance, the charges in the electrolytes, ions will be exchanged through the salt bridge.
- The voltmeter will measure the potential difference (voltage) between both electrodes, which indicates the rate of dissolution of the anode.
- To apply this concept to concrete and to interpret the results of corrosion potential, a reference electrode with a known potential is needed.
- Typically, for reinforced concrete applications, a copper/copper sulphate electrode ($Cu/CuSO_4$) or silver/silver chloride electrode (Ag/AgCl) is used for the half-cell reference.
- This reference electrode is connected to the other half-cell represented by the embedded rebar. By connecting that reference electrode to the reinforcing steel and placing the reference electrode on the surface of the concrete, it is possible to measure the potential difference between the two half-cells.
- ASTM C876 provides a guideline on how this measurement can be undertaken and the relationship between the measured potential values and the

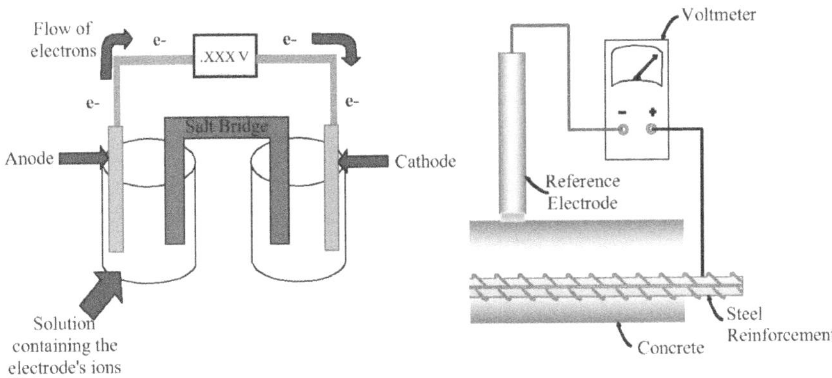


FIGURE 10.10 Half-cell potential measurement.

**TABLE 10.1**
**Relationship Between the Potential Values and Corrosion Probability**

| Measured Potential (mV CSE) | Probability of Steel Corrosion Activity |
|---|---|
| > −200 | Less than 10% corrosion is possible. |
| −200 to −350 | Corrosion activity of the reinforcing steel in the area is uncertain. |
| < −350 | There is a greater than 90% probability that reinforcing steel corrosion is occurring in that area at the time of measurement. |

*Source:* Adapted from ASTM C876.

corrosion probability. Interpretation of the result is qualitative and is based on the copper sulphate electrode (CSE).

- Table 10.1 shows the general interpretation guideline proposed by ASTM, where the measured potential ranges are categorized into three categories; more than 90% chance, less than 10% chance or an uncertain chance of corrosion.

### 10.3.3.1 Significance and Uses of Half-Cell Potential

- This test method is applicable to members regardless of their size or the depth of concrete cover over the reinforcing steel.
- This test method may be used at any time during the life of a concrete member.
- The results obtained using this test method shall not be considered as a means for estimating the structural properties of the steel or of the reinforced concrete member.
- The potential measurements should be interpreted by engineers or technical specialists experienced in the fields of concrete materials and corrosion testing.

FIGURE 10.11 Measurement during half-cell potential test.

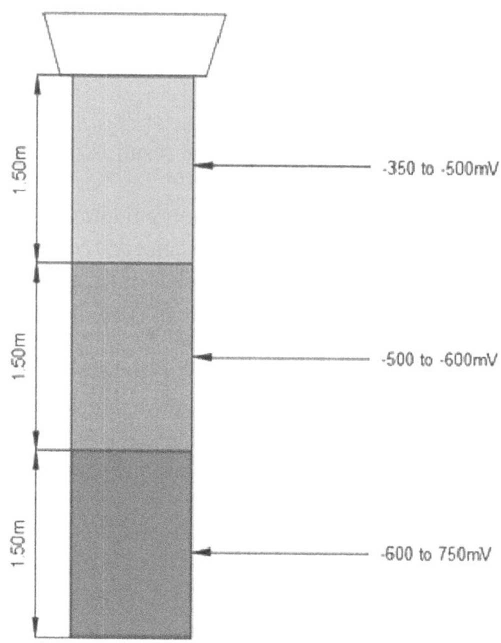


FIGURE 10.12 Measurement showing half-cell potential at pile location: the tidal zone is the most affected.

**TABLE 10.2**
**Half-Cell Potential Values in mV**

| Sl. No. | Description | Half-Cell Potential Values (mV) |
|---|---|---|
| 1 | Pile X-1 Top | −423, −459, −359, −356 |
| 2 | Pile X-1 Mid | −529, −557, −520, −519 |
| 3 | Pile X-1 Intertidal zone | −678, −671, −700, −662 |

## 10.4 CORE SAMPLE COLLECTION

### 10.4.1 General Description

Core sampling is a non-destructive or minimally invasive technique employed to extract cylindrical specimens from structural or geotechnical materials such as concrete and rock.

### 10.4.2 Objective

The objective of the core sample collection in this context includes:

- Assessing the material condition (e.g., corrosion, pitting, wall thinning)
- Determining residual strength of concrete
- Identifying chemical and physical degradation due to marine exposure
- Evaluating corrosion products, chlorides and sulphate penetration
- Supporting rehabilitation or replacement decisions

### 10.4.3 Methodology

#### 10.4.3.1 Equipment Used

- Core drilling machine with diamond or carbide-tipped core bit
- Portable water supply or coolant system (not clearly visible but typically used)
- Marking tools and measurement tapes
- Safety equipment

### 10.4.4 Procedure

- Marking the sampling location based on structural plan and corrosion extent
- A rebar locator shall be used to identify the rebars and to avoid the cutting of the same
- Drilling perpendicular to the member using rotary coring with continuous cooling to prevent thermal damage
- Extracting the core carefully to preserve the material integrity

- Labelling and sealing cores in moisture-proof containers to prevent further degradation
- Transporting samples to the laboratory for detailed analysis

### 10.4.5 Laboratory Testing and Analysis

Depending on the material type (likely steel or concrete encasement), the following tests are conducted:

For concrete cores:

- Compressive strength
- Carbonation depth and chloride and pH content
- Permeability and porosity
- Microcrack assessment using microscopic imaging

### 10.4.6 Interpretation of Results

**Indicative parameters:**

Loss of sectional area >25% may indicate structural danger.

Chloride ingress > threshold (~0.6 kg/m$^3$ by weight of concrete) indicates high corrosion risk.

Cracks or voids compromise durability and require remediation.

## 10.5 MATERIALS FOR REPAIR

Materials for the repair and rehabilitation of ocean structures are not under the recommendation of international codes. These codes only suggest repair procedures and desirable characteristics of materials for repair. Among several reasons for this limitation, the foremost is that the material choice for repair is case specific. The characteristics of a material chosen for repair should enable speedy construction and attain the desired strength at the earliest possible time. This is because the downtime available for repair of ocean structures is generally for a limited period; constraints may arise from the weather window or functional priorities. The material chosen for repair is not based on the existing design requirements but should also compensate for the degraded performance of materials that are deteriorated. Both the factors, namely strength and serviceability are required to be fulfilled.

**Advanced material**

- Epoxy resin/epoxy mortar (for shear connector fixing)
- Epoxy concrete (for filling cavities)
- Moisture-cured urethane coating for steel and concrete surfaces
- Silane siloxane coating for concrete surfaces
- Self-regulating anodes for corrosion prevention and monitoring

- Zinc-based coating for rebars (concreting to be carried out immediately after coating)
- Epoxy based bond coating between old concrete surfaces and new concrete surfaces. The new concrete shall be placed within 60 minutes or as per the manufacturer's specifications.
- Polymer modified cementitious products
- Micro-concrete jacketing
- Glass/carbon/aramid fibre reinforced polyester composites (carbon fibre or glass fibre for retrofitting)

## 10.6 METHODOLOGY

The methodology of the repair and rehabilitation (removal of damaged fenders, retrofitting of fender beams and installation of new fenders) is explained as a case study in the following sections.

### 10.6.1 Case Study I – Repair of Pile

Before finalizing the rehabilitation methodology, a full-fledged visual inspection shall be carried out to facilitate a site-specific methodology. The following sequence illustrates the sequence of rehabilitation of offshore pile.

**FIGURE 10.13** View of corrosion-affected pile with exposed reinforcement.

FIGURE 10.14 View of tentative hanging support arrangement.

#### 10.6.1.1 Stage 1: Provide Access Platform

The working platform above the ground/water level shall be erected using the friction and hanging support. The fixing and erection work shall be carried out with the help of floating pontoon and boats, etc. Figure 10.14 shows tentative arrangement of the working platform.

#### 10.6.1.2 Stage 2: Surface Preparations

The pile liner shall be cut and removed using the gas cutting equipment/electrically operated cutting grinder to expose the concrete surface to continue the rehabilitation works. Figure 10.15 shows a view of the pile after the cutting of the liner and chipping of cover concrete.

#### 10.6.1.3 Stage 3: Removal of Cover Concrete and Exposing the Reinforcement

The cover concrete shall be chipped and removed and the rear side of the steel bar shall be exposed using the mechanical arrangements and hand chiselling. It is recommended to use water jet for the removal of all loose particles and corrosion stains, which might be present due to wave action.

#### 10.6.1.4 Stage 4: Anchoring Shear Connector

In order to fix the shear connectors (secondary reinforcement) of 10 mm diameter, holes shall be drilled at the required locations as per the drawings. The pilot holes shall be drilled using a 10 mm diameter hammer drill bit. The holes shall be widened using the appropriate drill bit size of 16/18 mm diameter. Drilling shall be carried out without much vibration so as not to disturb adjacent concrete. The holes shall be roughened for better bonding.

FIGURE 10.15 View of exposed reinforcement.

FIGURE 10.16 Drilling and fixing of shear connectors using pneumatic drilling at low tide.

**FIGURE 10.17** Provision of additional reinforcement to the existing corroded reinforcement.

#### 10.6.1.5 Stage 5: Placing of Additional Reinforcement

Additional reinforcement of FE 500 rebars for main reinforcement including the stirrups wherever the size of the reinforcement has reduced by 15% from the original diameter shall be fabricated and placed in position as per the Good for Construction Drawing.

#### 10.6.1.6 Stage 6: Provision of Sacrificial Anode

All repairs to chloride-contaminated concrete shall be protected from future corrosion and incipient anode formation by providing sacrificial zinc anodes embedded in concrete. The sacrificial anode shall be sacrificial metal surrounded by a highly alkaline cementitious product. This has a sufficiently high pore solution of pH for corrosion of the anode to occur and for passive film formation on the anode. Sacrificial anode shall be fixed over the main reinforcement in the jacket portion. The number of sacrificial anodes shall be as per the drawing and the manufacturer's specifications. A monitoring box shall be provided as shown in Figure 10.25 for long term measurement.

#### 10.6.1.7 Stage 7: Provision of Anti-corrosive Treatment for Old and New Reinforcement

Anticorrosive coating to the reinforcement consists of providing and applying two coats of zinc-rich coating to the rebars. An unbroken 40 micron dft/coat coating shall be capable of providing active galvanic protection and avoiding the generation of incipient anodes in the immediately adjacent locations. It shall be of suitable viscosity to enable the coating to penetrate into imperfections and pits within the surface of the structural steel strictly following the manufacturer's instruction.

FIGURE 10.18 The provision of sacrificial anode in the additional reinforcement.

FIGURE 10.19 Protective coating on additional reinforcement and existing reinforcement and provision of epoxy bonding coat.

#### 10.6.1.8 Stage 8: Provision of Epoxy Bonding Agent

Before pouring the micro-concrete, the base and hardener of the epoxy jointing compound shall be mixed mechanically using a slow-speed heavy duty drilling machine. The same shall be applied over the prepared concrete surface using a good quality brush.

#### 10.6.1.9 Stage 9: Place Form Work/Shuttering

Slurry tight leak proof form work which should not deform due to pressure of micro-concrete shall be fabricated and erected in position as per the drawing. The formwork should be coated with mould-releasing agent. Proper supporting arrangement shall be made for keeping the shutter in correct line and length.

#### 10.6.1.10 Stage 10: Pouring of Micro-concrete

A sufficient quantity of micro-concrete shall be stocked along with 12 mm down-graded chips at the site to enable completion of pouring in a continuous operation.

**FIGURE 10.20** Provision of epoxy bonding agent.

**FIGURE 10.21** Shuttering arrangement.

**FIGURE 10.22** A tentative arrangement for pouring micro-concrete.

**FIGURE 10.23** A view of application of curing membrane after deshuttering.

An approved grout concrete mixer of slow-speed heavy duty drill shall be used for the mixing of the micro-concrete. The exact quantity of water as recommended by the manufacturers should be poured slowly in to the mixing vessel along with the micro-concrete and 12 mm cleaned chips (in the ratio of 1:0.50 or 0.75). The material shall be mixed thoroughly in forced action. The mixed material should be placed immediately in the formwork continuously without any gap through the PVC pipe fixed from the top of the deck.

### 10.6.1.11 Stage 11: Membrane Curing Agent

Immediately after the striking form work, a membrane curing agent shall be applied on the concrete surface.

FIGURE 10.24 View of tentative application of protective coating.

FIGURE 10.25 View of corrosion monitoring box.

### 10.6.1.12 Stage 12: Protective Coating

Two coats of ultraviolet-resistant breathable protective coating shall be applied to the pile in view of the marine environment along with one coat of compatible primer as per the manufacturer's specification.

### 10.6.1.13 Stage 13: Corrosion Monitoring Box and Monitoring

A corrosion monitoring box as shown in Figure 10.25 is then installed. The typical readings are shown in Table 10.3.

**TABLE 10.3**
**Half-Cell Potential Values**

| Before Repair | Immediately After Repair | After 6 Months | After 12 Months | After 18 Months |
|---|---|---|---|---|
| -457–422–412 | -004–007–008 | -052–047–038 | -098–065–092 | -085–085–095 |
| -419–398–412 | -009–007–008 | -024–032–040 | -076–083–073 | -095–098–099 |
| -395–401–389 | -005–006–009 | -021–029–035 | -076–080–082 | -085–089–092 |
| -425–436 418 | -004–008–009 | -029–027–022 | -060–065–068 | -075–075–072 |
| -412–389–395 | -004–006–006 | -034–042–049 | -072–072–082 | -075–082–093 |
| -375–385–399 | 008–011–009 | -034–044–050 | -074–065–073 | -080–080–087 |
| -385–400–408 | -006–009–008 | -042–037–034 | -077–085–090 | -090–090–105 |

The half-cell potential of about −400 reduces to about −10 immediately after repair. The half-cell potential increases to about −80 after 18 months if it further increases to −200 the sacrificial anodes shall be replaced.

## REVIEW QUESTIONS

1. What is meant by rehabilitation and retrofitting?
2. What is UPV? Write the principle of it.
3. How is the UPV test conducted? Explain the ways of measuring pulse velocity with a diagram.
4. Write the interpretation of a UPV test result as per the Indian Standard.
5. What are the applications for the UPV test?
6. Why does spalling of concrete occur? Explain with a sketch.
7. Describe in brief the half potential technique with a neat sketch.
8. How is the half-cell potential data result interpreted?
9. What are the various materials used for the repair of concrete?
10. How is the concrete structure monitored with sacrificial anodes?

## REFERENCES

**J. Blitz and G. Simpson (1996)** *Ultrasonic Methods of Non-Destructive Testing*, Chapman and Hall, London

**Srinivasan Chandrasekaran and Arvind Kumar Jain (2016)** *Materials for Ocean Structures from: Ocean Structures, Construction, Materials, and Operations*. Available at www.routledgehandbooks.com/doi/10.1201/9781315366692-4

**John W. Gaythwaite (2014)** *Design of Marine Facilities: Engineering for Port and Harbour Structures*, ASCE Press, ISBN (print): 9780784414309, ISBN (PDF): 9780784479933

Guide Book on Non-destructive Testing of Concrete Structures (2002) International Atomic Energy Agency, IAEA-TCS-17, ISSN: 1018-5518

**IS 13311 (Part 1) (1992)** Method of Non-destructive Testing of Concrete, Part 1: Ultrasonic Pulse Velocity

**Jin-A. Jeong and Chung-Kuk Jin (2013)** Three year performance of sacrificial anode cathodic protection system in the reinforced concrete bridge structures, *Advanced Materials Research*, Volumes 753–755, pp. 467–475, Online: 2013–08–30 © (2013) Trans Tech Publications, Switzerland. Available at https://doi.org/10.4028/www.scientific.net/AMR.753-755.467

**R. Sundaravadivelu, M.R. Pranesh, R. Ambalavanan, C.R. Balaji, and G. Kanniappan (2002)** *Performance of Underwater Grouting System for Rehabilitation of Marine Concrete Structures*, World Conference on Concrete Materials and Structures, 14th–16th May, Malaysia, pp. 161–172.

**H. Wall (2013)** Service Life Assessment of Harbour Structures – Case Studies of Chloride Ingress into Concrete and Sheet Piling Corrosion Rates, Division of Building Materials, LTH, Lund University.

# Index

Note: Page numbers in *italics* indicate figures and page numbers in **bold** indicate tables in the text.

For Product Safety Concerns and Information please contact our EU representative GPSR@taylorandfrancis.com Taylor & Francis Verlag GmbH, Kaufingerstraße 24, 80331 München, Germany